AF581847

ALOS-2/PALSAR-2 Calibration, Validation, Science and Applications

ALOS-2/PALSAR-2 Calibration, Validation, Science and Applications

Editors

Takeo Tadono
Masato Ohki

MDPI • Basel • Beijing • Wuhan • Barcelona • Belgrade • Manchester • Tokyo • Cluj • Tianjin

Editors
Takeo Tadono
Earth Observation
Research Center
Japan Aerospace
Exploration Agency
Tsukuba, Ibaraki
Japan

Masato Ohki
Earth Observation
Research Center
Japan Aerospace
Exploration Agency
Tsukuba, Ibaraki
Japan

Editorial Office
MDPI
St. Alban-Anlage 66
4052 Basel, Switzerland

This is a reprint of articles from the Special Issue published online in the open access journal *Remote Sensing* (ISSN 2072-4292) (available at: www.mdpi.com/journal/remotesensing/special_issues/alos_palsar).

For citation purposes, cite each article independently as indicated on the article page online and as indicated below:

LastName, A.A.; LastName, B.B.; LastName, C.C. Article Title. *Journal Name* **Year**, *Volume Number*, Page Range.

ISBN 978-3-0365-6106-6 (Hbk)
ISBN 978-3-0365-6105-9 (PDF)

Cover image courtesy of Japan Aerospace Exploration Agency (JAXA)

Contents

Preface to "ALOS-2/PALSAR-2 Calibration, Validation, Science and Applications"

Dear Readers,

The Advanced Land Observing Satellite-2 (ALOS-2, nicknamed "DAICHI-2") was launched on May 24th, 2014, which is a follow-on mission of L-band Synthetic Aperture Radars (SARs) by the ALOS "DAICHI" from 2006 to 2011, and the Japanese Earth Resources Satellite-1 (JERS-1, "FUYO-1") from 1992 to 1998. Thus, global coverage and almost three decades of L-band SAR data are currently available. In addition, ALOS-4 is under development and will be launched in the Japanese Fiscal Year 2023 as a successor to Japanese L-band SAR missions.

The mission objectives of ALOS-2 that were defined to fulfill social needs include the following: 1) disaster monitoring of damaged areas, both in considerable detail and when these areas may be large, 2) continuous updating of data archives related to national land and infrastructure information, 3) effective monitoring of cultivated areas, and 4) global monitoring of tropical rain forests to identify carbon sinks. The Phased Array-type L-band SAR-2 (PALSAR-2) mounted on ALOS-2 has capabilities of high-resolution, wide-swath width, and both right- and left-looking observations, and is now operating globally to achieve these objectives. The Japan Aerospace Exploration Agency (JAXA) is continuously conducting research announcement (RA) programs that provide opportunities to use PALSAR-2 and other satellite data to engage and enhance science and application activities worldwide.

This Special Issue has collected original manuscripts on calibration, validation, science, and applications based on PALSAR-2 data. We hope that it will serve as one of the guidelines for future satellite-borne SAR missions, as well as the utilization of L-band SAR data.

Takeo Tadono and Masato Ohki
Editors

Communication

Calibration and Validation of Polarimetric ALOS2-PALSAR2

Ridha Touzi [1,*], Masanobu Shimada [2,3] and Takeshi Motohka [3]

1 Canada Center for Remote Sensing, 560 Rochester Street, Ottawa, ON K1A 0E4, Canada
2 School and Science Engineering, Tokyo Denki University, Saitama 350-0394, Japan; shimada@g.dendai.ac.jp
3 Japan Aerospace Exploration Agency (JAXA), Tsukuba 305-8505, Japan; motooka.takeshi@jaxa.jp
* Correspondence: ridha.touzi@canada.ca

Abstract: PALSAR2 polarimetric distortion matrix is measured using corner reflectors deployed in the Amazonian forest. The Amazonian forest near the geomagnetic equator provides ideal sites for the assessment of L-band PALSAR2 antenna parameters, at free Faraday rotation. Corner reflectors (CRs) deployed at free Faraday rotation provide accurate estimation of antenna cross-talks in contrast to the biased measurements obtained with CRs deployed at significant Faraday rotation. The extended Freeman–Van Zyl calibration method introduced and validated for ALOS-PALSAR calibration is used for the assessment of PALSAR-2 calibration parameters. Six datasets collected over the Amazonian rainforests (with CRs) are used to assess PALSAR-2 distortion matrix for five beams (FP6-3 to FP6-7) with incidence angle varying from 25° to 40°. It is shown that the PALSAR2 antenna is highly isolated with very low cross-talks (lower than −40 dB). Finally, the impact of a significant Faraday rotation on antenna cross-talk measurements using CR is discussed.

Keywords: radar polarimetry; synthetic aperture radar; calibration; Faraday rotation

Citation: Touzi, R.; Shimada, M.; Motohka, T. Calibration and Validation of Polarimetric ALOS2-PALSAR2. *Remote Sens.* **2022**, *14*, 2452. https://doi.org/10.3390/rs14102452

Academic Editors: Takeo Tadono and Masato Ohki

Received: 22 February 2022
Accepted: 16 April 2022
Published: 20 May 2022

1. Introduction

ALOS2, which was launched on the 24th of May 2014, is equipped with a fully polarimetric L-band SAR, PALSAR2 [1–3]. Unlike ALOS-PALSAR, which used to collect polarimetric data at only one off-nadir angle (21.5°) [4], PALSAR-2 offers the possibility of providing polarimetric measurements at various beams (seven beams, with incidence angle varying from 25° to 40°) [1,2]. The active antenna uses 180 transmit-receive (T/R) modules; each T/R excites a single subarray [1]. The requirement on the overall antenna array cross-talk is low (better than −30 dB [1,2]), and this should permit the measurement of pure HV (at low Faraday conditions) [5,6]. However, different distortion matrices might be required for the extraction of pure HH, HV, VH, and VV from the PALSAR2 PLR measurements at the various beams. Calibration residual errors should be minimized since PALSAR-2 is expected to have a signal-to-noise ratio (S/N) 3-dB better than PALSAR. The latter was already operating at low noise floor (NESZ better than −34 dB [5]).

Like L-band JERS-1 SAR and ALOS PALSAR, PALSAR-2 measurements might be affected by Faraday rotation. In the past, the use of corner reflector (CR) measurements in the presence of Faraday rotation led to mixed conclusions regarding the actual isolation of the ALOS-PALSAR antenna [7–9]. PALSAR antenna measurements using CR deployed in Germany and Australia led to the conclusion that the PALSAR antenna is not highly isolated (−23 dB isolation) [7,8]. However, the use of the CR measurements collected in the Amazonian rainforests, at low Faraday rotation conditions, led to the conclusion that PALSAR antenna is highly isolated, with a cross-talk lower than −37 dB [3,5,6,9,10].

In Section 2, the lessons learned from PALSAR calibration are first presented. The sites of calibration in the Amazonian rainforests and the polarimetric PALSAR2 images used for the assessment of PALSAR2 calibration, are described in Section 3. The extended Freeman–VanZyl calibration method [6] is briefly presented in Section 4 and used for the assessment of polarimetric PALSAR2 transmit-receive distortion matrices. In Section 5, an estimation of PALSAR-2 distortion matrix is provided for five beams (FP6-3 to FP6-7) with incidence angle varying between 25° and 40°. Measurement of the Faraday rotation during PALSAR2 acquisition is conducted to confirm the legitimacy of the estimation of antenna cross-talks under negligible Faraday rotation. Finally, the impact of a significant Faraday rotation on CR-based measurements of antenna cross-talks is discussed using PALSAR2 images collected over the CCRS calibration site in Ottawa (Canada).

2. Lessons Learned from ALOS-PALSAR Calibration

ALOS-PALSAR, launched in 2006, was the first polarimetric L-band SAR satellite mission. The use of various calibration sites led to confusion that arose at the 2006 ALOS Calibration/Validation meeting regarding the actual PALSAR antenna isolation. The presence of Faraday rotation in addition to the uncertainty regarding the actual isolation of the PALSAR antenna, led to mixed conclusions regarding the actual isolation of the H-V PALSAR antenna [8,9,11].

The use of the CR measurements obtained at the DLR calibration site led to the following misleading conclusion regarding PALSAR antenna isolation; PALSAR antenna is not well isolated with significant antenna cross-talks varying between −18 dB and −23 dB [7,11]. Such significant cross-talk would have made the use of PALSAR HV in dual-pol (HH-HV) useless since the contamination of HV by the like polarization (HH and VV) cannot be corrected for, as discussed in [12]. In fact, PALSAR antenna is higly isolated and this permits an excellent exploitation of PALSAR dual-pol (HH-HV) systematic coverage for global forest mapping and monitoring [3,13,14].

Using CR deployed in the Amazonian rainforests, Touzi and Shimada [6] showed that the PALSAR antenna is highly isolated (−37 dB isolation) [9,10]. The thorough investigation they conducted in [6] using PALSAR data collected over various calibration sites (with CRs), in Japan (JAXA, Sweden (Chalmers University), Germany (DLR), and Canada (CCRS), in addition to the ones collected over the Amazonian rainforests, led to the following conclusions:

- PALSAR2 CR measurements in the Amazonian rainforests, collected at free Faraday rotations [15], revealed insignificant return at HV and VH CR measurements. These CR measurements integrated in two different calibration methods led to the conclusion that ALOS-PALSAR is highly isolated with a cross-talk lower than −37 dB [5,6];
- The presence of small but still significant Faraday rotation (2–3°), at the JAXA (Japan), CCRS, DLR and, Chalmers U. calibration sites, induces a significant CR return at the cross-polarization (cross-pol) HV and VH. The integration of these contaminated CR response in the conventional polarimetric calibration methods [16–18] led to erroneous (i.e., biased) estimates of the antenna cross-talks.

These lessons learned with the calibration of ALOS-PALSAR has encouraged JAXA to deploy CRs in the Amazonian rainforests. The latter should permit accurate assessment of ALOS2-PALSAR2 calibration parameters, as discussed in Section 3.

However, the impact of a significant Faraday rotation on CR response in addition to the antenna cross-talks related contamination was not thoroughly investigated in [6]. This is conducted in the following Section 4. The analytical response of CR is derived as a function of the Faraday rotation and antenna cross-talks, and the cross-pol (HV and VH) contamination with the like-pol (HH and VV) due to the Faraday rotation is demonstrated using PALSAR2 data collected over the CCRS calibration site in Ottawa.

3. The Extended Freeman-Van Zyl Calibration Method for Accurate Assessment of Antenna Cross-Talks

The van Zyl calibration, which offers a more convenient solution than the conventional method that requires the deployment of many reference point targets across the swath, has become the standard method for estimation and calibration of antenna distortion matrix variations with incidence angle [17–19]. However, the van Zyl algorithm is limited to symmetric SAR systems. This problem has been circumvented by Freeman [17] who introduced a symmetrization method to adapt the van Zyl calibration to non-symmetric SAR systems. The Freeman–van Zyl calibration technique [17], which symmetrizes the system prior to the estimation of the distortion matrix elements, uses the van Zyl iterative method [16] for antenna cross-talk estimation. Recently, the accuracy of the van Zyl method [16] has been questioned [6,18] for azimuthally symmetric targets of low HV return in comparison with HH, VV, and the HH-VV cross-correlation. The van Zyl algorithm [16] was reconsidered [6] and an extension of van Zyl algorithm was introduced to solve this problem. An entity RR was introduced to quantify the importance of the HV return in comparison with HH, VV, and the HH-VV cross-correlation. The van Zyl equations of [16] were expressed in terms of RR, and this led to the development of an iterative method for accurate estimation of antenna X-talk, at low and significant RR conditions, using azimuthally symmetric targets. The method was validated using PALSAR data collected over the Amazonian forest [6]. Errors higher than 10 dB can occur when RR is ignored. The iterative method corrects for these errors, and permits the demonstration of ALOS-PALSAR high isolation (better than −37 dB) [6].

The extended Freeman–van Zyl calibration method [6] includes two steps:

1. Data symmetrization using the Freeman symmetrization method [17];
2. Generation of the required calibration parameters and calibration using an extended version of van Zyl calibration method that leads to unbiased estimation of antenna X-talk for all azimuthally symmetric targets, including the ones with low HV return in comparison with HH, VV, and the HH-VV cross-correlation.

3.1. *The FREEMAN Symmetrization Method*

The "Uncalibrated" PALSAR2 data provided by JAXA were in fact calibrated for the transmit-receive antenna gain variations with incidence angle, as well as for slant range variations [2,20]. At free Faraday rotations, the following model can be used to express the voltage measurements, as a function of the illuminated target scattering matrix $[S]$ [17,21]:

$$[V] = \begin{bmatrix} 1 & \delta_1 \\ \delta_2 & F_1 \end{bmatrix}^T [S] \begin{bmatrix} 1 & \delta_3 \\ \delta_4 & F_2 \end{bmatrix} \quad (1)$$

where the measured voltage matrix $[V]$ is given by:

$$[V] = \begin{bmatrix} V_{hh} & V_{hv} \\ V_{vh} & V_{vv} \end{bmatrix} \quad (2)$$

and the actual target scattering matrix $[S]$ is given by:

$$[S] = \begin{bmatrix} S_{hh} & S_{hv} \\ S_{vh} & S_{vv} \end{bmatrix} \quad (3)$$

In Equation (1), $[.]^T$ denotes the matrix transpose; F_1 and F_2 are the channel imbalances between the H and V channels on receive and transmit, respectively. δ_3 and δ_1 are the cross-talks when a vertically polarized wave is transmitted and received, respectively. δ_4 and δ_2 are the cross-talks when a horizontally polarized wave is transmitted and received, respectively. The van Zyl calibration method assumes that the SAR system is symmetric; the transmitting and receiving distortion matrices are identical with $F_1 = F_2$, $\delta_1 = \delta_3$, and $\delta_2 = \delta_4$. Such assumptions may not be satisfied in general, and as result, the van Zyl calibration method might be of limited use in certain applications. Freeman introduced the symmetrization method to extend the use of the van Zyl algorithm to non-symmetric systems. After application of the Freeman symmetrization [17], the symmetrized measured voltage matrix $[V]_{sym}$ is related to the actual scattering matrix $[S]$ by the following equation (with $S_{hv} = S_{vh}$ under target reciprocity assumption):

$$[V]_{sym} = \begin{bmatrix} 1 & \Delta_1 \\ \Delta_2 & F \end{bmatrix}^T [S] \begin{bmatrix} 1 & \Delta_1 \\ \Delta_2 & F \end{bmatrix} \tag{4}$$

where $F = |F_2|$; and Δ_1 and Δ_2 are expressed in [17] as functions of δ_i ($i = 1,4$) and F_j ($j = 1,2$) [6]. The van Zyl algorithm can then be applied to the symmetrized system of Equation (4), as follows [16]:

1. Apply the van Zyl iterative process to derive Δ_1 and $\Delta_2' = \Delta_2/F$ using an azimuthally symmetric target;
2. Deduce F, F_1, and F_2 using the additional information provided by a trihedral.

3.2. Reconsideration of Van Zyl Algorithm

The accuracy of van Zyl's algorithm [16] has been questioned for azimuthally symmetric reference targets of low HV return in comparison with HH, VV, and their cross-correlation [6,18]. This issue was taken into account during the development of the calibration method specific for the Convair-580 SAR [19]. The Convair-580 SAR system, which uses different receiving configurations depending on the transmitted (horizontal or vertical) polarization, required a more complex calibration method than the ones developed for systems with one receiving configuration (such as the JPL AIRSAR system, for example). To quantify the residual relative error that would result if the entities $< |HH|^2 >$, $< |VV|^2 >$, and $< HHVV^* >$ were ignored during the cross-talk estimation process, the following entity RR (of Equation (23) in [19]) was introduced [6,19]:

$$\begin{aligned} RR = \; & |\Delta_1|^2 \cdot \frac{<|V_{hh}|^2>}{<|V'_{hv}|^2>} + |\Delta_2'|^2 \cdot \frac{<|V_{vv}|^2>}{<|V'_{hv}|^2>} + \\ & 2Real(\Delta_1 \Delta_2'^* \cdot \frac{<V_{hh} \cdot V_{vv}^*>}{<|V'_{hv}|^2>}) \end{aligned} \tag{5}$$

where $V'_{hv} = V_{hv} + C \cdot V_{vh}$ and C is a calibration constant [19]. The actual cross-polarized intensity mean $< |S_{hv}|^2 >$ is expressed as a function of the measured voltage and RR as:

$$< |S_{hv}|^2 > \; \simeq \; (1/F^2) \cdot < |V'_{hv}|^2 > \cdot (1 - RR) \tag{6}$$

For an azimuthally symmetric target of $< |S_{hv}|^2 >$ that is significant in comparison with $< |S_{hh}|^2 >$, $< |S_{vv}|^2 >$, $| < S_{hh} S_{hv}^* > |$, and $| < S_{vv} S_{hv}^* > |$, RR is close to zero, and the measured and actual cross-polarized intensity means are identical (modulo a multiplicative coefficient). Since the calibration of the Convair-580 X-band polarimetric SAR becomes too complex if RR is not negligible [19], only azimuthally symmetric targets with RR close to zero are used [19]. PALSAR2, like PALSAR and most of the conventional polarimetric SARs, requires the use of a calibration model much simpler than the Convair-580 SAR (whose receiver is adapted to the transmitted polarization [19]). The van Zyl model can then be retained for PALSAR2 calibration, and the van Zyl equations [16] are expressed as a function of RR, which takes into account all the required terms for unbiased estimation of antenna X-talks [6].

3.3. Assessment of Antenna Cross-Talks Using an Iterative Method

Van Zyl's Equations in Terms of RR for Azimuthally Symmetric Target

Under the assumption of an azimuthally symmetric target, the following equations are derived from [16] as a function of RR and the cross-talk estimate Δ_1 and Δ_2' [6]:

$$\begin{aligned} < V'_{hv} V_{hh}^* > \simeq \; & \Delta_1 < |V_{hh}|^2 > + \Delta_2' \cdot < V_{vv} V_{hh}^* > + \\ & 2\Delta_2'^* \cdot < |V'_{hv}|^2 > \cdot (1 - RR) \end{aligned} \tag{7}$$

$$< V'_{hv} V^*_{vv} > \simeq \Delta_1 \cdot < V_{hh} V^*_{vv} > + \Delta'_2 < |V_{vv}|^2 > + 2\Delta^*_1 \cdot < |V'_{hv}|^2 > \cdot (1 - RR) \quad (8)$$

The two unknowns Δ_1 and Δ'_2 can be determined by solving the two equations above as a function of the voltages V_{hh}, V'_{hv}, V_{vv}, and their cross-correlations measured over the azimuthally symmetrical reference target [6]. The two unknowns are determined using an analytical equation or an iterative method depending on the RR values:

1. Case 1: RR close to zero. The solutions are obtained by solving an analytical equation of 1st order. This analytical method is named Method 1;
2. Case 2: RR of significant value (not negligible): In this case, an additional equation is needed to solve for the two unknowns. An iterative method similar to the one suggested by van Zyl can be applied to solve the problem.

The iterative method is applied as follows:

1. Derive Δ_1 and Δ'_2 using Method 1, under the approximation that $RR = 0$;
2. Compute the corresponding RR by inserting the two solutions Δ_1 and Δ'_2 in Equation (5);
3. Insert RR in Equations (7) and (8) and compute the residual errors Test1 and Test2 as the difference between the left and right terms of the two equations above (7) and (8);
4. If RR is close to zero or if RR is larger than the one computed in the previous iteration, or Test1 and Test2 are too low, stop the iterative process, and retain the solutions for Δ_1 and Δ'_2;
5. If not, update Equation (6) of S_{hv} with the actual RR value and apply Method 1 again with the new S_{hv} estimate;
6. Repeat 1–3 until condition 4 is satisfied.

After the estimation of Δ_1 and Δ'_2, all the required calibration parameters Δ_1, Δ_2, F_1, and F_2 can be deduced using the additional equations provided by the corner reflector, as shown in [16].

In the following, the iterative method described above is adopted for the assessment of PALSAR2 system parameters using datasets collected over the Amazonian rainforest sites (with CRs), at free Faraday rotation conditions.

4. Assessment of PALSAR2 System Parameters at Low Faraday Rotation Conditions Using Amazonian Rainforests

4.1. Calibration Sites

The Amazonian rainforest near the geomagnetic equator, at free Faraday rotation [15], has been used as the ideal site for the assessment and calibration of L-band ALOS-PALSAR and PALSAR2 [3,5,6,22]. CRs deployed at free Faraday rotation provide accurate estimation of antenna cross-talks in contrast to the biased measurements obtained with CRs deployed at significant Faraday rotation, as shown in [6]. In 2014, five CRs were deployed by JAXA in the Amazonian rainforests [23], and used in support of polarimetric PALSAR2 calibration. Six polarimetric images collected over the Amazonian rainforests (near Rio Branco, Brazil) at various incidence angles are used in the following to assess PALSAR-2 distortion matrix for five beams (FP6-3 to FP6-7) with incidence angle varying from 25° to 40°. The CR measurements are also used to confirm the very low Faraday rotation conditions during PALSAR2 data acquisition.

4.2. Assessment of Polarimetric PALSAR2 Distortion Matrix

Figures 1–4 present the response (in range) of a CR covered with the FP6-3 polarimetric mode on the 8th of August, 2014. According to Figures 3 and 4, there is no CR (HV and VH) return at the HH and VV peak location. This shows that the PALSAR2 cross-talk is very low, and the ALOS2 acquisition took place at very low Faraday rotation conditions. The various PALSAR2 modes considered herein, the dates of the acquisition, and the incidence angle (in degree) at the CR location are indicated in Table 1. Channel imbalance (F_1 and F_2) (magnitude and phase) are also given in Table 1. The following observations can be noted:

- F_1 and F_2 vary (slightly) with incidence angle, as might be expected. The magnitude of the ratio F_1/F_2 varies between 1.021 and 1.044; which corresponds to a variation in intensity ratio within 0.2 dB;
- Channel imbalance are very stable (in time) for the same mode, as can be noted for the 2 acquisitions in FP6-7 (last rows of Table 1);
- Channel imbalance phase varies with incidence angle, as might be expected. However, it remains stable (in time) for the same mode at different acquisitions, as can be noted for the two acquisitions in FP6-7.

The extended Freeman–van Zyl iterative method is applied for the generation of PALSAR2 antenna cross-talks. Δ_1 and Δ_2 are given (in dB) in Table 2. The residual error RR is computed using Equation (5). The final error RRf computed at the last iteration corresponds to the residual error. All the cross-talks given in the table were obtained with a very low residual error (RRf lower than -43 dB).

Table 1. PALSAR2 channel imbalance (magnitude and phase (in degree)).

Mode	Date	CR Inc	$\|f_1\|$	ϕ_{f1}	$\|f_2\|$	ϕ_{f2}
FP6-3	8−08	28	1.06	−1.1	1.26	−28.0
FP6-4	8−22	31	1.06	−22.1	1.01	−25.6
FP6-5	9−05	35	1.03	−5.4	1.01	−28.2
FP6-6	9−19	37	1.06	−23.6	1.04	−27.4
FP6-7	8−13	39	1.05	−3.3	1.02	−25.7
FP6-7	8−27	39	1.04	−3.7	1.01	−25.7

Table 2. PALSAR2 Antenna Cross-Talks (in dB) and Faraday Rotation (in degree).

Mode	Date	Δ_1	Δ_2	Ω
FP6-3	8−08	−40	−45	−0.12
FP6-4	8−22	−44	−41	−0.19
FP6-5	9−05	−41	−41	−0.24
FP6-6	9−19	−45	−42	−0.15
FP6-7	8−13	−40	−40	−0.17
FP6-7	8−27	−40	−44	−0.16

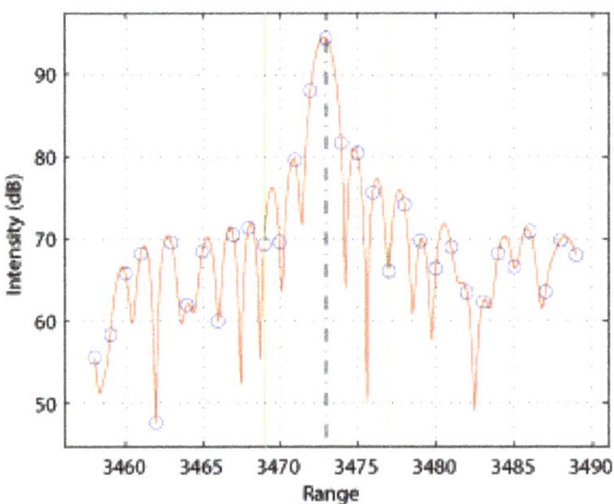

Figure 1. Corner reflector reflector HH response (in range expressed in pixel numbers).

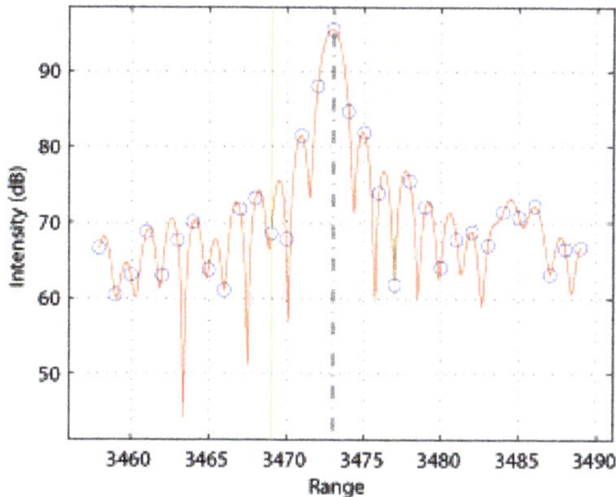

Figure 2. Corner reflector VV response (in range).

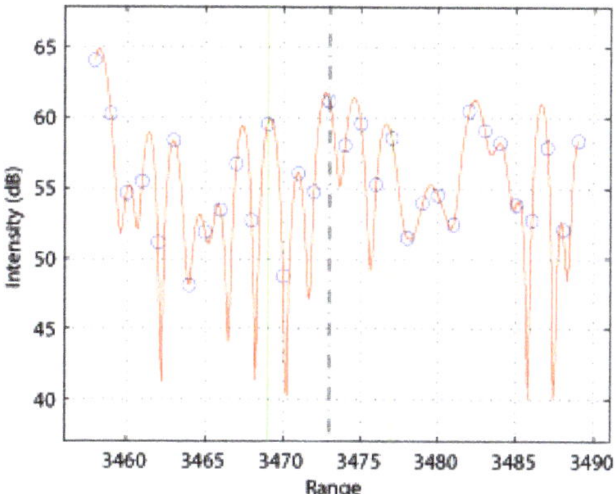

Figure 3. Corner reflector HV response (in range).

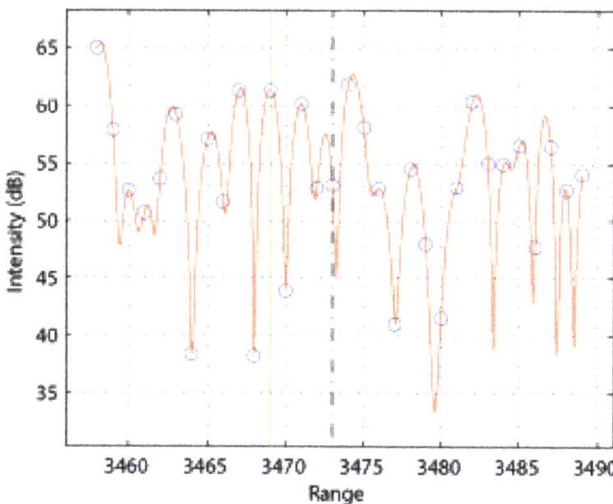

Figure 4. Corner reflector VH response (in range).

As can be noted in Table 2, the PALSAR2 antenna is highly isolated with cross-talk lower than −40 dB. The CR-measurements permit demonstrating the very high isolation of the PALSAR2 antenna: cross-talks lower than −40 dB [24]. Similar results were obtained in [3,22,25] using different calibration approaches.

4.3. Measurement of the Actual Faraday Rotation during PALSAR2 Acquisition

Many studies have shown that Faraday rotation is low near the equator [15,26]. These results are confirmed in the following for the five PALSAR2 images collected over the JAXA sites of calibration in the Amazonian rainforests. Several methods can be used to measure the Faraday rotation [26–28]. We have adopted herein the Bickel and Bates method [29] which calculates the Faraday rotation at the circular basis as follows:

$$\begin{bmatrix} Z_{11} & Z_{12} \\ Z_{21} & Z_{22} \end{bmatrix}^T = \begin{bmatrix} 1 & j \\ j & 1 \end{bmatrix}^T [S]_\Omega \begin{bmatrix} 1 & j \\ j & 1 \end{bmatrix} \tag{9}$$

where $[S]_\Omega$ is given as a function of the Faraday rotation angle Ω and the scattering matrix [S] of Equation (3) by:

$$[S]_\Omega = \begin{bmatrix} \cos\Omega & -\sin\Omega \\ \sin\Omega & \cos\Omega \end{bmatrix} [S] \begin{bmatrix} \cos\Omega & -\sin\Omega \\ \sin\Omega & \cos\Omega \end{bmatrix} \tag{10}$$

Ω is given by [27,29]:

$$\Omega = < \frac{1}{4} \cdot arg(Z_{12} Z_{21}^*) > \tag{11}$$

where $arg(Z)$ is the argument of the complex Z, and $<>$ denotes spatial averaging.

Equation (11) is used to measure the Faraday rotation for the five PALSAR images. The measurements obtained at the CRs are given in Table 2. As expected, the Faraday rotation angle is very low (lower than 0.24°) during the 5 PALSAR2 acquisitions. This confirms the accuracy of the PALSAR2 antenna distortion matrix measurements obtained above, under the assumption of negligible Faraday rotation.

5. Impact of Significant Faraday Rotation on PALSAR2 Polarimetric Distortion Matrix Measurement

The model of Equation (1) can be extended to take into account the Faraday rotation. Channel imbalances can also be separated from the polarimetric distortion matrices as done in [21]. This leads to the following equation [21,30]:

$$[V] = \begin{bmatrix} 1 & 0 \\ 0 & F_1 \end{bmatrix} \begin{bmatrix} 1 & \delta_2 \\ \delta_1' & 1 \end{bmatrix} [S]_\Omega \begin{bmatrix} 1 & \delta_3' \\ \delta_4 & 1 \end{bmatrix} \begin{bmatrix} 1 & 0 \\ 0 & F_2 \end{bmatrix} \tag{12}$$

with $\delta_1' = \delta_1/F_1$, $\delta_3' = \delta_3/F_2$, and $[S]_\Omega$ is given by Equation (10) as a function of the scattering matrix $[S]$ and the Faraday rotation angle Ω. It is worth noting that Equation (12), which is an extension of the van Zyl calibration model that takes into account the Faraday rotation, is expressed as a function of the conventional scattering matrix $[S]$ in contrast to the Freeman model [30] expressed as a function of the transposes of $[S]$ and $[V]$ matrices.

The CR response combined with the channel imbalance phase difference, which can be derived using the Zebker method [31], permits the measurement of the channel imbalance F_1 and F_2 ([6,17,31]). After the channel imbalance correction, the measured voltages can be expressed (using Equation (12)) as a function of the antenna cross-talks and $[S]_\Omega$ matrix elements in the following equation:

$$\begin{bmatrix} V_{hh} \\ V_{hv} \\ V_{vh} \\ V_{vv} \end{bmatrix} = \begin{bmatrix} S_{hh}^\Omega + \delta_4 S_{hv}^\Omega + \delta_2 S_{vh}^\Omega \\ \delta_3' S_{hh}^\Omega + S_{hv}^\Omega + \delta_2 S_{vv}^\Omega \\ \delta_1' S_{hh}^\Omega + S_{hv}^\Omega + \delta_4 S_{vv}^\Omega \\ \delta_1' S_{hv}^\Omega + \delta_3' S_{vh}^\Omega + S_{vv}^\Omega \end{bmatrix} \tag{13}$$

where

$$\begin{bmatrix} S_{hh}^{\Omega} \\ S_{hv}^{\Omega} \\ S_{vh}^{\Omega} \\ S_{vv}^{\Omega} \end{bmatrix} = \begin{bmatrix} \cos^2\Omega \cdot S_{hh} - \sin^2\Omega \cdot S_{vv} \\ S_{hv} - \sin\Omega\cos\Omega \cdot (S_{hh} + S_{vv}) \\ S_{hv} + \sin\Omega\cos\Omega \cdot (S_{hh} + S_{vv}) \\ -\sin^2\Omega \cdot S_{hh} + \cos^2\Omega \cdot S_{vv} \end{bmatrix} \tag{14}$$

For a CR, $S_{hh} = S_{vv} = K$ and $S_{hv} = S_{vh} = 0$, and Equation (13) can be used to derive the CR-measured voltage vector as a function of the Faraday rotation angle Ω and the antenna cross-talks $(\delta_1', \delta_2, \delta_3', \delta_4)$:

$$\begin{bmatrix} V_{hh}^{CR} \\ V_{hv}^{CR} \\ V_{vh}^{CR} \\ V_{vv}^{CR} \end{bmatrix} = K \begin{bmatrix} \cos 2\Omega + \sin 2\Omega(\delta_2 - \delta_4) \\ \cos 2\Omega(\delta_2 + \delta_3') - \sin 2\Omega \\ \cos 2\Omega(\delta_1' + \delta_4) + \sin 2\Omega \\ \cos 2\Omega - \sin 2\Omega(\delta_1' - \delta_3') \end{bmatrix} \tag{15}$$

In 2014, several PALSAR2 images were collected over the CCRS calibration site in Ottawa. The PALSAR2 image collected on the 9th of September is used herein. Figures 5–8 present the polarimetric response of the CCRS (2.5 m CR) at HH, VV, HV, and VH. In contrast to the low return at cross-pol of the JAXA CR deployed in the Amazonian rainforests (Figures 3 and 4), the Ottawa CR presents a significant return at HV and VH polarization as seen in Figures 7 and 8.

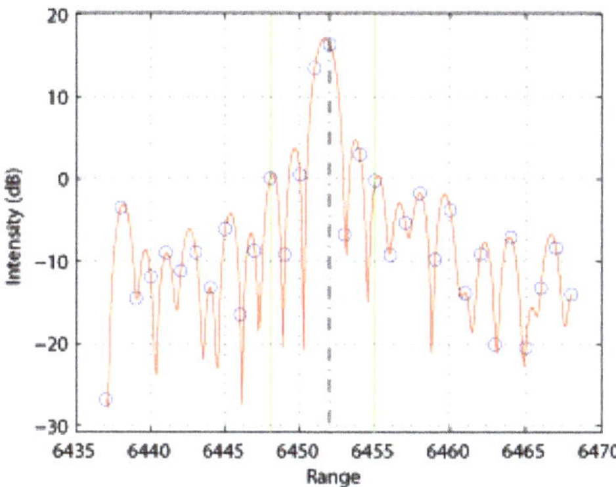

Figure 5. Corner reflector HH response (in range expressed in pixel numbers).

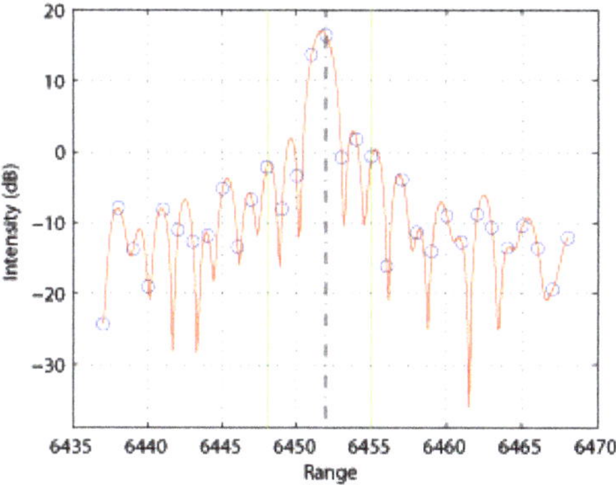

Figure 6. Corner reflector VV response (Ottawa).

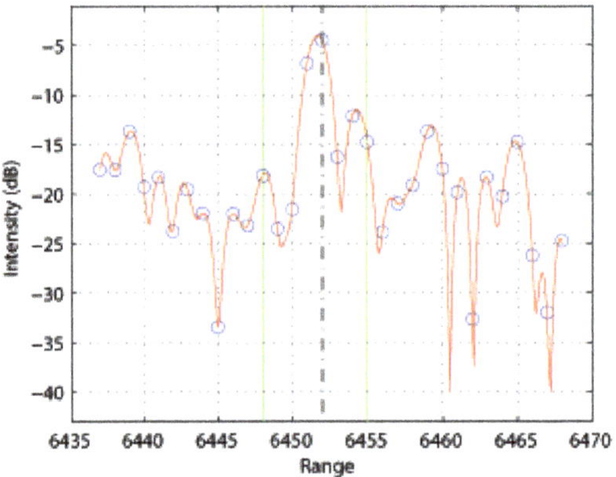

Figure 7. Corner reflector HV response (Ottawa).

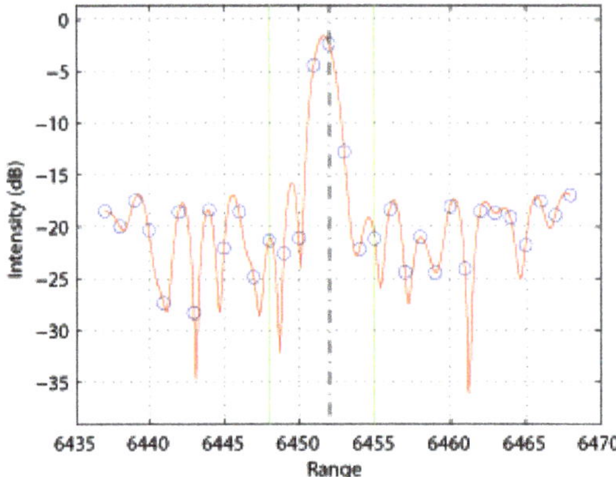

Figure 8. Corner reflector VH response (Ottawa).

The results obtained in the previous section show that the PALSAR2 antenna is highly isolated with cross-talks lower than −40 dB. The analysis of the CR voltage Equation (15) with negligible cross-talks (cross-talks lower than −40 dB) leads to the following conclusions:

1. The ratios of the cross-like polarization voltage, $V_{hv}^{CR}/V_{hh}^{CR} = -20.03$ dB and $V_{vh}^{CR}/V_{vv}^{CR} = -19.95$ dB do not correspond to the actual antenna cross-talk contamination of the cross-pol (HV and VH) by the like polarization (HH and VV), as could be misinterpreted if the Faraday rotation contamination is ignored ([7]). In fact, these cross-like polarization voltage ratios lead to an estimation of the Faraday rotation angle Ω for a highly isolated antenna, as can be shown using Equation (15):

$$\begin{bmatrix} \frac{V_{hv}^{CR}}{V_{hh}^{CR}} \\ \frac{V_{vh}^{CR}}{V_{vv}^{CR}} \end{bmatrix} = K \cdot \begin{bmatrix} -\tan 2\Omega \\ \tan 2\Omega \end{bmatrix} \tag{16}$$

   Equation (16) is used to estimate the Faraday rotation during PALSAR2 acquisitions. The results obtained are similar (within 0.2) with the ones obtained with the Bickel and Bates method (Equation (11)); $\Omega = 2.8°$ with V_{hv}^{CR}/V_{hh}^{CR}, $\Omega = 2.9°$ with V_{vh}^{CR}/V_{vv}^{CR} in comparison with $\Omega = 3.1°$ obtained over a forested area using Equation (11).

2. The sum of the CR cross-pol voltages cancels the Faraday rotation contamination.

This can be shown using the following equation derived from (15):

$$(V_{hv}^{CR} + V_{vh}^{CR})/K = (\delta_1' + \delta_2 + \delta_3' + \delta_4)\cos 2\Omega \quad (17)$$

Since the cross-talks are negligible, $(V_{hv}^{CR} + V_{vh}^{CR})$ should be close to zero. Figure 9 presents the CR response of the averaged PALSAR2 image $(hv + vh)/2$. The significant peaks that occur at HV and VH images of Figures 6 and 7 vanish in Figure 9. At the HH and VV peak location, the intensity of the averaged cross-pol return vanishes in the CR surrounding clutter of radar backscattering (−18.23 dB) which is much lower than HH and VV retro-diffusion (about 15 dB).

In summary, the analytical presentation of the CR voltage as a function of the antenna cross-talks and Faraday rotation leads to the conclusion that the HV and VH contaminations, which can be measured from SAR images collected with a highly isolated antenna, are mainly due to the actual Faraday rotation. In the case that the antenna is not highly isolated, Equation (13) obtained after the channel imbalance correction (using the CR) shows that additional reference point targets (of scattering response different from the CR) should be used for the measurement of the 5 unknowns; the 4 antenna cross-talks and the Faraday rotation.

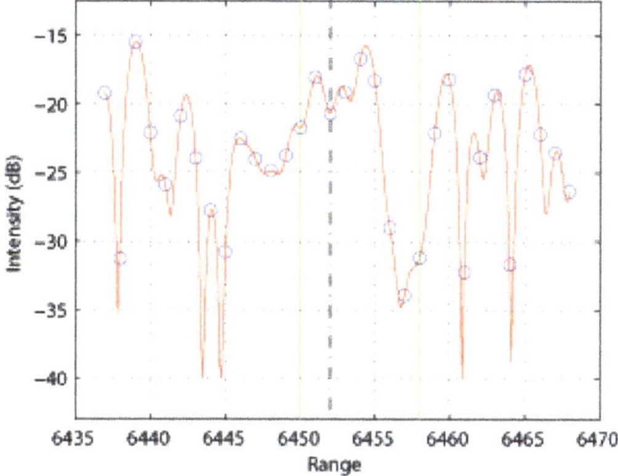

Figure 9. Corner reflector (HV + VH)/2 response (Ottawa).

6. Conclusions

In summary, the PALSAR antenna is highly isolated. Both the antenna subarray and T/R are highly isolated, and as a result, the global antenna is highly isolated for all the modes considered. The excellent performances of the polarimetric PALSAR2 in terms of NESZ (better than −37 dB ([1,3]) combined with the high antenna isolation permit the demonstration of the unique L-band long-penetration SAR capabilities at low incidence angles (25°) for subsurface peatland hydrology monitoring and discontinuous permafrost mapping [32,33]. The high isolation of the polarimetric ALOS2-PALSAR2 permits a simplification of the assessment and calibration of the ALOS2-Compact experimental mode, as discussed in [34].

Author Contributions: Concepts and Methodology, R.T. and M.S.; ALOS2-PALSAR2 raw data processing and Calibration, M.S. and T.M.; writing—original draft preparation, R.T. and M.S.; writing—review and editing, R.T., M.S. and T.M. All authors have read and agreed to the published version of the manuscript.

Funding: This research received no external funding.

Data Availability Statement: The ALOS-2 original data are provided by JAXA under ALOS-2 Calibration Validation and Science Team (CVST) and the Research Announcement on the Earth Observations (EO-RA).

Acknowledgments: The authors would like to thank the anonymous reviewers for their helpful comments and suggestions. We would like to express our special thanks to JAXA for data observation and provisions. S. Nedelcu from CCRS is thanked for his help in PALSAR2 data processing.

Conflicts of Interest: The authors declare no conflict of interest.

References

1. Okada, Y.; Hamasaki, T.; Tsuji, M.; Iwamoto, M.; Hariu, K.; Kankaku, Y.; Suzuki, S.; Osawa, Y. Hardware performance of L-band SAR system onboard ALOS2. In Proceedings of the IGARSS 2011, Vancouver, BC, Canada, 24–29 July 2011.
2. Shimada, M. PALSAR-2 beam calibration. In Proceedings of the ALOS2 Cal-Val Meeting CVST2, Tsukuba, Japan, 25–26 November 2013.
3. Shimada, M. *Imaging from Spaceborne and Airborne SARs, Calibration, and Applications*; Taylor and Francis, CRC Press: Boca Raton, FL, USA, 2018.
4. Ito, N.; Hamazaki, T.; Tomioka, K. ALOS/PALSAR chracteristics and status. In Proceedings of the CEOS SAR Workshop Proceedings, Tokyo, Japan, 2–5 April 2001; pp. 191–194.
5. Shimada, M.; Isoguchi, O.; Tadono, T.; Isono, K. PALSAR Radiometric and Geometric Calibration. *IEEE Trans. Geosci. Rem. Sens.* **2009**, *47*, 3951–3959. [CrossRef]
6. Touzi, R.; Shimada, M. polarimetric PALSAR Calibration. *IEEE Trans. Geosci. Rem. Sens.* **2009**, *47*, 3951–3959. [CrossRef]
7. Papathanassiou, K.P.; Zandona-Schneider, R. ALOS PALSAR- PolInSAR data analysis: First results. In Proceedings of the Pol-InSAR 2007, Frascatti, Italy, 22–26 January 2007.
8. Cloude, S. PALSAR- PolInSAR calibration results for Australian test sites polarimetric evaluation. In Proceedings of the ALOS Cal-Val Meeting CVST5, Tokyo, Japan, 19 October 2006.
9. Touzi, R.; Shimada, M.; Nedelcu, S.; Hawkins, R.K.; Murnagham, K. Calibration of PALSAR polarimetric modes: SAR system model assessment using the Amazonian and Ottawa calibration site data. In Proceedings of the ALOS Cal-Val Meeting CVST5, Tokyo, Japan, 19 October 2006.
10. Touzi, R.; Shimada, M. Assessment and Calibration of polarimetric PALSAR. In Proceedings of the PolinSAR'09 Workshop Proceedings, Frascatti, Italy, 26–30 January 2009.
11. Papathanassiou, K.P.; Zandona-Schneider, R. ALOS PALSAR- PolInSAR data analysis: First results. In Proceedings of the ALOS Cal-Val Meeting CVST5, Tokyo, Japan, 19 October 2006.
12. Touzi, R.; Vachon, P.W.; Wolfe, J. Requirement on antenna isolation for operational use of future C-band SAR Constellations missions in maritime surveillance. *IEEE Trans. Geosci. Rem. Sens. Lett.* **2010**, *7*, 861–865. [CrossRef]
13. Shimada, M.; Itoh, T.; Motooka, T.; Watanabe, M.; Shiraishi, T.; Thapa, R.; Lucas, R. New global forest/non-forest maps from ALOS PALSAR data (2007–2010). *Remote. Sens. Environ.* **2014**, *155*, 13–31. [CrossRef]
14. Rosenqvist, A.; Shimada, M.; Suzuki, S.; Ohgushi, F.; Tadono, T.; Watanabe, M.; Tsuzuku, K.; Watanabe, T.; Kamijo, S.; Aoki, E. Operational performance of the ALOS global systematic acquisition strategy and observation plans for ALOS-2 PALSAR-2. *Remote. Sens. Environ.* **2014**, *155*, 3–12. [CrossRef]
15. Wright, P.; Quegan, S.; Wheadon, N.; Hall, D. Faraday rotation effects on L-band spaceborne SAR data. *IEEE Trans. Geosci. Rem. Sens.* **2003**, *41*, 2735–2744. [CrossRef]
16. Zyl, J.J.V. Calibration of polarimetric radar images using only image parameters and trihedral corner reflectors responses. *IEEE Trans. Geosci. Rem. Sens.* **1990**, *28*, 337–348.
17. Freeman, A.; Zyl, J.J.V.; Klein, J.D.; Zebker, H.A.; Shen, Y. Calibration of Stokes and scattering matrix format polarimetric SAR data. *IEEE Trans. Geosci. Rem. Sens.* **1992**, *30*, 531–539. [CrossRef]
18. Quegan, S.; Rhodes, I. Statistical models for polarimetric data: Consequences, testing and validity. *Int. J. Rem. Sens.* **1995**, *16*, 1183–1210. [CrossRef]
19. Touzi, R.; Livingstone, C.E.; Lafontaine, J.R.C.; Lukowski, T.I. Consideration of antenna gain and phase patterns for calibration of polarimetric SAR data. *IEEE Trans. Geosci. Rem. Sens.* **1993**, *31*, 1132–1145. [CrossRef]
20. Shimada, M. PALSAR-2 initial calibration results. In Proceedings of the ALOS2 Cal-Val Meeting CVST4, Tsukuba, Japan, 20 November 2014.
21. Zyl, J.J.V.; Zebker, H.A. Imaging Radar Polarimetry. In *Progress in Electromagnetics Research, PIER 3*; Kong, J.A., Ed.; Elsevier: New York, NY, USA, 1990; Chapter 5.
22. Shimada, M.; Watanabe, M.; Motooka, T.; Kankaku, Y.; Suzuki, S. Calibration and validation of PALSAR2. In Proceedings of the 2015 International Geoscience and Remote Sensing Symposium (IGARSS'15), Milano, Italy, 26–31 July 2015.
23. Sano, E.E. CALVAL of ALOS-2 multi-polarized and time series for monitoring deforestation in the Brazilian Amazon. In Proceedings of the ALOS2 Cal-Val Meeting CVST3, Tsukuba, Japan, 25–26 October 2014.
24. Touzi, R.; Shimada, M. Assessment and Calibration of polarimetric PALSAR-2. In Proceedings of the 2015 International Geoscience and Remote Sensing Symposium (IGARSS'15), Milano, Italy, 26–31 July 2015.

25. Moriyama, T. Polarimetric calibration of PALSAR2. In Proceedings of the 2015 International Geoscience and Remote Sensing Symposium (IGARSS'15), Milano, Italy, 26–31 July 2015.
26. Meyer, F.J.; Nicoll, J.B. Prediction, detection and correction of Faraday rotation in full-polarimetric L-band SAR data. *IEEE Trans. Geosci. Rem. Sens.* **2008**, *46*, 3076–3086. [CrossRef]
27. Freeman, A. SIR-C/X data quality and calibration results. *IEEE Trans. Geosci. Rem. Sens.* **1995**, *33*, 848–857. [CrossRef]
28. Sandberg, G.; Eriksson, L.E.B.; Ulander, L.M.H. Measurement of Farady rotation using polarimetric PALSAR images. *IEEE Geosci. Rem. Sens. Lett.* **2009**, *6*, 142–146. [CrossRef]
29. Bickel, S.H.; Bates, R.H. Effects of magneto-ionic on the propagation of the polarization scattering matrix. *Proc. IRE* **1965**, *53*, 1089–1091. [CrossRef]
30. Freeman, A. Calibration of linearly polarized SAR data subject to Faraday rotation. *IEEE Trans. Geosci. Rem. Sens.* **2004**, *in press*. [CrossRef]
31. Zebker, H.A.; Lou, Y. Phase calibration of imaging radar polarimeter stokes matrices. *IEEE Trans. Geosci. Rem. Sens.* **1990**, *28*, 246–252. [CrossRef]
32. Touzi, R.; Omari, K.; Sleep, B.; Jiao, X. Scattered and received wave polarization optimization for enhanced peatland classification and fire damage assessment using polarimetric PALSAR. *IEEE J. Sel. Top. Appl. Earth Obs. Remote. Sens.* **2018**, *11*, 4452–4477. [CrossRef]
33. Touzi, R.; Pawley, S.; Hosseini, M.; Jiao, X. Polarimetric L-band PALSAR2 for Discontinuous Permafrost Mapping In Peatland Regions. In Proceedings of the 2015 International Geoscience and Remote Sensing Symposium (IGARSS'19), Yokohama, Japan, 28 July–2 August 2019.
34. Touzi, R.; Shimada, M.; Motohka, T.; Nedelcu, S. Assessment of ALOS-2 Compact non circularity using the Amazonian rainforests. *IEEE Trans. Geosci. Rem. Sens.* **2020**, *58*, 7472–7482. [CrossRef]

Article

Performance Study of Landslide Detection Using Multi-Temporal SAR Images

Yunung Nina Lin [1,*], Yi-Ching Chen [1], Yu-Ting Kuo [2] and Wei-An Chao [3]

[1] Institute of Earth Sciences, Academia Sinica, Taipei 11529, Taiwan; yiching@earth.sinica.edu.tw
[2] Department of Earth and Environmental Sciences, National Chung Cheng University, Chiayi 62102, Taiwan; yutingkuo@ccu.edu.tw
[3] Department of Civil Engineering, National Yang Ming Chiao Tung University, Hsinchu 30010, Taiwan; vvnchao@nycu.edu.tw
* Correspondence: ninalin@earth.sinica.edu.tw

Abstract: This study addresses one of the most commonly-asked questions in synthetic aperture radar (SAR)-based landslide detection: How the choice of datatypes affects the detection performance. In two examples, the 2018 Hokkaido landslides in Japan and the 2017 Putanpunas landslide in Taiwan, we utilize the Growing Split-Based Approach to obtain Bayesian probability maps for such a performance evaluation. Our result shows that the high-resolution, full-polarimetric data offers superior detection capability for landslides in forest areas, followed by single-polarimetric datasets of high spatial resolutions at various radar wavelengths. The medium-resolution single-polarimetric data have comparable performance if the landslide occupies a large area and occurs on bare surfaces, but the detection capability decays significantly for small landslides in forest areas. Our result also indicates that large local incidence angles may not necessarily hinder landslide detection, while areas of small local incidence angles may coincide with layover zones, making the data unusable for detection. The best area under curve value among all datatypes is 0.77, suggesting that the performance of SAR-based landslide detection is limited. The limitation may result from radar wave's sensitivity to multiple physical factors, including changes in land cover types, local topography, surface roughness and soil moistures.

Keywords: SAR-based landslide detection; Growing Split-Based Approach (GSBA); Hokkaido landslide; Putanpunas landslide; SAR polarimetry; model-free 3-component decomposition for full polarimetric data (MF3CF)

Citation: Lin, Y.N.; Chen, Y.-C.; Kuo, Y.-T.; Chao, W.-A. Performance Study of Landslide Detection Using Multi-Temporal SAR Images. *Remote Sens.* **2022**, *14*, 2444. https://doi.org/10.3390/rs14102444

Academic Editors: Takeo Tadono and Masato Ohki

Received: 31 March 2022
Accepted: 17 May 2022
Published: 19 May 2022

1. Introduction

According to the Global Fatal Landslide Database, Asia suffers the greatest impact of fatal landslides among all continents [1,2]. This fact has to do with the physiographical environment of this region, including active tectonics, frequent typhoons/tropical cyclones, as well as socioeconomic factors, such as rapid economic growth, human population increase, habitat expansion and even loose regulations [1,3]. Among all questions associated with landslides, where and when they occur and how big they are remain the first information people demand to know. From a rapid response perspective, landslide sizes and locations are the key information needed by the ground crews to ensure the safety of human lives and the transportation of aids and supplies. From a policy-making perspective, long-term spatiotemporal evolution of landslide hotspots impacts the formulation of management strategies and mitigation plans [4]. From a scientific perspective, landslide volumes and frequencies may provide information about rock strength and denudation rates [5,6], the influence of climate changes [2], and the interaction among the lithosphere, hydrosphere and biosphere [7,8].

Given the large-area imaging capability from the sky, remote sensing has been the most widely-used tool in landslide detection during the past decades. For example, based on

Formosat-2 satellite images, Lin et al. (2017) found that about 70% of the mountainous area in Taiwan has experienced at least 1 landslide within the decade between 2003 and 2012 [4]. In Japan, the National Research Institute for Earth Science and Disaster Prevention (NIED) also conducts regular landslide mappings based on aerial photos (https://www.bosai.go.jp/e/research/database/earth-and-sand.html, accessed on 10 March 2022; a study using the national landslide database can be found in [8]). The optical image-based landslide mapping has high quality and accuracy, although the availability usually depends on the weather condition and the source of sun lights after a landslide occurs. In some cases, the latency between the event and the first usable image can be weeks. This is where the synthetic aperture radar (SAR)-based mapping may serve as an alternative solution particularly during the early phase of the hazards. Some recent examples include the earthquake-triggered landslides after the 1999 Chi-Chi earthquake in Taiwan [9], the 2015 Gorkha earthquake in Nepal [10,11], the 2015 Mt. Kinabalu earthquake in Malaysia [12], the 2018 Lombok earthquakes in Indonesia [10], and the 2018 Hokkaido Eastern Iburi earthquake in Japan [10,13–17]. Examples of rainfall-triggered landslides include the 2009 Typhoon Morakot in Taiwan [18], the 2011 Typhoon Talas in Honshu, Japan [19], the 2015 heavy rain in Chin State, Myanmar [20], and the 2017 heavy rain in Kyushu, Japan [16].

In general, SAR-based change detection can be classified into two categories–the coherent change detection (CCD) and incoherent change detection (ICD)–depending on whether interferometric phase is used [15]. Speaking of change detection for landslides, CCD includes the pre-failure landslide monitoring by using interferometric phase time-series, as well as the post-failure landslide detection by comparing the change of the interferometric phase qualities (interferometric coherence). In the post-failure landslide detection, a landslide patch is usually depicted by low coherence due to the changes in surface geometry, roughness and dielectric properties. It works particularly well in regions with intermediate-to-high pre-event coherence [10,12,15]. However, in places where phase decorrelation occurs constantly, CCD may fail to provide accurate landslide information. One such a place is the forest, where volumetric decorrelation and temporal decorrelation prevail due to frequent changes in vegetation structures and dielectric properties, as well as the disturbance by winds, rains, water vapors and other atmospheric conditions [21]. It has been shown that coherence-based landslide detection methods may yield unreliable results in the forest areas where the pre-event coherence is too low [10,15].

On the other hand, ICD compares SAR backscattering amplitudes or intensities before and after the landslide event. Similar to interferometric coherence, changes in intensity are also associated with variations in surface geometry, roughness and dielectric properties. Backscattering intensities are, however, less sensitive to atmospheric conditions, and they also appear in a wider range of values that allow a better separation of major changes (such as from vegetation to bare surface) from minor variations (such as tree growth). In addition, most ICD methods are relatively easy to implement as they only involve image-wise calculations (except for the intensity correlation method in [15,19,20]). ICD has therefore been adopted more widely for landslide detection than CCD, especially for forest areas [13,15,17,19,22].

Single-polarimetric (single-pol) SAR data is by far the most commonly-used datatype in ICD. Having said that, multi-polarimetric (multi-pol) datasets and their decomposition parameters can also be used in ICD. SAR polarimetry, or PolSAR, takes into account of both the intensities and phases from the same image epoch acquired at different transmitting-receiving polarizations (HH, HV, VV and VH). Polarimetric decomposition then recombines the complex scattering coefficients to extract parameters that can directly infer physical properties of the scatterers. These decomposition parameters, even from a single image epoch, have been proved to be efficient in differentiating land cover types including landslides [9,16,23,24]. Some studies also use PolSAR datasets and decomposition parameters in the dual-temporal (1 pre-event and 1 post-event image) change detection to improve the result accuracy and to gain physical insights [16,18,25]. In comparison, PolSAR's detection capability in a multi-temporal context ($\geq$2 pre-event images and 1 post-event image) has not received much attention yet and deserves more exploration.

This study attempts to incorporate the multi-temporal PolSAR decomposition parameters into SAR-based landslide detection in forest areas, and compares the performance with those from single-pol datasets. This comparison is out of a practical consideration: Given more and more public and private players in SAR space missions, we expect the list of SAR sensors to increase and the overpass latency to shorten substantially in the near future. Soon SAR-based landslide detection may be not so much limited by data availability, and hence our knowledge about how different data properties affect the detection performance will be pivotal. Such knowledge will help researchers and government agencies make decisions before choosing the best dataset for landslide mapping. It will also provide information about the uncertainties and limitations especially for a rapid-response product.

The goal of this study is therefore to evaluate the influence of different SAR data properties on landslide detection, including radar wavelengths, spatial resolutions, polarizations and viewing geometry. Different datatypes, or combinations of the properties listed above, are produced from some of the most commonly-used sensors (the L-band ALOS-2, C-band Sentinel-1 and X-band COSMO-SkyMed) to facilitate this performance study. To unify the comparison basis, we carry out change detection using a newly-designed algorithm Growing Split-Based Approach (GSBA). GSBA takes in a SAR-derived value, either a backscattering intensity or decomposition parameter, normalized over its time-series variations and computes the Bayesian probability of landslides given that value. We compare the detection performance in two landslide cases, the earthquake-triggered landslides due to the 2018 Hokkaido Eastern Iburi Earthquake in Japan, and the rainfall-triggered landslide caused by the 2017 heavy rain in the catchment of Putanpunas River, southern Taiwan. With results from these two cases, we discuss how different data properties affect the landslide detection, and what may limit the detection efficacy.

2. SAR Data Processing

To achieve our objectives, we need to produce multiple datatypes from the same SAR data (see Table 1 for the list of SAR data used in this study). Next we describe the processing flows for two major datatype categories and the generation of Z-score maps, which are the input to the GSBA algorithm.

Table 1. List of SAR data used in this study.

Sensor & Track [*1]	Pre-Event Epochs	Post-Event Epoch	Average Look Angle (θ)	Mode and Resolution [*2]	Wavelength	Polarization [*3]
Hokkaido Landslides (Japan), 2018-09-06, earthquake-triggered						
ALOS-2 A122	2018-08-25 2017-08-26 2016-08-27	2018-09-08	30°	High-Sensitive 6 m (HR)	L-band 22.9 cm	Full-pol HH, HV, VV, VH
S-1 A68	2018-09-01 2018-08-20 2018 08 08	2018-09-13	39°	Interferometric Wide 15 m (MR)	C-band 5.6 cm	Dual-pol VV, VH
CSK A	2018-06-04 2017-07-16	2018-09-08	37°	StripMap 3 m (UHR)	X-band 3.1 cm	Single-pol HH
Putanpunas Landslide (southern Taiwan), 2017-06-07, rainfall-triggered						
ALOS-2 A137	2016-12-22 2016-08-18 2016-06-09 2016-04-14 2016-03-03	2017-08-03	33°	ScanSAR 60 m (LR)	L-band 22.9 cm	Dual-pol HH, HV (HV-mode is missing on the post-event epoch)
ALOS-2 D27	2017-05-21 2017-04-23 2017-01-01 2016-12-04 2016-10-09	2017-07-02	44°	ScanSAR 60 m (LR)	L-band 22.9 cm	Dual-pol HH, HV

Table 1. *Cont.*

Sensor & Track *1	Pre-Event Epochs	Post-Event Epoch	Average Look Angle (θ)	Mode and Resolution *2	Wavelength	Polarization *3
S-1 A69	2017-05-27 2017-05-15 2017-05-03 2017-04-21 2017-04-09	2017-06-08	35°	Interferometric Wide 15 m (MR)	C-band 5.6 cm	Dual-pol VV, VH
S-1 D105	2017-05-29 2017-05-17 2017-05-05 2017-04-23 2017-04-11	2017-06-10	38°	Interferometric Wide 15 m (MR)	C-band 5.6 cm	Dual-pol VV,VH
CSK D	2017-06-01 2017-05-24 2017-05-08 2017-04-22 2017-04-14	2017-06-09	27°	StripMap 3 m (UHR)	X-band 3.1 cm	Single-pol HH

*1 S-1 = Sentinel-1; CSK = COSMO-SkyMed; A = ascending; D = descending. *2 UHR = ultra-high resolution; HR = high resolution; MR = medium resolution; LR = low resolution. *3 Full-pol = full-polarimetric; Dual-pol = dual-polarimetric; Single-pol = single-polarimetric. H = horizontally polarized; V = vertically polarized. The first letter in the combination stands for the polarization of the transmitted wave, and the second is for the received wave.

2.1. Single-Polarization: Backscattering Coefficient (σ^0)

Backscattering coefficient σ^0 is the normalized measure of the radar signal's strength reflected by a distributed target. From the single-look complex (SLC) stack, the general processing steps to obtain σ^0 include radiometric calibration [26], speckle noise attenuation [27,28], multilooking and geocoding (Figure 1). For Sentinel-1 (S-1) data, thermal noise removal is carried out concurrently with radiometric calibration before converting the digital numbers to σ^0 [29]. We only process the co-polarized σ^0 stacks (HH or VV) for their higher sensitivity to surface-related scattering [30]. All data processing is carried out using the graph processing tool (gpt) in SeNtinel Application Platform (SNAP) and the InSAR Scientific Computing Environment (ISCE, for ALOS-2 ScanSAR data only) built in a high-performance computing cluster. Two auxiliary datasets, local incidence angle (LIA) and layover-shadow mask, are also generated during data processing [31]. LIA is the angle between the radar incidence direction and the slope normal vector. Small LIAs indicate slopes facing the satellite, while large LIAs indicate either slopes facing the satellite but significantly deviating from the line-of-sight (LOS) direction, or slopes facing away from the satellite.

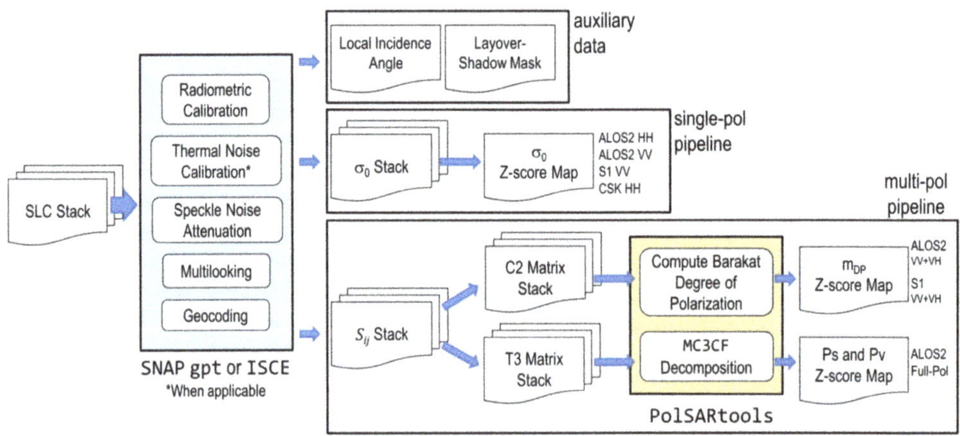

Figure 1. Processing flows for different SAR datatypes used in this study.

2.2. Multi-Polarization: Degree of Polarization (m_{DP}) and Scattering Powers

For a dual-polarimetric SAR data, we follow the same steps as those in the σ^0 processing flow to generate geocoded complex scattering coefficients S_{ij} $(i,\ j = H \text{ or } V)$ (Figure 1). We then construct the 2×2 covariance matrix $\mathbf{C_2}$ [32],

$$\mathbf{C_2} = \langle \mathbf{k} \cdot \mathbf{k}^{*T} \rangle = \begin{bmatrix} C_{11} & C_{12} \\ C_{21} & C_{22} \end{bmatrix} = \begin{bmatrix} \langle |S_{ii}|^2 \rangle & \langle S_{ii} S_{ij}^* \rangle \\ \langle S_{ij} S_{ii}^* \rangle & \langle |S_{ij}|^2 \rangle \end{bmatrix} \tag{1}$$

where $\langle \cdot \rangle$ indicates spatial ensemble averaging (using a 5×5 window in our case) and $\mathbf{k}$ is the target vector,

$$\mathbf{k} = \begin{bmatrix} S_{ii} \\ S_{ij} \end{bmatrix} \tag{2}$$

From $\mathbf{C_2}$ we can derive the 2D Barakat degree of polarization m_{DP}, which is defined as the ratio between the intensity of the polarized portion to that of the total intensity [33],

$$m_{\text{DP}} = \sqrt{1 - \frac{4 \times \det(\mathbf{C_2})}{(\text{Tr}(\mathbf{C_2}))^2}} \tag{3}$$

m_{DP} represents the anisotropy from polarization structures, whereas the scattering randomness, $1 - m_{\text{DP}}\beta$ (β is a measure of the relative dominance of polarized scattering), is considered as the dual-pol radar vegetation index [34].

For the full-pol SAR data (ALOS-2 A122 in Table 1), we form the 3×3 coherency matrix $\mathbf{T_3}$ [30]:

$$\mathbf{T_3} = \langle \underline{\mathbf{k}} \cdot \underline{\mathbf{k}}^{*T} \rangle = \begin{bmatrix} T_{11} & T_{12} & T_{13} \\ T_{21} & T_{22} & T_{23} \\ T_{31} & T_{32} & T_{33} \end{bmatrix}$$
$$= \frac{1}{2} \begin{bmatrix} \langle |S_{HH} + S_{VV}|^2 \rangle & \langle (S_{HH} + S_{VV})(S_{HH} - S_{VV})^* \rangle & 2\langle (S_{HH} + S_{VV}) S_{HV}^* \rangle \\ \langle (S_{HH} - S_{VV})(S_{HH} + S_{VV})^* \rangle & \langle |S_{HH} - S_{VV}|^2 \rangle & 2\langle (S_{HH} - S_{VV}) S_{HV}^* \rangle \\ 2\langle S_{HV}(S_{HH} + S_{VV})^* \rangle & 2\langle S_{HV}(S_{HH} - S_{VV})^* \rangle & 4\langle |S_{HV}|^2 \rangle \end{bmatrix} \tag{4}$$

where the target vector $\mathbf{k}$ is defined as [30]

$$\mathbf{k} = \frac{1}{\sqrt{2}} \begin{bmatrix} S_{HH} + S_{VV} \\ S_{HH} - S_{VV} \\ 2S_{HV} \end{bmatrix} \tag{5}$$

From $\mathbf{T_3}$, we carry out a Model-Free 3-Component decomposition for Full-pol data (MF3CF) which jointly considers the Barakat degree of polarization and the received wave information to allow the estimation of scattering powers without any assumption of scattering models [35]. The odd-bounce surface scattering power P_s, even-bounce scattering power P_d and the diffused (volumetric) scattering power P_v can be estimated as [35]:

$$\begin{cases} P_s = \frac{m_{\text{FP}}\text{Span}}{2}(1 + \sin 2\theta_{\text{FP}}) \\ P_d = \frac{m_{\text{FP}}\text{Span}}{2}(1 - \sin 2\theta_{\text{FP}}) \\ P_v = \text{Span}(1 - m_{\text{FP}}) \end{cases} \tag{6}$$

where m_{FP} is the 3D Barakat degree of polarization,

$$m_{\text{FP}} = \sqrt{1 - \frac{27 \times \det(\mathbf{T_3})}{(\text{Tr}(\mathbf{T_3}))^3}} \tag{7}$$

$\text{Span} = T_{11} + T_{22} + T_{33}$, and the scattering type parameter θ_{FP} is defined as,

$$\tan\theta_{\text{FP}} = \frac{m_{\text{FP}}\text{Span}(T_{11} - T_{22} - T_{33})}{T_{11}(T_{22} + T_{33}) + m_{\text{FP}}^2\text{Span}^2} \tag{8}$$

The calculation of 2D degree of polarization and scattering powers is carried out in PolSAR-tools available at https://github.com/Narayana-Rao/PolSAR-tools (accessed on 10 March 2022).

2.3. Generating Z-Score Maps

To standardize the change detection flow for various data values (σ^0, m_{FP} and scattering powers), we adopt a dimensionless Z-score (Z) of the post-event value normalized by the statistical information obtained from its pre-event time-series. It is calculated as [36],

$$Z = \frac{y_{post} - \overline{y}_{pre}}{\sigma_{pre}} \tag{9}$$

where y_{post} is the data value on the post-event epoch, and $[\overline{y}_{pre}, \sigma_{pre}]$ are the mean and standard deviation calculated from the pre-event time-series. The Z-score value represents the difference between a pixel's change value and its background mean value in the unit of background standard deviations. The background mean and standard deviation are derived from multiple pre-event epochs, and therefore may avoid the issues associated with a single biased pre-event image, a consideration commonly seen in a dual-temporal change detection scheme [37]. It also allows a more robust detection of minor changes if the pre-event time-series contains relatively stable values [36].

Among different datatypes, however, we caution that the Z-score map for CSK data in the Hokkaido case (Table 1) may not be as accurate due to the insufficient number of pre-event images–no acquisition exists before 16 July 2017. The temporal standard deviations thus derived tend to be too large and yield Z-score values smaller than those generated from other datasets. To work around this problem, we estimate the spatial standard variation within a 21 × 21 window centered at each pixel, and use it as σ_{pre} in Equation (9) when the value is smaller than the temporal standard deviation. The Z-score values thus generated are visually more comparable to other Z-score maps. The detection results, however, may still be less accurate because of incomplete pre-event information.

In addition to Z-score maps generated from the aforementioned datasets, we further generate a Z-score map that combines the positive Z-score values in P_s (Z_{P_s}) and negative values in P_v (Z_{P_v}),

$$Z_{P_c} = \begin{cases} Z_{P_v} \text{ if } Z_{P_v} < 0 \text{ and } |Z_{P_v}| > |Z_{P_s}| \\ Z_{P_s} \text{ otherwise} \end{cases} \tag{10}$$

where Z_{P_c} stands for the combined Z-score map. The necessity and performance of such a combined Z-score map will be further demonstrated in Section 4. Next, we describe how these Z-score maps are used in the GSBA change-detection algorithm.

3. Change Detection Method

The method proposed in this study is a variation of the split-based approach (SBA), which was first proposed for SAR-based flood mapping [38]. SBA is designed for the generalization of a mapping algorithm regardless of the image's spatial resolution, swath size and the target's geospatial distribution. The basic idea is to separate the image into multiple, non-overlapping tiles (also called splits). The tiles are then checked one by one to identify the existence of a certain proxies that signify the changes. Some suggested proxies include standard deviation [38], coefficient of variation [39,40], and the ratio between the tile mean and the global mean [40]. Tiles with proxies above a given threshold are selected to jointly determine the global threshold either via a non-parametric approach such as the

Otsu method [41] or the KI algorithm [42], or through a parameterized fitting to the data histogram in order to determine the best cutoff point [43,44].

One variation of SBA is the hierarchical split-based approach (HSBA) [45]. Different from the conventional SBA which adopts a uniform tile size [38,40,46,47], HSBA adaptively and hierarchically splits the image to variable sizes until a bimodal histogram (representing the change and non-change classes) can be identified in the tile or when a minimal tile size is reached. This way it avoids the need of a pre-defined tile size, which in some cases may compromise the detection if the change area within a tile is extremely large or small. The issue with HSBA is that the splitting process is nonlinear-every split depends on the result of the previous split, and hence the computation time can be long when the image is large.

Instead of the top-down splitting strategy of HSBA, here we propose a bottom-up approach called Growing Split-Based Approach (GSBA) (Figure 2). We initialize the image splitting as in the conventional SBA. Once the tiles with changes are identified, we "grow" a patch within each tile cluster until the maximum patch area with a consistent bimodal histogram is reached. This growing step produces patches of different areas that mimic the variable tile size obtained by HSBA. The second variation in GSBA is that rather than obtaining a global threshold from the patches to generate a binary map, we calculate the Bayesian probability of changes instead [48]. Given the probability map, we can obtain a binary change map at a given cutoff probability (by default 0.5).

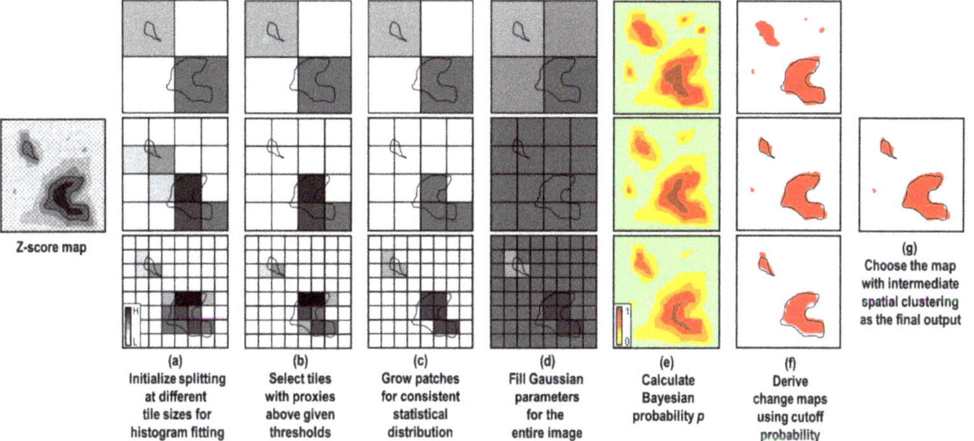

Figure 2. Growing Split-Based Approach (GSBA) workflow.

The last variation in GSBA is that instead of using a single tile size, the flow described above is repeated at different tile sizes. This practice is to acknowledge the observation that depending on the size and spatial distribution of changes, a single arbitrary tile size may in some cases produce results that fall into a local minimum or maximum of change area. With multiple binary maps generated at different tile sizes, we can select the one with intermediate spatial clustering (using Ripley's K, see step (g) in Appendix A) as the final output. In other cases where different tile sizes produce similar binary maps, this practice also offers reassurance regarding the robustness of the detection output. The processing flow in GSBA is linear and can be fully parallelized, so the increase in computation time can be minor in a multi-processor computing system.

To avoid distraction from the main focus of this paper, details of the GSBA algorithm are given in the Appendix A. The output Bayesian probability map is used in the following Receiver Operating Characteristics (ROC) curve analysis. To generate the ROC curves, we calculate the true positive rate (TPR, or recall) and false positive rate (FPR) at different cutoff probabilities. They are defined as follows:

$$\begin{cases} \mathrm{TPR} = \mathrm{TP}/(\mathrm{TP}+\mathrm{FN}) \\ \mathrm{FPR} = \mathrm{FP}/(\mathrm{FP}+\mathrm{TN}) \end{cases} \tag{11}$$

where [T, F] stand for true and false, [P, N] stand for positive and negative, and any two-letter combination means the number of such pixels identified through validation. We further analyze the area under curve (AUC) to evaluate the overall detection performance. This continuous tracking of trade-off effects between TPR and FPR allows users to have a complete and visual overview of the detection performance [15]. To compare with other studies using the same landslide cases, we also report the overall accuracy (OA) by using the final binary maps from GSBA. The overall accuracy is defined as:

$$\mathrm{OA} = (\mathrm{TP} + \mathrm{TN})/(\mathrm{TP} + \mathrm{TN} + \mathrm{FP} + \mathrm{FN}) \tag{12}$$

Note that these metrics are calculated at each datatype's spatial resolutions, which means, the validation dataset is resampled to match the resolution of the SAR images.

4. Results

4.1. Case Study 1: Earthquake-Triggered Hokkaido Landslides in Japan

Widely-distributed landslides occurred due to seismic shaking during the 6 September 2018 Mw 6.6 Hokkaido Eastern Iburi Earthquake (Figure 3). After the earthquake, the Geospatial Information Authority of Japan (GSI) acquired aerial photos on 6 and 11 September over the landslide areas and identified more than 3300 landslide patches manually (https://www.gsi.go.jp/BOUSAI/H30-hokkaidoiburi-east-earthquake-index.html#1, accessed on 10 March 2022). Visual comparison between the aerial photos taken on 6 and 11 September shows that only minor changes occurred between these two dates, and hence the majority of the landslides have existed since the earthquake.

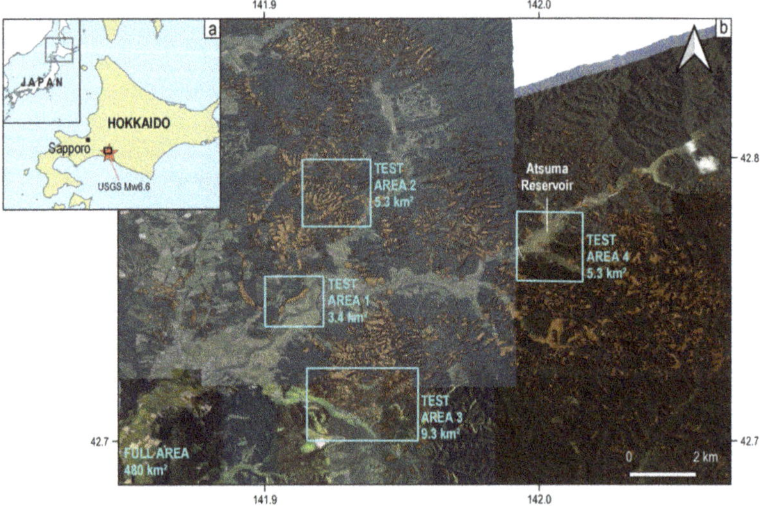

Figure 3. Distribution of earthquake-triggered landslides in Hokkaido, after the 6 September 2018 Mw. 6.6 Hokkaido Easter Iburi Earthquake. Red star in (**a**) represents the epicenter; black box represents the full extent of AOI in (**b**). Background images are aerial photos taken on 6 and 11 September by GSI. Orange polygons are the manually identified landslide patches based on these aerial photos (source: https://www.gsi.go.jp/common/000204728.zip, accessed on 10 March 2022).

We define a large area of interest (AOI) of 480 km^2 that covers most of the landslide patches (Figure 3). Within the large AOI we further define four test areas, three of which are the same as those chosen by Jung and Yun (2020) [15]. This choice allows us to compare the effect of different spatial resolutions-they adopt the ALOS-2 PALSAR-2 Ultra-Fine HH-pol SAR data of 3-m resolution, while the data used in this study is the 6-m High-Sensitive full-pol SAR. In addition, we define a fourth test area around the Atsuma Reservoir in order to examine the effect of water bodies during landslide detection.

4.1.1. Qualitative Comparison

From the three sensors used in this case study (Table 1), we produce the following seven SAR datatypes and one combined Z-score map to carry out the performance comparison:

(1) High-res L-band HH-pol σ^0
(2) High-res L-band VV-pol σ^0
(3) Medium-res C-band VV-pol σ^0
(4) Ultra-high-res X-band HH-pol σ^0
(5) High-res L-band dual-pol (VV + VH) m_{DP}
(6) Medium-res C-band dual-pol (VV + VH) m_{DP}
(7) High-res L-band full-pol P_s
(8) High-res L-band full-pol Z_{P_c} (denoted as P_c datatype hereafter)

In Figure 4 we choose Test Area 2 to demonstrate the differences among some of the selected datatypes. Both the high-res L-band HH-pol σ^0 and the ultra-high-res X-band HH-pol σ^0 provide sharp outlines for landslides compared to the multi-pol datatypes (m_{DP} and P_s). This difference is caused by the spatial ensemble averaging carried out on the polarimetric datasets. The C-band VV-pol σ^0, acquired at a lower spatial resolution, does not capture the landslide boundaries as clearly. In addition, many pixels in the landslide patches contain low Z-score values, suggesting that these pixels are indistinguishable from non-landslides. The same phenomenon is also observed more in the X-band σ^0 than in the L-band σ^0, implying that shorter radar wavelengths may not perform as efficient in landslide detection within forest areas, possibly due to their higher sensitivity to small-scale changes in vegetation during the pre-event periods.

Next, we examine the effect of viewing geometry on different polarimetric combinations in the L-band datatypes. The P_s datatype shows strong positive Z-scores for the landslides on the slopes with small LIAs. On the slopes with large LIAs, the landslide patches contain low to nearly zero P_s Z-scores (pointed by yellow arrows in Figure 4). The same is also seen on the m_{DP} Z-score map. In comparison, the L-band σ^0 may still show clear and predominantly negative Z-score values on the slopes with large LIAs. We investigate the Z-score maps from other scattering powers and find that, instead of a positive increase in P_s Z-score, landslide patches with larger LIAs tend to show a stronger decrease in P_v Z-score (yellow arrows in Figure 5). This different dependency on LIA between P_s and P_v is also observed in other landslide cases [23]. Numerical simulation in [49] offers some possible explanation to this phenomenon, in which backscattering energy due to direct-ground reflection and crown-ground interaction decreases with LIA, while the energy due to direct-crown backscattering increases. In other words, landslides on the slopes with larger LIAs are sensed as "loses in tree crowns" instead of "increases in bare surfaces". That means P_s or P_v each only carries half of the information about landslides. It is therefore necessary to consider both values when carrying out landslide detection, such as by combining them into a joint Z-score map Z_{P_c} (Equation (10) and Figure 5).

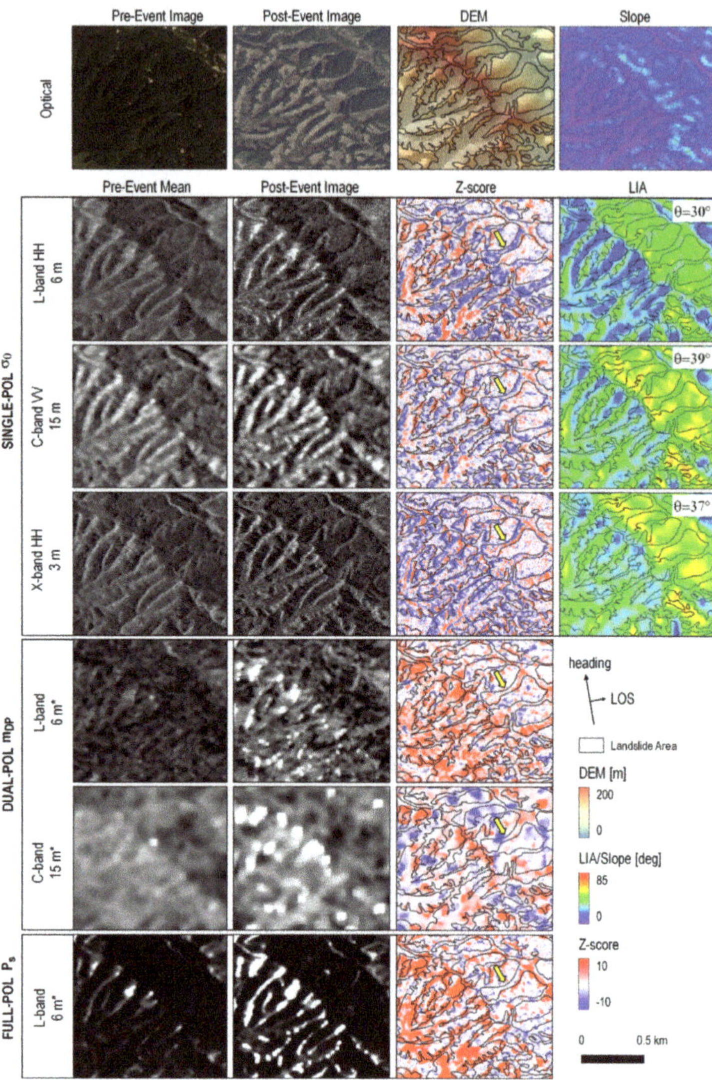

Figure 4. A qualitative comparison among different datatypes for Hokkaido Test Area 2 (see Table 1 for data sources). The value 3 m, 6 m and 15 m are the spatial resolutions, with the asterisk (*) indicating images processed by taking spatial ensemble averaging. DEM: digital elevation model. LIA: local incidence angle. The LIA maps for the multi-pol datatypes are the same as those in the single-pol. Θ: satellite look angle. Yellow arrows on Z-score maps indicate one particular landslide that is well captured by σ^0 but not by m_{DP} and P_s.

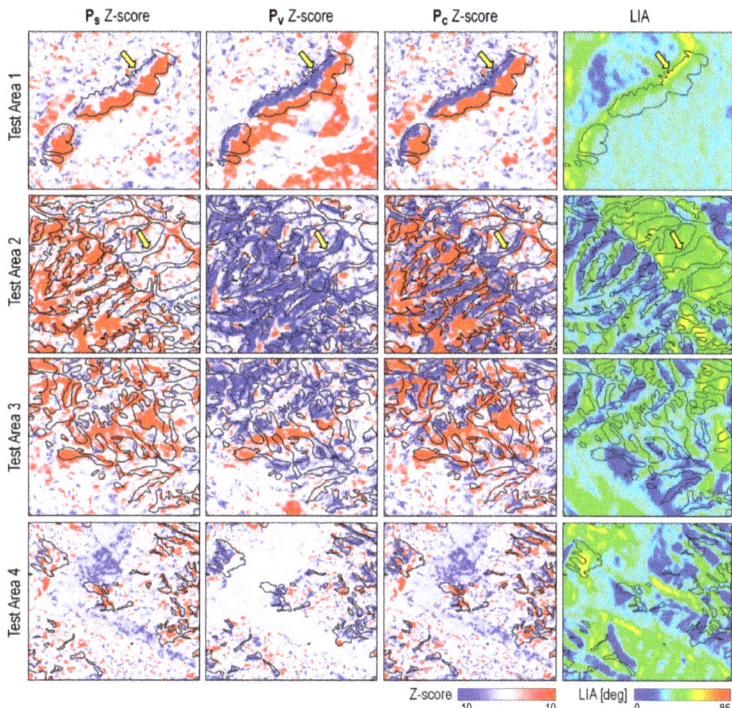

Figure 5. Comparison of Z-score maps generated from the odd-bounce surface scattering power (P_s) and the diffused (volumetric) scattering power (P_v), and the combined Z-score map (P_c Z-score). Yellow arrows indicate places with low Z-score amplitudes in P_s but high amplitude in P_v, a phenomenon associated with the local incidence angles (LIAs). Yellow arrows indicate landslide patches better depicted by combining the Z-score values from P_s and P_v.

The aforementioned complementary effect between P_s or P_v is not seen between the dual-pol m_{DP} and the dual-pol radar vegetation index [34]. In fact, when comparing the Z-score maps generated from these dual-pol parameters, one appears as a sign-flipped image of the other–pixels in the landslide patches display similar amplitudes but opposite signs on the two Z-score maps. This is probably due to the incomplete information of scattering properties in dual polarization. We therefore do not attempt to combine these two parameters and remain with the m_{DP}-only Z-score map. Next we will look into the quantitative comparison of detection performance among the eight listed datatypes.

4.1.2. Quantitative Comparison

Figure 6 shows the Bayesian probability maps calculated from different datatypes. Overall, the full-pol P_c depicts the most complete shape of the landslide bodies, followed by the single-pol σ^0 at different radar wavelengths. The dual-pol datatypes detect only few landslides. The X-band σ^0 captures an additional high-probability patch in Test Area 1 with sharp outlines. This patch is possibly related to crops, and the false detection is likely associated with the lower number of pre-event images. In areas where water bodies exist (Test Area 4), the single-pol σ^0 may detect changes associated with both the landslides and the water bodies, while the multi-pol datatypes are relatively free of such confusions.

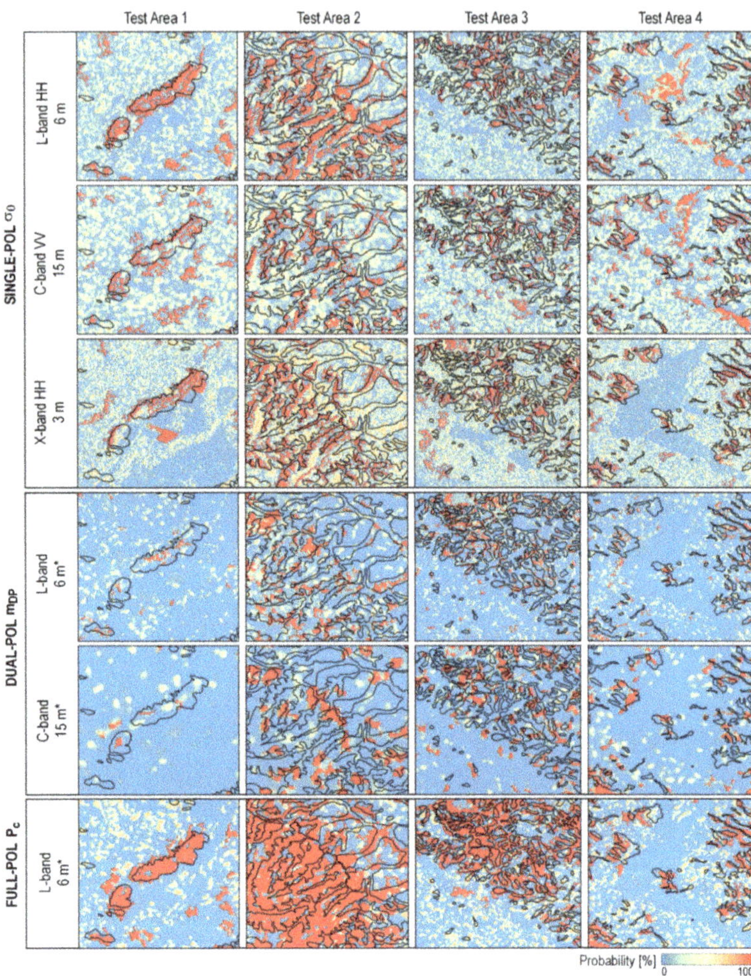

Figure 6. Bayesian probability maps obtained from different datatypes in the Hokkaido landslide case. The asterisk (*) indicates the image is processed by taking spatial ensemble averaging using a 5×5 window.

The performance of each datatype is shown in the ROC curves (Figure 7). In Test Area 1, the ROC curves for the L-band σ^0 and P_c show a steep increase of TPR at low FPR, yielding AUC values as high as 0.83 to 0.86 (Figure 7f). These values are comparable to the detection results based on the ALOS-2 Ultra-Fine HH-pol multi-temporal intensities of 3-m spatial resolution (AUC = 0.79) [15]. The AUC for X-band σ^0 and L-band P_s rank the second and third at 0.77 and 0.73. The L-band m_{DP}, C-band m_{DP} and C-band σ^0 perform poorly, yielding AUC values lower than 0.7.

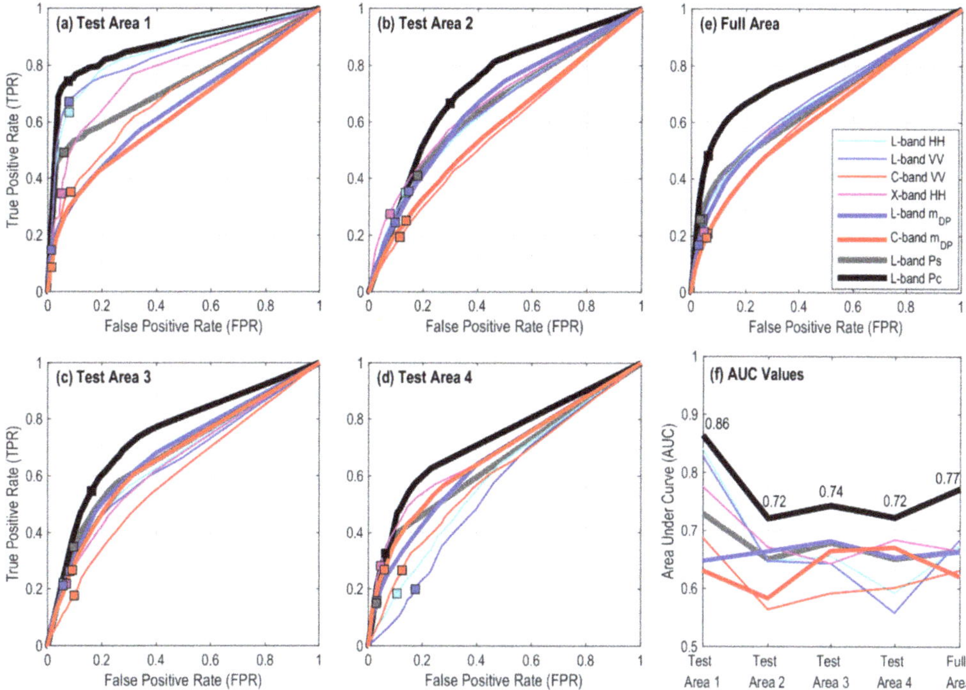

Figure 7. Receiver operating characteristics (ROC) curves and area under curve (AUC) for (**a**–**d**) the four test areas and (**e**) the full area of the Hokkaido AOI (Figure 3). The numbers shown in (**f**) the AUC plot are the values from the full-pol P_c. Squares of different colors indicate the positions of the GSBA binary maps generated at a cutoff probability of 0.5.

The landslide patches in Test Area 2, 3 and 4 result in slow-rising ROC curves and low AUC values in almost all datatypes (Figure 7b–d,f). Among them, the full-pol P_c still gives the highest and more stable AUC values, above 0.7 for all three test areas. All the other datatypes yield AUC values lower than 0.7. The C-band σ^0 yields the lowest AUC in Test Area 2 and 3–lower than 0.6. In Test Area 4, the L-band σ^0 gives the lowest AUC values as a result of false detection over the reservoir lake (Figures 3 and 6).

For the full area, the L-band full-pol P_c has superior performance in landslide detection with AUC = 0.77 (Figure 7f). The comparison between P_c and P_s-only AUC values again confirms that the combination of P_s and P_v is necessary for landslide detection. The full-area AUC value for the high-res L-band σ^0 and ultra-high-res X-band σ^0 ranks the second and third, higher than the medium-res C-band σ^0. We can therefore confirm that spatial resolutions plays a role as important as radar wavelengths in SAR-based landslide detection. The performance of the dual-pol m_{DP} is least favored because of its lower full-area AUC and its tendency to detect fewer landslides.

In Figure 7a–e, we also mark the positions of the binary change maps determined by GSBA on each ROC curve. These solutions are located close to the turning points of the curves, but slightly leaning towards the lower-FPR end. These positions suggest that the default cutoff probability of 0.5 tends to create conservative change maps with a higher positive likelihood ratio (TPR-to-FPR ratio).

4.2. Case Study 2: Rainfall-Triggered Putanpunas Landslide in Southern Taiwan

In the second case, we look into a different scenario–a rainfall-triggered landslide. During the first 4 days in June 2017, a total of nearly 900 mm of precipitation occurred in the upstream of the Laonong River (according to the records from weather station C0V210, available at https://e-service.cwb.gov.tw/HistoryDataQuery/index.jsp, accessed on 10 March 2022) (Figure 8). Three days later on 7 June 2017, a landslide-associated earthquake (landquake) was detected through the Real-time Landquake Monitoring System (RLAMS, http://collab.cv.nctu.edu.tw/older/catalog_20171231.html, accessed on 10 March 2022) [50]. Given the order of occurrence between the torrential rain and the landquake, we attribute this event to a rainfall-triggered landslide.

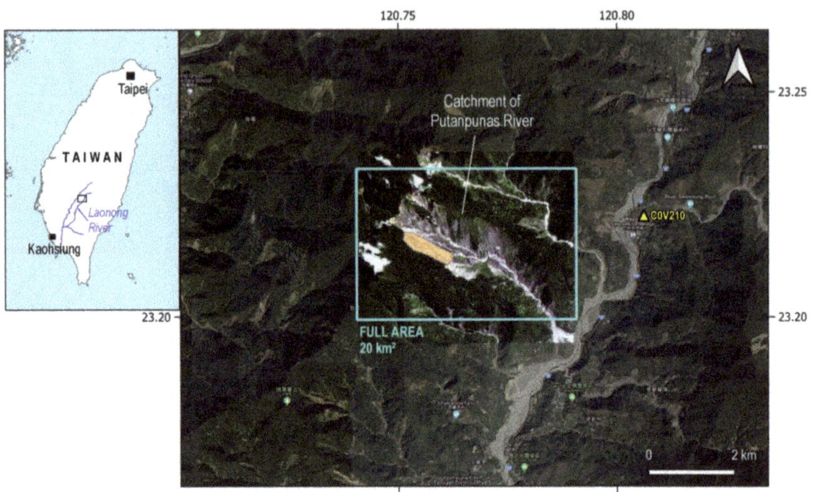

Figure 8. Basemap of the Putanpunas River catchment and the landslide on 7 June 2017. Blue rectangle represents the AOI. Orange polygon: the landslide patch from manual mapping. C0V210 is the weather station. Source of image: SPOT-6/7 acquired on 5 July 2017 overlaid on Google Earth Pro Image © 2022 CNES/Airbus.

To determine the landslide area associated with the 7 June 2017 landquake, we use SPOT-6/7 images acquired on 18 April 2017 and 5 July 2017 to carry out manual detection. Within a 15-km radius (approximately the estimation error of landquake relocation) of the epicenter, we locate a new sliding patch in the existing landslide area of the Putanpunas River catchment. Actually, there have been repeating landslides in this catchment since typhoon Morakot in the year of 2009 [18,51], making this catchment one of the most actively evolving landslide area in southern Taiwan. Different from the multiple small-scale landslides triggered by the Hokkaido Eastern Iburi Earthquake, this rainfall-triggered Putanpunas landslide contains a single patch of a relatively large area, up to 400,000 m^2 (Figure 8). We caution that the actual landslide patch may differ from what we map here due to the longer latency between the event date and the post-event optical image.

4.2.1. Qualitative Comparison

In this case study, we produce the following five datatypes for comparison (note that different viewing geometry is involved):

(1) Low-res L-band HH-pol σ^0, ascending
(2) Low-res L-band HH-pol σ^0, descending
(3) Medium-res C-band VV-pol σ^0, ascending
(4 Medium-res C-band VV-pol σ^0, descending
(5) Ultra-high-res X-band, HH-pol σ^0, descending

Some datasets allow dual-pol polarimetric combinations, such as HH-HV for ALOS-2 track D27 and VV-VH for both Sentinel-1 tracks (Table 1). However, in the Hokkaido case we have demonstrated that the use of dual-pol m_{DP} datatype tends to capture fewer landslides, so we decide not to adopt the dual-pol datatype in this case study. Here we wish to focus on the detection performance among the σ^0 datatypes of different wavelengths, spatial resolutions and viewing geometry.

Figure 9 shows how different σ^0 datatypes compare visually. The ascending tracks show more prominent landslide signatures on the Z-score maps compared to those from the descending tracks, regardless of the wavelengths and spatial resolutions. The intriguing point is that the viewing geometry from ascending tracks yield larger LIAs compared to those from descending tracks. In addition, we find that the viewing geometry from descending tracks produces large layover zones that tend to overlap with the slopes of small LIAs. Among all sensors, the X-band CSK descending track produces the largest layover area, which also corresponds to the smallest look angle (27°) among all three sensors. Within the layover zones, landslide-related changes may still be recorded but are mixed with energies from multiple ground targets of the same range distance, leading to brighter pixels and stretched patterns after geocoding (Figure 10). We also notice that the current layover-shadow mask does not necessarily mask out all layover areas (Figure 10), which is possibly caused by errors in DEM or limitations in the SAR geometric distortion simulation [52].

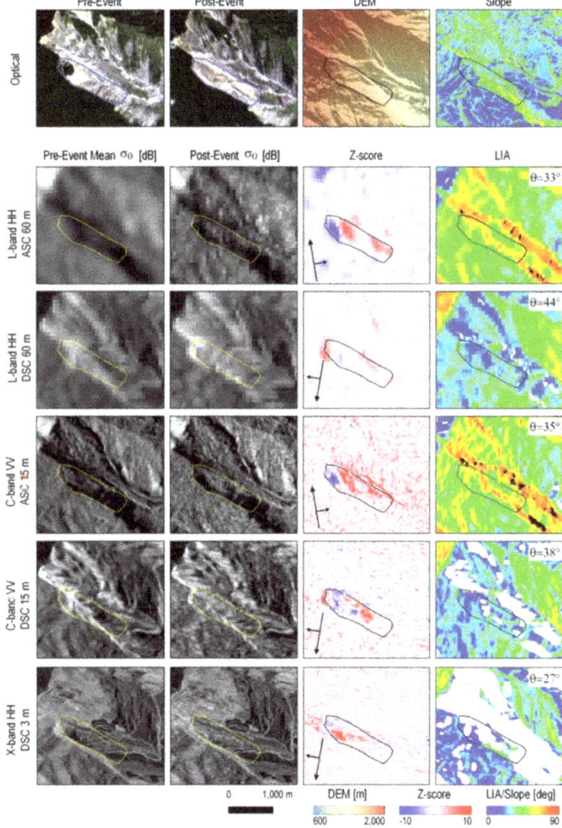

Figure 9. A qualitative comparison among different datatypes for the rainfall-triggered Putanpunas landslide on 7 June 2017 in southern Taiwan. White and black pixels on the LIA maps are layover and shadow zones, respectively. Refer to Figure 4 captions for more information.

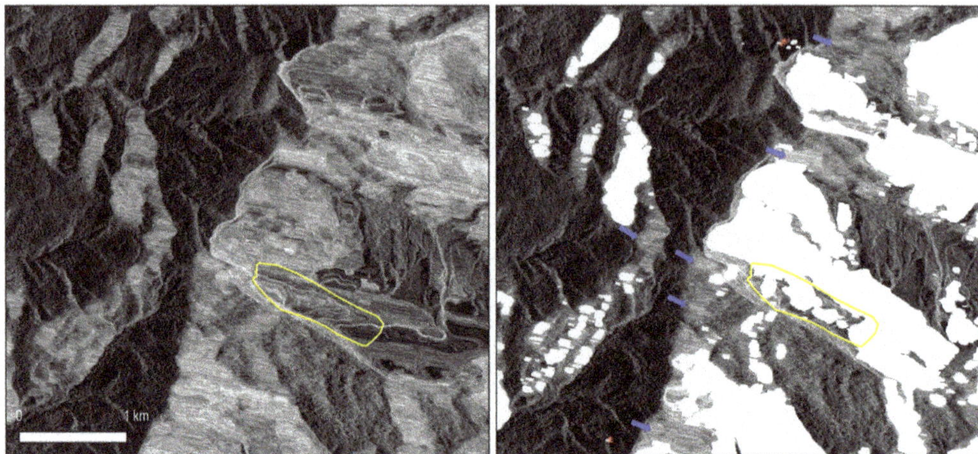

Figure 10. (**Left**) Slopes within the layover zones appear as brighter and stretched pixels in the post-event X-band CSK σ^0 image. In comparison, slopes in non-layover areas show typical speckle textures. Yellow polygon is the 7 June 2017 landslide patch. (**Right**) After applying the layover-shadow mask (white and red pixels), some severely stretched patterns remain unmasked (blue arrows) possibly due to errors in DEM or incorrect prediction in the SAR geometric distortion simulation.

We have to emphasize that the same layover-shadow masks are also calculated and applied on the SAR images in the previous Hokkaido landslide case. However, given the smaller slope angles (Figure 4), the layover and shadow effects are nearly negligible. In comparison, the Putanpunas River catchment has a large topographic relief and hence greater slope angles (Figure 9), resulting in a larger portion of unusable data within the layover zone.

4.2.2. Quantitative Comparison

Figure 11 shows the Bayesian probability, ROC curves and AUC values for different datatypes. The ascending tracks show better performance than descending tracks, with the highest AUC of 0.78 from the medium-res C-band VV σ^0. The ascending L-band HH σ^0, albeit its low spatial resolution (60 m), still yields an AUC value of 0.71. On the other hand, the descending tracks do not detect as many changes, with an AUC value of 0.65 for the medium-res C-band VV σ^0 and 0.64 for the ultra-high-res X-band HH σ^0. No change is detected from the descending L-band HH σ^0. We should point out that despite the similar AUC values from the descending C-band VV σ^0 and the X-band HH σ^0, their effective area (unmasked area) is different–more than 50% of the catchment is under the layover-shadow mask of the X-band data.

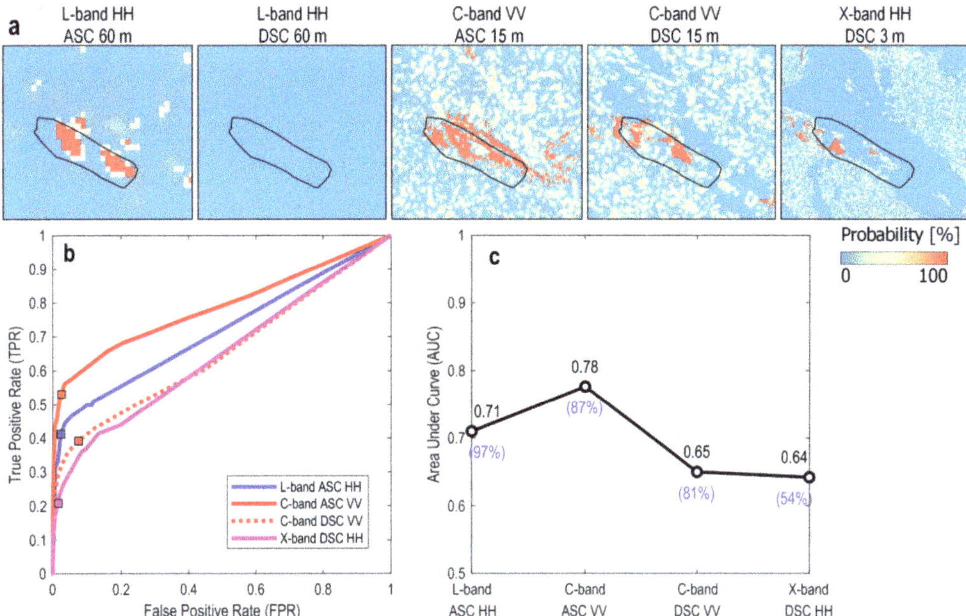

Figure 11. (**a**) Bayesian probability maps obtained from different σ^0 datatypes for the Putanpunas landslide on 7 June 2017. (**b**) ROC curves and (**c**) AUC values for different datatypes. Blue numbers in parentheses are the percentage of effective area within the AOI.

5. Discussion

5.1. How Data Properties Affect Detection Performance

In Table 2 we summarize the performance of different datatypes, including the full-area AUC and OA. Note that AUC does not represent the performance of any particular detection outcome but a full spectrum of outcomes, whereas OA is calculated at a specific selection (e.g., at the cutoff probability of 0.5) to mimic the operator's choice. The OA values among different datatypes in the Hokkaido case, however, are all very similar and cannot reflect their actual relative performance. This similarity results from the large number of non-landslide pixels (TN in Equation (12)) when computing the OA values. To allow a better judgement of the relative performance at a specific landslide mapping outcome, we compute the full-area TPR at a fixed FPR of 0.1 ($TPR_{FPR\,=\,0.1}$). This value indicates the ratio of detectable landslides at the cost of 10% false positive detection. In the Hokkaido case, we further normalize the $TPR_{FPR\,=\,0.1}$ values by using the best $TPR_{FPR\,=\,0.1}$ from P_c (Table 2). With the AUC and $TPR_{FPR\,=\,0.1}$ values, we discuss how the following data properties affect the performance of SAR-based landslide detection.

Table 2. Summary of full-area AUC, OA, TPR at PRF = 0.1 and the ratio of effective area (A_e) for the two case studies.

Event and AOI Area	Sensor-Track	Datatype *1		AUC	OA	$TPR_{FPR = 0.1}$ *2		A_e *3 Ratio
Hokkaido 480 km^2	ALOS-2 A122	Single-pol HR L-band	HH σ^0	0.67	0.89	0.37	(0.66)	0.99
		Single-pol HR L-band	VV σ^0	0.68	0.89	0.39	(0.70)	0.99
		Dual-pol HR L-band	VV + VH m_{DP}	0.65	0.89	0.42	(0.75)	0.99
		Full-pol HR L-band	P_s	0.66	0.90	0.40	(0.71)	0.99
		Full-pol HR L-band	P_c Z-score	0.77	0.90	0.56	(1.00)	0.99
	S-1 A68	Single-pol MR C-band	VV σ^0	0.63	0.87	0.27	(0.48)	0.99
		Dual-pol MR C-band	VV + VH m_{DP}	0.62	0.87	0.27	(0.48)	0.99
	CSK-A	Single-pol UHR X-band	HH σ^0	0.67	0.88	0.34	(0.61)	0.99
Putanpunas 20 km^2	ALOS-2 A137	Single-pol LR L-band	HH σ^0	0.71	0.98	0.50		0.97
	ALOS-2 D27	Single-pol LR L-band	HH σ^0	-	-	-		0.97
	S-1 A69	Single-pol MR C-band	VV σ^0	0.78	0.92	0.61		0.87
	S-1 D105	Single-pol MR C-band	VV σ^0	0.65	0.90	0.41		0.81
	CSK-D	Single-pol UHR X-band	HH σ^0	0.64	0.98	0.37		0.54

*1 UHR: ultra-high resolution; HR: high resolution; MR: medium resolution; LR: low resolution. *2 The TPR value at FPR = 0.1. Values in parentheses represent the normalized value by the best performance. *3 The ratio between the effective area (area outside the layover-shadow mask) and the full AOI.

Radar wavelengths and spatial resolutions. In the Hokkaido case, the L-band datatypes have better AUC and $TPR_{FPR = 0.1}$ values than the other two radar wavelengths. This result seems to suggest that longer wavelengths work better in landslide detection. However, spatial resolutions can be an equally important factor. This inference is made from the poor performance of the medium-res C-band single-pol datatype–its result is worse than that from the X-band single-pol data at a higher spatial resolution. At the same time, the medium-res C-band single-pol data seems to perform relatively well in the Putanpunas case. This better performance is probably associated with the geometry of the landslide patches (small and distributed landslides in Hokkaido vs. one single large patch in Putanpunas), and the fact that the Putanpunas landslide is a repeated landslide on a bare surface instead of on a forested land (Figure 8).

Polarizations. In the Hokkaido case, the high-res L-band full-pol data can offer the best landslide detection capability in forest areas. It can even avoid false detection over water bodies. Even though the landslide boundaries become slightly blurry compared to those detected by using single-pol datatypes, the amount of information contained in a full-pol dataset and the detection performance thereof is indeed unparalleled. The L-band dual-pol datatype, despite a slightly better $TPR_{FPR = 0.1}$, gives an AUC value lower than those from many single-pol datatypes. It also tends to detect fewer landslide patches. As dual polarization will be a major observation mode for some upcoming missions such as

NISAR, more efforts are needed to explore a better utilization strategy for dual-pol data in landslide mapping.

Viewing Geometry. In the Hokkaido case, we do not see a notable difference in LIA between the detected and the missed pixels (Figure 12a). This similarity in LIA distributions suggests that large LIAs do not necessarily hinder landslide detection. In the Putanpunas case, landslides are even better detected on datatypes with larger LIAs (Figure 9). On the other hand, slopes with small LIAs may coincide with overlay zones, which are also seen in the Putanpunas case. The smaller (steeper) the satellite look angle (θ), the larger the overlay zone and the smaller the effective area. The percentage of effective area drops from 97% at θ = 44° (ALOS-2 D27) to 54% at θ = 27° (CSK-D) (Figure 9 and Table 2). As the area of layover and shadow depends on the specific topography of a region, a DEM-based SAR geometric distortion simulation should be executed before planning for a mission or submitting a tasking request in order to determine the best viewing geometry [52].

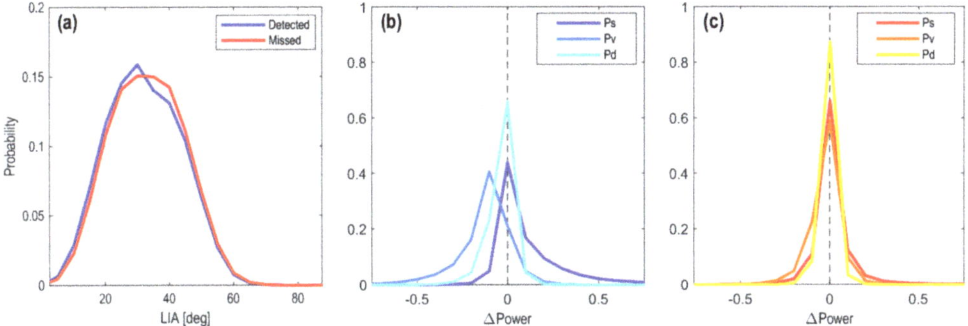

Figure 12. Normalized histogram for (**a**) the LIA and the post-to-pre-event difference in scattering powers (ΔPower) for the (**b**) detected and (**c**) missed pixels in the L-band quad-pol datatype (detected and missed pixels are based on the GSBA binary map generated with the P_c Z-score map). The histograms are computed over the entire Hokkaido AOI.

5.2. Limitations in SAR-Based Landslide Detection

In the Hokkaido case, the best full-area AUC is 0.77 and the best OA is 0.90 (Table 2). These values are close to the results produced by using the ultra-high-res L-band HH-pol multi-temporal intensity with different detection algorithms [15]. The similarity in limited performance from different studies suggests that there may exist some factors that prevent the SAR-based landslide mapping from achieving the high accuracy attained in optical image-based mapping [53]. We plot the histogram of the post-to-pre-event difference in scattering powers (ΔPower) for the detected and the missed pixels, respectively, within the Hokkaido AOI (Figure 12b,c). While the histograms for the detected pixels are clearly skewed, the histograms for the missed pixels are centered and symmetric around zero, indicating no significant difference in any of the scattering powers before and after the landslide. This phenomenon is bewildering and needs some explanation.

The first possibility is the change of local topography. From a broader scale, there does not seem to be a systematic difference in LIA between the detected and the missed pixels (Figure 12a). However, the LIA is calculated using the pre-event global 30-m DEM [54]. The topography must have changed locally after the landslide. In fact, according to the post-event DEM obtained by airborne laser survey, the surface morphology within the landslide patches has changed substantially [55]. Features such as scarps and crown cracks can reach meters tall with very steep (nearly vertical) facets [55]. Field photos also reveal that large boulders, huge piles of dead trees and meter-scale surface undulations exist on the ground [55,56]. The chaotic distribution of these features may result in scattering properties considerably deviating from what is expected from a pure land cover-type change (forest to bare surface).

Another equally important factor to consider is the spatial variation of water contents. Several studies show that the water contents can vary remarkably in the Hokkaido area, from 30% to 280% among different geological materials sampled at the same site [57,58]. As these materials are spread out during the landslide, the randomness in surface soil moisture within the landslide patches increases. To sum up, soil moistures, local topography, surface roughness and land cover changes together form a complicated backscattering field in the landslide patches, leading to clear changes of scattering powers in some places and no clear changes in other places. We may even hypothesize that it is radar wave's sensitivity to multiple physical properties that limits its performance in landslide detection. This hypothesis needs further validations though, such as through numerical simulation of 3D backscattering fields.

The increased randomness in backscattering fields within a landslide patch may explain why intensity correlation method can yield better detection results than those from pixel-by-pixel detection methods [15,19]. Intensity correlation is calculated based on a number of pixels within a moving window, which offers contextual information about the objects and their changes. Most important of all, it can potentially average out the randomness in the complicated scattering field within a landslide patch, leading to a smoother and less noisy detection result.

6. Conclusions

By applying a newly-designed change detection algorithm Growing Split-Based Approach (GSBA) on two different landslide cases, this study examines how different SAR data properties affect the performance of landslide detection. Our result shows that the high-resolution, full-polarimetric SAR datatype has unparalleled performance in landslide detection over forest areas. Single polarimetric datasets of high or ultra-high spatial resolution rank the next, regardless of their radar wavelengths. This result suggests that high spatial resolution is critical especially for detecting small and distributed landslides in forest areas. Datatypes of medium or low spatial resolution work better in detecting large landslide patches over bare surfaces; their detection performance decays significantly over small landslides in forest areas. Dual polarimetric datatypes have the worst performance among all; a better utilization strategy may be needed for their use in landslide detection. Different viewing geometry mainly impacts the effective detection area by creating layover and shadow zones. This problem is more severe in areas of large slopes ($\geq$40–50°), in which a steep viewing angle (<30°) may render half of the image unusable for landslide detection. SAR geometric distortion simulation is recommended before planning for a mission or submitting a tasking request for landslide mapping purposes. Given the limited performance of SAR-based landslide detection (both in this study and in previous studies) as compared to that from optical image-based landslide detection, we propose that other confounding factors, including but not limited to local topography, surface roughness and soil moistures, are all contributing to the randomness in the backscattering field and hinder the detection of land cover changes. Such limitations need to be properly acknowledged when adopting SAR-based landslide mapping for emergency responses or post-hazard assessments.

Author Contributions: Conceptualization, Y.N.L.; data curation, Y.-C.C., Y.-T.K. and W.-A.C.; formal analysis, Y.N.L.; methodology, Y.N.L. and Y.-C.C.; project administration, Y.N.L.; software, Y.N.L. and Y.-C.C.; validation, Y.-C.C. and Y.-T.K.; writing—original draft, Y.N.L. and Y.-C.C.; writing—review and editing, Y.-T.K. and W.-A.C. All authors have read and agreed to the published version of the manuscript.

Funding: This research was supported by research grant AS-TP-108-M08-1 from Academia Sinica, Taiwan and MOST 110-2119-M-001-006 from the Ministry of Science and Technology, Taiwan to Y.N.L.

Data Availability Statement: ALOS-2 data between 2016 and 2018 used in this study are provided by JAXA via the EO-RA2 research program (PI number: ER2A2N006). Copernicus Sentinel data between 2016 and 2018 used in this study are processed by ESA and retrieved from ASF DAAC. The COSMO-SkyMed data are purchased from e-GEOS. The airborne optical images for the Hokkaido landslides are available through the Geospatial Information Authority of Japan.

Acknowledgments: We thank Hsin Tung for handling the COSMO-SkyMed dataset. We also thank two anonymous reviewers for their constructive suggestions which improve the quality of this paper. This work is IES paper number 2408.

Conflicts of Interest: The authors declare no conflict of interest.

Appendix A. GSBA Algorithm

The following description corresponds to the steps illustrated in Figure 2.

(a) Initialize splitting and histogram fitting

We first split the image into multiple tiles. After splitting, we fit the data histogram $h(z)$ within each tile with a model histogram $g(z)$ of tri-modal Gaussian distribution (modified after the bimodal Gaussian distribution in [48]):

$$g(z) = \sum_{i=1}^{3} G_i = \sum_{i=1}^{3} A_i \exp\left[-\frac{1}{2}\frac{(z-m_i)^2}{s_i^2}\right] \tag{A1}$$

where $g(z)$ is the modeled histogram discretized at z bin centers, $[A_i,\ m_i,\ s_i]$ is the amplitude, mean and standard deviation for each of the i-th Gaussian modes. Mode 1 (G_1) and mode 3 (G_3) imply negative and positive changes in Z-score values, respectively. Mode 2 (G_2) has a mean value close to zero for the unchanged class (Figure A1). In the rest of this paper, we use Gaussian parameters to refer to $[A_i,\ m_i,\ s_i]$ for the three modes. The curve-fitting optimization is carried out by using the fast nonlinear solver of the Levenberg-Marquardt algorithm [59].

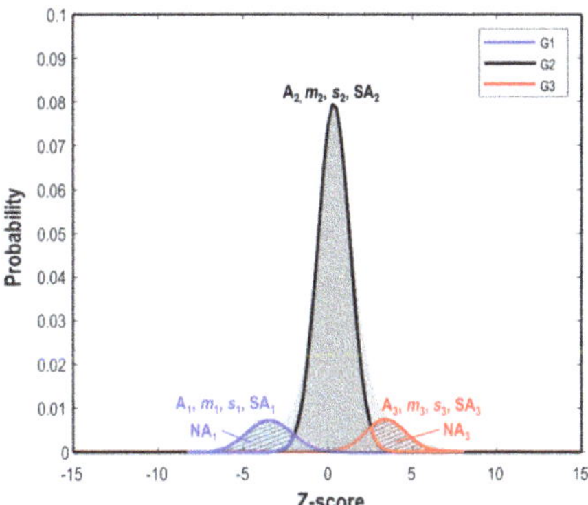

Figure A1. Tri-model Gaussian distribution. Mode 1 (G_1) and Mode 3 (G_3) are for the changes with decreased and increased Z-score values, whereas Mode 2 (G_2) is for the unchanged class. SA is surface area under the colored curve. NA is the non-overlapping area, marked by slash stripes.

(b) Select tiles using thresholds

The tiles are selected based on the following proxies:

i. Ashman D coefficient (AD). It represents the separation between two modes. The value is defined as [60]:

$$\text{AD}_i = \sqrt{2}\frac{|m_i - m_2|}{\sqrt{\left(s_i^2 + s_2^2\right)}} \tag{A2}$$

ii. Bhattacharyya coefficient (BC). It represents the goodness of fit in terms of probability. It is defined as [61]:

$$\mathrm{BC} = \sum_k \sqrt{\frac{h(z_k)}{\sum_k h(z)}} \sqrt{\frac{g(z_k)}{\sum_k g(z_k)}} \quad \text{(A3)}$$

where k stands for the k-th histogram bin.

iii. Surface ratio (SR). It represents the significance of the changes in terms of probability compared to the unchanged class. It is defined as [45]:

$$\mathrm{SR}_i = \frac{\min(SA_i, SA_2)}{\max(SA_i, SA_2)} \quad \text{(A4)}$$

where SA stands for the surface area for each mode (Figure A1).

iv. Non-overlapping Ratio (NR). It represents the significance of the changes in terms of the cumulative probability that is not overlapped with the unchanged class (G_2). It is defined as:

$$\mathrm{NR}_i = \frac{NA_i}{SA_i} \quad \text{(A5)}$$

The first three proxies are also used by HSBA [45], while the last one (NR) is an additional proxy implemented in GSBA. The main purpose of NR is to weed out the tiles where the G_2 mode has a wide distribution that overlap significantly with the other two modes. Except for BC, the other three proxies are calculated for G_1 (i = 1) and G_3 (i = 3) mode separately. Table A1 shows the empirical tile selection thresholds used for landslide selection in this study.

Table A1. Tile selection thresholds for landslide detection.

Ashman D Coeff. (AD)	>1.9
Bhattacharyya Coeff. (BC)	>0.98
Surface Ratio (SR)	>0.05
Non-overlapping Ratio (NR)	>0.4

(c) Grow patches for consistent statistical distribution

Next we try to merge the tiles in order to obtain more robust Gaussian parameters from clustered changes. Within each tile cluster, GSBA first chooses a random seed tile and identifies the neighboring tiles around it. The first round of histogram fitting (Equation (A1)) is carried out on the joint histogram between the seed and each of the neighboring tiles. The neighboring tiles with proxies above the thresholds shown in Table A1 are selected as the next round of seeds, and new neighboring tiles are identified around them. This process is repeated until all the tiles in the cluster are touched, or until the growing can no longer propagates onwards.

To avoid the situation where the first seed tile is significantly different from the rest of the tiles in the cluster, this growing process will be repeated a few times from a few randomly-selected seed tiles. The resultant patch with the largest number of merged tiles will be adopted. Through this growing process, the merged tiles (or a patch) contain consistent statistical distribution.

(d) Fill Gaussian parameters

After tile growing, we could have moved on to estimate the Bayesian probabilities by using the Gaussian parameters averaged over all patches. However, we found that in the case of landslide detection, where changes can be affected by spatially-varying factors such as local incidence angles [23], a global set of Gaussian parameters may not be the best solution. So instead, we keep the Gaussian parameters unchanged in the patches, and fill only the area outside the patches with the global average.

(e) Calculate Bayesian probability

The Bayesian probability is defined as [48]:

$$p(X|Z) = \frac{p(Z|X)p(X)}{p(Z|X)p(X) + p(Z|\overline{X})p(\overline{X})} \quad (A6)$$

where $X, \overline{X}$ stands for changes and non-changes, and $p(X|Z)$ is the Bayesian probability of changes given the Z-score value of the pixel. The prior probabilities $p(X)$ and $p(\overline{X})$ are both set to 0.5 following the suggestions in [48]. We assume that the conditional probability for the changes $p(Z|X)$ has a Gaussian probability density function (PDF) with parameters of $[\widetilde{A}_1, m_1, s_1]$ for negative Z-scores or $[\widetilde{A}_3, m_3, s_3]$ for positive Z-scores:

$$p(Z|X) = \begin{cases} \frac{\widetilde{A}_1}{\sqrt{2\pi}\, s_1} \exp\left[-\frac{1}{2}\frac{(Z-m_1)^2}{(s_1)^2}\right], \ \widetilde{A}_1 = \frac{A_1}{A_1+A_2}\ if\ Z < 0 \\ \frac{\widetilde{A}_3}{\sqrt{2\pi}\, s_3} \exp\left[-\frac{1}{2}\frac{(Z-m_3)^2}{(s_3)^2}\right], \ \widetilde{A}_3 = \frac{A_3}{A_3+A_2}\ if\ Z > 0 \end{cases} \quad (A7)$$

and the conditional probability for the non-changes $p(Z|\overline{X})$ also has a Gaussian PDF with parameters of $\left[\widetilde{A}_2, m_2,\ s_2\right]$:

$$p(Z|\overline{X}) = \frac{\widetilde{A}_2}{\sqrt{2\pi}\, s_2} \exp\left[-\frac{1}{2}\frac{(Z-m_2)^2}{(s_2)^2}\right], \ \widetilde{A}_2 = \begin{cases} \frac{A_2}{A_1+A_2}\ if\ Z < 0 \\ \frac{A_2}{A_3+A_2}\ if\ Z > 0 \end{cases} \quad (A8)$$

(f) Derive binary change maps

This step is relatively straightforward. By default, we adopt a cutoff probability of 0.5 on the Bayesian probability map in order to obtain the binary map.

(g) Choose the final change map

Step (a) to (f) will be repeated at multiple tile sizes. The list of tile sizes are determined based on the image dimensions and the preferred number of test sets. Currently we limit the tile size to be between 10 and 500 pixels, and the default number of test sets is between 4 and 8. After step (f), a binary change map will be generated at each tile size. To determine which one is the final output, GSBA calculates Ripley's K (K_r) for each map to estimate the spatial randomness of the change points [62,63]:

$$\begin{aligned} K_r &= \tfrac{A}{n^2} \textstyle\sum_i^n \sum_{j \neq i}^n I_r\left(\left|x_i - x_j\right| \le r\right) \\ I_r &= \begin{cases} 1\ if\ \left|x_i - x_j\right| \le r \\ 0\ if\ \left|x_i - x_j\right| > r \end{cases} \end{aligned} \quad (A9)$$

where r is a pre-defined distance of interest, x_i and x_j are the positions of any two change pixels, n is the total number of change pixels, and A is the image area. Ripley's K is a geospatial index to tell if points are dispersive or clustered in space. When the point distribution is close to complete spatial randomness, K_r will be close to πr^2. The higher the value, the more clustered the points. For a more efficient calculation of K_r, we down-sample the change map to 100 m × 100 m resolution, and set r = 100 m. After computing the K_r value for all binary maps, we choose the one with intermediate K_r value as our final change map.

References

1. Froude, M.J.; Petley, D.N. Global fatal landslide occurrence from 2004 to 2016. *Nat. Hazards Earth Syst. Sci.* **2018**, *18*, 2161–2181. [CrossRef]
2. Petley, D.N. On the impact of climate change and population growth on the occurrence of fatal landslides in South, East and SE Asia. *Q. J. Eng. Geol. Hydrogeol.* **2010**, *43*, 487–496. [CrossRef]
3. Lin, Y.N.; Park, E.; Wang, Y.; Quek, Y.P.; Lim, J.; Alcantara, E.; Loc, H.H. The 2020 Hpakant Jade Mine Disaster, Myanmar: A multi-sensor investigation for slope failure. *ISPRS J. Photogramm. Remote Sens.* **2021**, *177*, 291–305. [CrossRef]
4. Lin, S.C.; Ke, M.C.; Lo, C.M. Evolution of landslide hotspots in Taiwan. *Landslides* **2017**, *14*, 1491–1501. [CrossRef]

5. Hovius, N.; Stark, C.P.; Chu, H.-T.; Lin, J.-C. Supply and Removal of Sediment in a Landslide-Dominated Mountain Belt: Central Range, Taiwan. *J. Geol.* **2000**, *108*, 73–89. [CrossRef]
6. Dadson, S.J.; Hovius, N.; Chen, H.; Dade, W.B.; Hsieh, M.-L.; Willett, S.D.; Hu, J.-C.; Horng, M.-J.; Chen, M.-C.; Stark, C.P.; et al. Links between erosion, runoff variability and seismicity in the Taiwan orogen. *Nature* **2003**, *426*, 648. [CrossRef]
7. Geertsema, M.; Pojar, J.J. Influence of landslides on biophysical diversity—A perspective from British Columbia. *Geomorphology* **2007**, *89*, 55–69. [CrossRef]
8. Natsuki, S.; Toshihiko, S. Distribution and Development Processes of Wetlands on Landslides in the Hachimantai Volcanic Group, NE Japan. *Geogr. Rev. Jpn. Ser. B* **2015**, *87*, 103–114. [CrossRef]
9. Czuchlewski, K.R.; Weissel, J.K.; Kim, Y. Polarimetric synthetic aperture radar study of the Tsaoling landslide generated by the 1999 Chi-Chi earthquake, Taiwan. *J. Geophys. Res. Earth Surf.* **2003**, *108*, 1–11. [CrossRef]
10. Burrows, K.; Walters, R.J.; Milledge, D.; Densmore, A.L. A systematic exploration of satellite radar coherence methods for rapid landslide detection. *Nat. Hazards Earth Syst. Sci.* **2020**, *20*, 3197–3214. [CrossRef]
11. Yun, S.-H.; Hudnut, K.; Owen, S.; Webb, F.; Simons, M.; Sacco, P.; Gurrola, E.; Manipon, G.; Liang, C.; Fielding, E.; et al. Rapid Damage Mapping for the 2015 Mw 7.8 Gorkha Earthquake Using Synthetic Aperture Radar Data from COSMO-SkyMed and ALOS-2 Satellites. *Seismol. Res. Lett.* **2015**, *86*, 1549–1556. [CrossRef]
12. Wang, Y.; Wei, S.; Wang, X.; Lindsey, E.O.; Tongkul, F.; Tapponnier, P.; Bradley, K.; Chan, C.-H.; Hill, E.M.; Sieh, K. The 2015 M w 6.0 Mt. Kinabalu earthquake: An infrequent fault rupture within the Crocker fault system of East Malaysia. *Geosci. Lett.* **2017**, *4*, 6. [CrossRef]
13. Aimaiti, Y.; Liu, W.; Yamazaki, F.; Maruyama, Y. Earthquake-Induced Landslide Mapping for the 2018 Hokkaido Eastern Iburi Earthquake Using PALSAR-2 Data. *Remote Sens.* **2019**, *11*, 2351. [CrossRef]
14. Fujiwara, S.; Nakano, T.; Morishita, Y.; Kobayashi, T.; Yarai, H.; Une, H.; Hayashi, K. Detection and interpretation of local surface deformation from the 2018 Hokkaido Eastern Iburi Earthquake using ALOS-2 SAR data. *Earth Planets Space* **2019**, *71*, 64. [CrossRef]
15. Jung, J.; Yun, S.-H. Evaluation of Coherent and Incoherent Landslide Detection Methods Based on Synthetic Aperture Radar for Rapid Response: A Case Study for the 2018 Hokkaido Landslides. *Remote Sens.* **2020**, *12*, 265. [CrossRef]
16. Ohki, M.; Abe, T.; Tadono, T.; Shimada, M. Landslide detection in mountainous forest areas using polarimetry and interferometric coherence. *Earth Planets Space* **2020**, *72*, 67. [CrossRef]
17. Ge, P.; Gokon, H.; Meguro, K.; Koshimura, S. Study on the Intensity and Coherence Information of High-Resolution ALOS-2 SAR Images for Rapid Massive Landslide Mapping at a Pixel Level. *Remote Sens.* **2019**, *11*, 2808. [CrossRef]
18. Lin, S.-Y.; Lin, C.-W.; van Gasselt, S. Processing Framework for Landslide Detection Based on Synthetic Aperture Radar (SAR) Intensity-Image Analysis. *Remote Sens.* **2021**, *13*, 644. [CrossRef]
19. Konishi, T.; Suga, Y. Landslide detection using COSMO-SkyMed images: A case study of a landslide event on Kii Peninsula, Japan. *Eur. J. Remote Sens.* **2018**, *51*, 205–221. [CrossRef]
20. Mondini, C.A. Measures of Spatial Autocorrelation Changes in Multitemporal SAR Images for Event Landslides Detection. *Remote Sens.* **2017**, *9*, 554. [CrossRef]
21. Jung, J.; Yun, S.; Kim, D.; Lavalle, M. Damage-mapping algorithm based on coherence model using multitemporal polarimetric—Interferometric SAR data. *IEEE Trans. Geosci. Remote Sens.* **2018**, *56*, 1520–1532. [CrossRef]
22. Mondini, A.C.; Santangelo, M.; Rocchetti, M.; Rossetto, E.; Manconi, A.; Monserrat, O. Sentinel-1 SAR Amplitude Imagery for Rapid Landslide Detection. *Remote Sens.* **2019**, *11*, 760. [CrossRef]
23. Shibayama, T.; Yamaguchi, Y.; Yamada, H. Polarimetric Scattering Properties of Landslides in Forested Areas and the Dependence on the Local Incidence Angle. *Remote Sens.* **2015**, *7*, 15424–15442. [CrossRef]
24. Plank, S.; Twele, A.; Martinis, S. Landslide Mapping in Vegetated Areas Using Change Detection Based on Optical and Polarimetric SAR Data. *Remote Sens.* **2016**, *8*, 307. [CrossRef]
25. Niu, C.; Zhang, H.; Liu, W.; Li, R.; Hu, T. Using a fully polarimetric SAR to detect landslide in complex surroundings: Case study of 2015 Shenzhen landslide. *ISPRS J. Photogramm. Remote Sens.* **2021**, *174*, 56–67. [CrossRef]
26. Miranda, N.; Meadows, P.J. *Radiometric Calibration of S-1 Level-1 Products Generated by the S-1 IPF*; ESA-EOPG-CSCOP-TN-0002; European Space Agency: Paris, France, 2015.
27. Jong-Sen, L.; Papathanassiou, K.P.; Ainsworth, T.L.; Grunes, M.R.; Reigber, A. A new technique for noise filtering of SAR interferometric phase images. *IEEE Trans. Geosci. Remote Sens.* **1998**, *36*, 1456–1465. [CrossRef]
28. Lee, J.-S.; Pottier, E. *Polarimetric Radar Imaging: From Basics to Applications*; CRC Press: Boca Raton, FL, USA, 2009.
29. Piantanida, R.; Miranda, N.; Franceschi, N.; Meadows, P. *Thermal Denoising of Products Generated by the S-1 IPF*; S-1 Mission Performance Centre: Ramonville-Saint-Agne, France, 2017.
30. Cloude, S.R.; Papathanassiou, K.P. Polarimetric SAR interferometry. *IEEE Trans. Geosci. Remote Sens.* **1998**, *36*, 1551–1565. [CrossRef]
31. Schreier, G. *SAR Geocoding: Data and Systems*; Schreier, G., Ed.; Wichmann: Karlsruhe, Germany, 1993.
32. Cloude, S. The dual polarization entropy/alpha decomposition: A PALSAR case study. *Sci. Appl. SAR Polarim. Polarim. Interferom.* **2007**, *644*, 2.
33. Barakat, R. Degree of polarization and the principal idempotents of the coherency matrix. *Opt. Commun.* **1977**, *23*, 147–150. [CrossRef]

34. Mandal, D.; Kumar, V.; Ratha, D.; Dey, S.; Bhattacharya, A.; Lopez-Sanchez, J.M.; McNairn, H.; Rao, Y.S. Dual polarimetric radar vegetation index for crop growth monitoring using sentinel-1 SAR data. *Remote Sens. Environ.* **2020**, *247*, 111954. [CrossRef]
35. Dey, S.; Bhattacharya, A.; Ratha, D.; Mandal, D.; Frery, A.C. Target Characterization and Scattering Power Decomposition for Full and Compact Polarimetric SAR Data. *IEEE Trans. Geosci. Remote Sens.* **2021**, *59*, 3981–3998. [CrossRef]
36. Lin, N.Y.; Yun, S.-H.; Bhardwaj, A.; Hill, M.E. Urban Flood Detection with Sentinel-1 Multi-Temporal Synthetic Aperture Radar (SAR) Observations in a Bayesian Framework: A Case Study for Hurricane Matthew. *Remote Sens.* **2019**, *11*, 1778. [CrossRef]
37. Hostache, R.; Matgen, P.; Wagner, W. Change detection approaches for flood extent mapping: How to select the most adequate reference image from online archives? *Int. J. Appl. Earth Obs. Geoinf.* **2012**, *19*, 205–213. [CrossRef]
38. Bovolo, F.; Bruzzone, L. A Split-Based Approach to Unsupervised Change Detection in Large-Size Multitemporal Images: Application to Tsunami-Damage Assessment. *IEEE Trans. Geosci. Remote Sens.* **2007**, *45*, 1658–1670. [CrossRef]
39. Bovolo, F.; Bruzzone, L. A detail-preserving scale-driven approach to change detection in multitemporal SAR images. *IEEE Trans. Geosci. Remote Sens.* **2005**, *43*, 2963–2972. [CrossRef]
40. Martinis, S.; Twele, A.; Voigt, S. Towards operational near real-time flood detection using a split-based automatic thresholding procedure on high resolution TerraSAR-X data. *Nat. Hazards Earth Syst. Sci.* **2009**, *9*, 303–314. [CrossRef]
41. Otsu, N. A Threshold Selection Method from Gray-Level Histograms. *IEEE Trans. Syst. Man Cybern.* **1979**, *9*, 62–66. [CrossRef]
42. Kittler, J.; Illingworth, J. Minimum error thresholding. *Pattern Recognit.* **1986**, *19*, 41–47. [CrossRef]
43. Young, T.Y.; Coraluppi, G. Stochastic estimation of a mixture of normal density functions using an information criterion. *IEEE Trans. Inf. Theory* **1970**, *16*, 258–263. [CrossRef]
44. Chow, C.K.; Kaneko, T. Automatic boundary detection of the left ventricle from cineangiograms. *Comput. Biomed. Res.* **1972**, *5*, 388–410. [CrossRef]
45. Chini, M.; Hostache, R.; Giustarini, L.; Matgen, P. A hierarchical split-based approach for parametric thresholding of SAR images: Flood inundation as a test case. *IEEE Trans. Geosci. Remote Sens.* **2017**, *55*, 6975–6988. [CrossRef]
46. Pulvirenti, L.; Marzano, F.S.; Pierdicca, N.; Mori, S.; Chini, M. Discrimination of Water Surfaces, Heavy Rainfall, and Wet Snow Using COSMO-SkyMed Observations of Severe Weather Events. *IEEE Trans. Geosci. Remote Sens.* **2014**, *52*, 858–869. [CrossRef]
47. Martinis, S.; Kersten, J.; Twele, A. A fully automated TerraSAR-X based flood service. *ISPRS J. Photogramm. Remote Sens.* **2015**, *104*, 203–212. [CrossRef]
48. Giustarini, L.; Hostache, R.; Kavetski, D.; Chini, M.; Corato, G.; Schlaffer, S.; Matgen, P. Probabilistic Flood Mapping Using Synthetic Aperture Radar Data. *IEEE Trans. Geosci. Remote Sens.* **2016**, *54*, 6958–6969. [CrossRef]
49. Park, S.; Moon, W.M.; Pottier, E. Assessment of Scattering Mechanism of Polarimetric SAR Signal from Mountainous Forest Areas. *IEEE Trans. Geosci. Remote Sens.* **2012**, *50*, 4711–4719. [CrossRef]
50. Chao, W.-A.; Wu, Y.-M.; Zhao, L.; Chen, H.; Chen, Y.-G.; Chang, J.-M.; Lin, C.-M. A first near real-time seismology-based landquake monitoring system. *Sci. Rep.* **2017**, *7*, 43510. [CrossRef]
51. Lo, C.-M. Evolution of deep-seated landslide at Putanpunas stream, Taiwan. *Geomat. Nat. Hazards Risk* **2017**, *8*, 1204–1224. [CrossRef]
52. Chen, X.; Sun, Q.; Hu, J. Generation of Complete SAR Geometric Distortion Maps Based on DEM and Neighbor Gradient Algorithm. *Appl. Sci.* **2018**, *8*, 2206. [CrossRef]
53. Comert, R. Investigation of the Effect of the Dataset Size and Type in the Earthquake-Triggered Landslides Mapping: A Case Study for the 2018 Hokkaido Iburu Landslides. *Front. Earth Sci.* **2021**, *9*, 23. [CrossRef]
54. Farr, T.G.; Rosen, P.A.; Caro, E.; Crippen, R.; Duren, R.; Hensley, S.; Kobrick, M.; Paller, M.; Rodriguez, E.; Roth, L.; et al. The Shuttle Radar Topography Mission. *Rev. Geophys.* **2007**, *45*, 1–33. [CrossRef]
55. Ito, Y.; Yamazaki, S.; Kurahashi, T. Geological features of landslides caused by the 2018 Hokkaido Eastern Iburi Earthquake in Japan. In *Characterization of Modern and Historical Seismic-Tsunamic Events and Their Global—Societal Impacts*; Dilek, Y., Ogawa, Y., Okubo, Y., Eds.; Special Publications; The Geological Society: London, UK, 2020; Volume 501, pp. 171–183.
56. Kawamura, S.; Kawajiri, S.; Hirose, W.; Watanabe, T. Slope failures/landslides over a wide area in the 2018 Hokkaido Eastern Iburi earthquake. *Soils Found.* **2019**, *59*, 2376–2395. [CrossRef]
57. Kameda, J.; Kamiya, H.; Masumoto, H.; Morisaki, T.; Hiratsuka, T., Inaoi, C. Fluidized landslides triggered by the liquefaction of subsurface volcanic deposits during the 2018 Iburi-Tobu earthquake, Hokkaido. *Sci. Rep.* **2019**, *9*, 13119. [CrossRef] [PubMed]
58. Zhou, H.; Che, A.; Wang, L.; Wang, L. Investigation and mechanism analysis of disasters under Hokkaido Eastern Iburi earthquake. *Geomat. Nat. Hazards Risk* **2021**, *12*, 1–28. [CrossRef]
59. Marquardt, D. An Algorithm for Least-Squares Estimation of Nonlinear Parameters. *J. Soc. Ind. Appl. Math.* **1963**, *11*, 431–441. [CrossRef]
60. Ashman, K.M.; Bird, C.M.; Zepf, S.E. Detecting Bimodality in Astronomical Datasets. *Astron. J.* **1994**, *108*, 2348. [CrossRef]
61. Bhattacharyya, A. On a Measure of Divergence between Two Multinomial Populations. *Sankhyā Indian J. Stat.* **1946**, *7*, 401–406.
62. Ripley, B.D. *Spatial Statistics*; Wiley: Hoboken, NJ, USA, 1981. [CrossRef]
63. Dixon, P.M. Ripley's K function. In *Encyclopedia of Environmetrics*; El-Shaarawi, A.H., Piegorsch, W.W., Eds.; John Wiley & Sons Ltd.: Hoboken, NJ, USA, 2002; Volume 3, pp. 1796–1803.

MDPI

Article

Comparative Study on Potential Landslide Identification with ALOS-2 and Sentinel-1A Data in Heavy Forest Reach, Upstream of the Jinsha River

Chen Cao, Kuanxing Zhu, Tianhao Song, Ji Bai, Wen Zhang *, Jianping Chen and Shengyuan Song 

College of Construction Engineering, Jilin University, Changchun 130026, China; ccao@jlu.edu.cn (C.C.); zhukx21@mails.jlu.edu.cn (K.Z.); songth21@mails.jlu.edu.cn (T.S.); baiji2418@mails.jlu.edu.cn (J.B.); chenjp@jlu.edu.cn (J.C.); songshengyuan@jlu.edu.cn (S.S.)

* Correspondence: zhang_wen@jlu.edu.cn

Abstract: Many SAR satellites such as the ALOS-2 satellite and Sentinel-1A satellite can be used in Interferometric Synthetic Aperture Radar (InSAR) to identify landslides. As their wavelengths are different, they can perform differently in the same area. In this study, we selected the alpine canyon heavy forest area of the Baishugong–Shangjiangxiang section of the Jinsha River with a strong uplift of faults and folds as the study area. The Small Baseline Subset (SBAS)–InSAR was used for landslide identification to compare the reliability and applicability of L-band ALOS-2 data and C-band Sentinel-1A data. In total, 13 potential landslides were identified, of which 12 potential landslides were identified by ALOS-2 data, two landslides were identified by Sentinel-1A data, and the Kongzhigong (KZG) landslide was identified by both datasets. Then, the field investigation was used to verify the identification results and analyze the genetic mechanism of four typical landslides. Both the Duila (DL) and KZG landslides are bedding slip, while the Jirenhe (JRH) and Maopo (MP) landslides are creep–pull failure. Then, the difference between ALOS-2 and Sentinel-1A data on KZG landslide was compared. A total of 35,961 deformation points on the KZG landslide were obtained using ALOS-2 data, which are relatively dense. Meanwhile, a total of 7715 deformation points were obtained by Sentinel-1A data, which are relatively scattered and seriously lacking, especially in areas with dense vegetation coverage. Comparing the advantages of ALOS-2 and Sentinel-1A data and the identification results of potential landslides, the reliability and applicability of ALOS-2 data in the identification of potential landslides in areas with dense vegetation cover and complex geological conditions were confirmed from the aspects of vegetation cover, topography, field investigation, and comparative analysis of typical landslides.

Keywords: ALOS-2; Sentinel-1A; SBAS-InSAR; heavy forest area; potential landslide identification

Citation: Cao, C.; Zhu, K.; Song, T.; Bai, J.; Zhang, W.; Chen, J.; Song, S. Comparative Study on Potential Landslide Identification with ALOS-2 and Sentinel-1A Data in Heavy Forest Reach, Upstream of the Jinsha River. *Remote Sens.* **2022**, *14*, 1962. https://doi.org/10.3390/rs14091962

Academic Editors: Takeo Tadono and Masato Ohki

Received: 20 February 2022
Accepted: 17 April 2022
Published: 19 April 2022

1. Introduction

Landslides are common geological disasters in mountainous areas, causing serious casualties and economic losses [1,2]. The Jinsha River Basin has dense vegetation coverage, complex geological conditions, and many deep faults. Strong tectonic activities have led to the rapid uplift of the reach, with a rate of up to 5 mm/y [3,4], resulting in frequent disasters such as landslides and debris flows [5,6]. These disasters are likely to cause blocked rivers, resulting in landslide-dammed-lake outburst floods, which seriously threaten the safety of downstream residents [7,8]. For example, the Baige landslide and river blocking occurred twice in the upper reaches of the Jinsha River, destroying hundreds of houses and causing direct economy losses of about USD 963.5 million [7]. In addition, there are many paleo-landslide deposits along the Jinsha River [9,10], which are potential hazards for residents and buildings in the downstream. Therefore, it is of great significance to identify potential landslides along the Jinsha River.

Accurate identification of potential landslides is the key to effective disaster prevention and reduction [11]. Therefore, identification and verification methods are particularly important, as they can effectively reduce the uncertainty of results. The traditional landslide identification methods are mainly field investigation and monitoring, including the macro characteristics of landslide deformation, global navigation satellite systems (GNSS), leveling measurements, and so on, and the accuracy can reach the centimeter-to-millimeter level [12]. However, these monitoring methods are point-measurement-based [13]. When the scope of the study area is large and the geological conditions are complex, there are some problems, such as being difficult for geologists and instruments to reach, the omission of geological disasters, and being relatively time consuming. Landslide susceptibility mapping is an effective method to predict landslides, and the use of a combined model can reduce the uncertainty of results [14–16]. The prediction result of this method is usually larger than the range of a landslide, and there will also be problems regarding the time-consuming and laborious nature of field investigation. With the development of remote sensing technology, optical remote sensing interpretation and InSAR technology have effectively solved these problems. However, there are various problems in a wide range of optical remote sensing images, such as untimely updating and it being difficult to determine the macro quantitative changes. InSAR technology has the advantages of all-weather and all-day, meaning that it is widely used. It was used to obtain deformation information as early as 1969. Then, with the continuous maturation and improvement in technology, time series InSAR technology has been proposed and applied in many fields and all over the world, such as landslide monitoring and identification [17,18], surface deformation monitoring in mined-out areas [13], land subsidence [19], earthquakes [20], glacier movement [21], and so on. When combined with optical images and field investigation, it can effectively determine and verify the accuracy of the results [12,22].

InSAR technology is the product of the combination of microwave imaging and electromagnetic wave interferometry, which uses the principle of interferometry at the macroscopic level and can monitor the surface deformation through SAR satellite images [23]. However, satellites in different bands and parameters have different penetration capabilities, resulting in different information quality. Radar satellites still in service are mainly divided into the X-, C-, and L-bands [24]. X-band SAR data have the characteristics of high resolution and short wavelength, with a wavelength of about 3.1 cm [25]. Representative satellites are COSMO-SkyMed and TerraSAR launched in June 2007. Affected by short wavelength, the influence of atmospheric phase screens (APS) on these satellites is more serious [26]. The wavelength of C-band is about 5.6 cm, between X-band and L-band, and the representative satellites are Sentinel-1 and Radarsat-2 [27]. The Sentinel-1 satellite data have the characteristics of high spatial resolution and being free, which means that they are widely used in deformation monitoring of geological hazards [12,28]. The L-band satellite has a wavelength of about 23.5 cm, with strong penetration capability, meaning that SAR images have a high coherence in the vegetation-covered areas. The representative satellites are ALOS and ALOS-2, which were launched by the Japan Aerospace Exploration Agency (JAXA) in 2006 and 2014, respectively [29]. In most studies, both ALOS-2 and Sentinel-1 data were used to obtain the land subsidence [30] and soil moisture [31,32], monitor landslides [33,34], and so on. The applicability of the two datasets was compared and analyzed, and better results were achieved [33,35]. In this study, ALOS-2 and Sentinel-1A data were used to obtain surface deformation and identify landslides.

In the alpine and canyon areas with high vegetation coverage, the long-term phase change map obtained by the ALOS-2 data is clearer, while the phase change map of the Sentinel-1A data is better in the exposed rock and soil area. A large number of vegetation and other factors will reduce the data coherence, and then produce noise points that can easily be mistaken for the deformation area, which makes the latter landslide field survey more ineffective [36]. The L-band ALOS-2 data provides better spatial coverage of landslide movements than the C-band Sentinel-1 data, especially in rural areas along lake shores [32]. Compared to the displacement signal amplitudes measured by Sentinel-1, the ALOS-2 data have higher values due to the high surface penetration of the L-band. Sentinel-1 showed better results on bare soil surfaces, while ALOS-2 was more sensitive on vegetation-covered surfaces [24,31,37]. Therefore, the accuracy of identification results can be improved by using InSAR technology to process multi-sensor satellite data [33].

The study aimed to identify potential landslides in the Baishugong–Shangjiangxiang section of the Jinsha River using ALOS-2 and Sentinel-1A data and analyze the applicability of the data. The study was conducted in three stages: (a) we used SBAS-InSAR technology to obtain the phase change and surface deformation rate of long-term series, and combined it with multi-phase Google Earth images to identify potential landslides to reduce the uncertainty of results; (b) the identification results were verified and the genetic mechanism of typical landslides was analyzed through field investigation; (c) the deformation characteristics of the KZG landslide were analyzed based on the results of ALOS-2 and Sentinel-1A data. Finally, the reliability and applicability of ALOS-2 and Sentinel-1A data were comprehensively analyzed and discussed.

2. Materials and Methods

2.1. Study Area

The study area is located in the Baishugong–Shangjiangxiang section of the upper reaches of the Jinsha River, Diqing Tibetan Autonomous Prefecture, Yunnan Province, China, on the southeast edge of the Qinghai Tibet Plateau (Figure 1a). The study area covers an area of about 1171 km^2, with developed vegetation, which is a heavy forest area. Under the strong uplifting action, the river is deeply incised to form a V-shaped deep and steep valley [3], and the erosion, denudation, and glacial erosion make it an alpine and canyon landform. The elevation of the study area is 1862~4502 m, and the height difference is 2640 m.

Affected by multi-stage movement, the geological structure is complex, and deep and large faults (Jinsha River East branch fault zone, Zhongdian-Longpan-Qiaohou fault zone, Daju-Lijiang fault zone, Xiaojinhe-Lijiang fault zone, Jinsha River fault zone, Heqing-Eryuan fault zone, Weixi-Qiaohou fault zone) and folds (Songpan-Ganzi geosyncline fold system of class I tectonic unit) have developed in a large area. Plate movement caused the fault zone to move and slip, forming large-scale thrust nappe and translational shear or strike slip in the region, resulting in discontinuity or loss of structural units. The exposed strata in the area from Cenozoic to Paleozoic are Quaternary (Q), Tertiary (E), Triassic (T), Permian (P), Carboniferous (C), Devonian (D), and Cambrian (∈). The quaternary stratum is dominated by sediments, and the Tertiary, Triassic, and Permian lithology is mainly sedimentary rocks, such as conglomerate, sandstone and limestone; the lithology of the Carboniferous system is mainly limestone; the lithology of the Devonian and Cambrian is mainly metamorphic rocks, such as schist and phyllite.

The study area is mainly characterized by a monsoon climate. Affected by the southwest monsoon and southeast monsoon, the rainfall is concentrated, and the rainy season is from May to October. The annual average rainfall is 954 mm and the average evaporation is 2179 mm. The annual average runoff is 1360 m^3/s and the annual average temperature is 12.6 °C. Small earthquakes are frequent in the region, with peak seismic acceleration of 0.20 g, characteristic period of seismic response spectrum of 0.40 s, and seismic intensity of VIII. Strong tectonic activity, continuous river undercutting erosion, long-term weathering, a complex high ground stress field, free surface unloading, and other factors in the study area are the fundamental factors leading to the instability of slopes on both banks of the river, resulting in frequent landslides and other disasters [6].

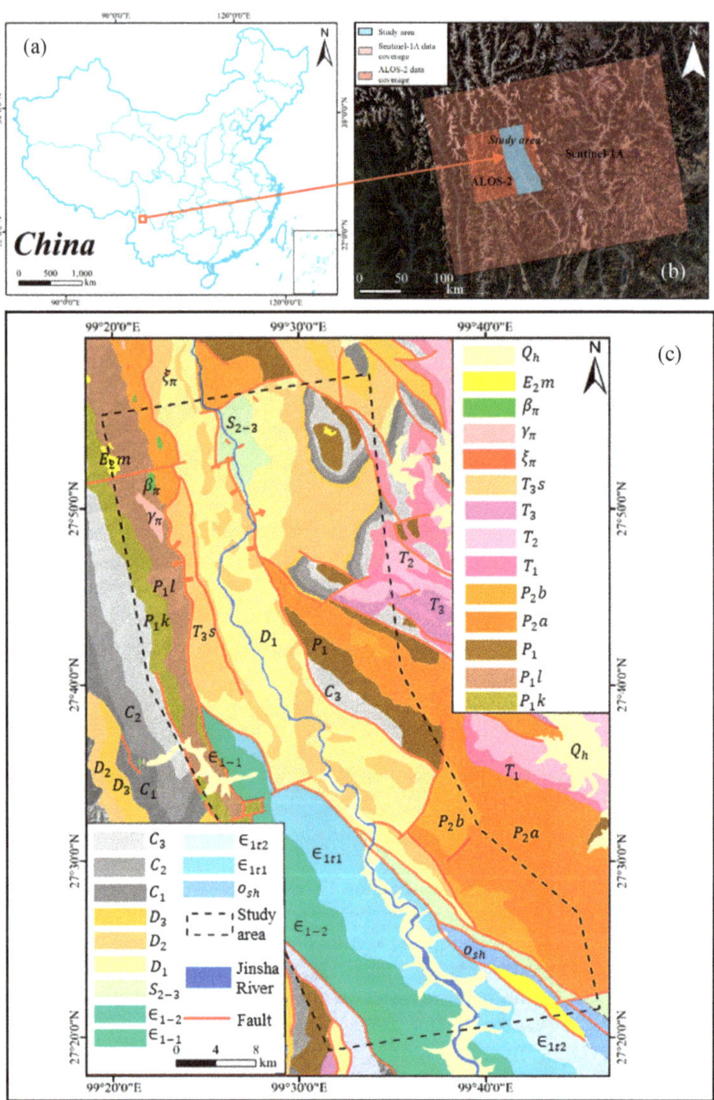

Figure 1. (**a**) The geographical location of the study area; (**b**) the coverage of the ALOS-2 and Sentinel-1A data; (**c**) the geological map of the study area.

2.2. Data Acquisition

The ascending data of ALOS-2 and Sentinel-1A satellites and digital elevation model (DEM) data were used for SBAS-InSAR technology processing. The ALOS-2 satellite images were provided by the Japan Aerospace Exploration Agency (JAXA) and launched in May 2014. Sentinel-1A satellite images were provided by the European Space Agency. It was launched in October 2014. The DEM used in the research was obtained by the Shuttle Radar Topography Mission (SRTM) sensor, with a spatial resolution of 30 m.

ALOS-2 is the only L-band SAR satellite in operation, with a wavelength of about 23.5 cm and a frequency of 1.2 GHz. The ALOS-2 satellite can obtain observation data without the influence of climate conditions and time. The L-band can penetrate vegetation, and it can be better used to monitor surface deformation in heavy forest areas. The coverage of the data is shown in Figure 1b, and the type is L1.1. The time span is from 6 October 2014 to 25 May 2020, with a total of 15 images. The resolution is 10 m and the incident angle is 36.28°. Affected by satellite shooting, the data time interval is not periodic.

Sentinel-1A is a C-band radar satellite with a wavelength of about 5.6 cm, which has certain limitations in densely vegetated areas. It can penetrate clouds and is not affected by weather and climate. It can be used to study the change in flood areas, landslides, and forest fire monitoring, and so on. The coverage of the image is shown in Figure 1b, and the data type is L1.1, slant-distance single look complex (SLC) mode. The time span is from 18 March 2017 to 21 November 2020, with a total of 29 images and a time interval of 48 days. The imaging mode is interferometric wide (IW) swath, the polarization mode is VV, the average incident angle is 33.91°, and the ground resolution is 5 × 20 m. The specific information relating to satellite data is shown in Table 1.

Table 1. Basic information of the ALOS-2 and Sentinel-1A images.

Satellite	ALOS-2	Sentinel-1A
Orbital direction	Ascending	Ascending
Temporal coverage	6 October 2014–25 May 2020	18 March 2017–21 November 2020
Level	L1.1	L1.1
Band	L-band	C-band
Wavelength	23.5 cm	5.6 cm
Resolution	10 m	5 × 20 m
Average angle of incidence	36.28°	33.91°
Polarization	HH	VV

2.3. SBAS-InSAR Technology

SBAS-InSAR technology was proposed in 2002, overcoming the limitations of spatio-temporal incoherence of SAR data and vulnerability to atmospheric effects, and generates more continuous ground phase change and deformation data in time and space [38,39], with monitoring accuracy up to the millimeter level. The GAMMA software was used to process ALOS-2 and Sentinel-1A data by SBAS-InSAR technology to obtain the surface deformation phase change and deformation rate of the Baishugong–Shangjiangxiang section. The main steps included: geocoding, image registration, generating connection diagrams, differential interference processing, removing the atmospheric error and elevation residual, estimating the deformation rate by means of the singular value decomposition (SVD) method, and reverse geocoding. The flow chart is shown in Figure 2c.

Geocoding and image registration constituted the preprocessing process of SAR data. The purpose was to match the data of the same satellite to the same common reference system. Then, according to the interval of SAR data, in order to ensure the coherence of the data, the time and spatial baseline of ALOS-2 data were set to 900 d and 420 m, respectively, and 60 interferometric pairs were obtained. The time and spatial baseline of Sentinel-1A data were set to 150 d and 240 m, respectively, and 81 interferometric pairs were obtained. The baseline connection diagrams are shown in Figure 2a,b, respectively. According to the engineering geological conditions of the study area, the adaptive filtering method was selected, and the unwrapping coefficient was set to 0.2 for processing, which was used to remove the interference of factors such as atmosphere, terrain, vegetation, and diffuse reflection of ground objects, improve the accuracy of phase map and velocity, and then ensure the accuracy of landslide identification results. The processing of removing atmospheric error and elevation residual sought to analyze the time series of the interference processed data and estimate the deformation rate and the phase change map of the study area combined with the SVD method. The color change in the phase diagram represents the change in phase, and the color difference represents the phase difference. When the phase difference is large, it represents that there is the deformation of this place that experiences a large change, which can be identified as a potential landslide area. Finally, through reverse geocoding, the results are encoded into the cartographic coordinate system, and then the deformation results with geographical coordinates are obtained, which are projected onto the map to further study the landslide.

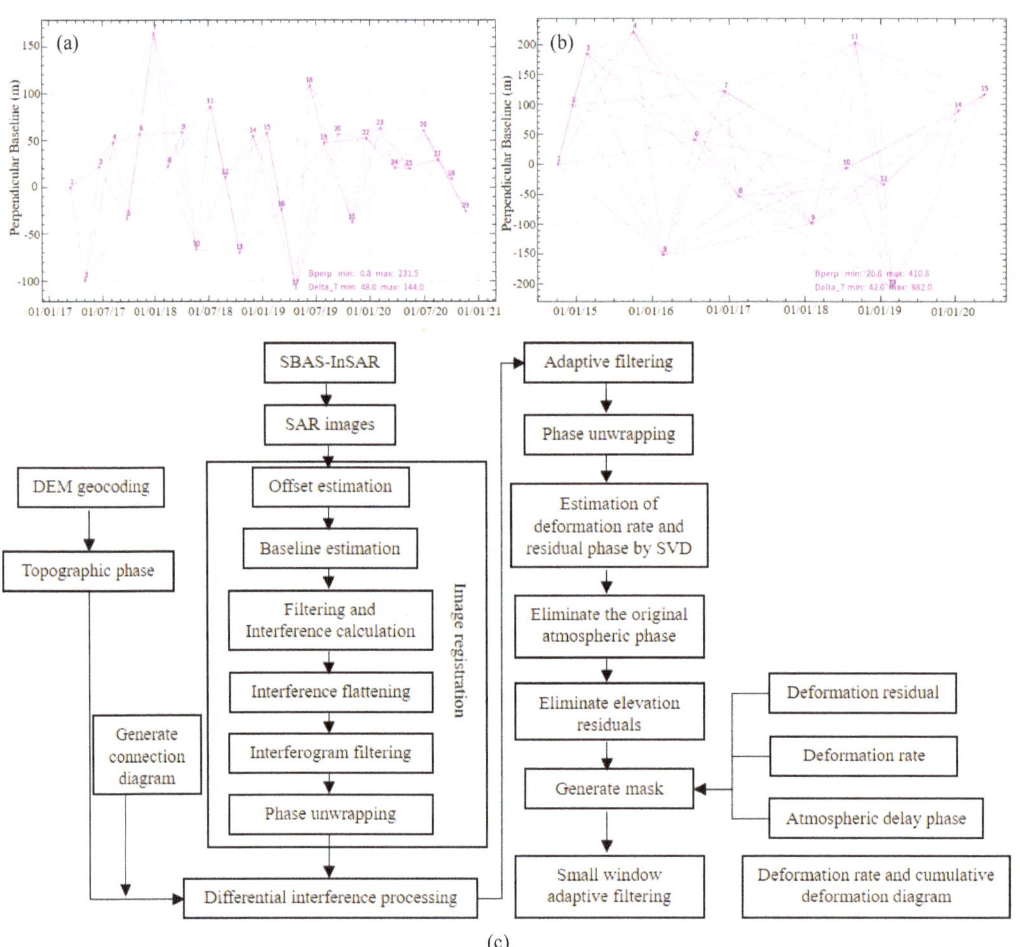

Figure 2. (**a**) Time−baseline of interferometric pairs of Sentinel-1A images; (**b**) time−baseline of interferometric pairs of ALOS-2 images; (**c**) SBAS-InSAR processing flow chart.

2.4. Normalized Difference Vegetation Index (NDVI)

The upper reaches of Jinsha River are heavily forested with dense vegetation. Due to the long river reach in the study area, there are differences in vegetation coverage at different locations, and the microwave signal strength and penetration ability of satellite data at different wavelengths are different. In this study, we used the normalized difference vegetation index (NDVI) to obtain the overall vegetation coverage of the study area, and then compare and analyze the applicability of ALOS-2 and Sentinel-1A data in the study area. NDVI was obtained by processing Landsat 8 satellite data with ArcGIS software. The data were accessed from http://www.gscloud.cn for free, (accessed on 9 April 2020). The calculation formula is as follows:

$$\text{NDVI} = (\text{IR} - \text{R})/(\text{IR} + \text{R}) \quad (1)$$

where, R and IR are the energy reflected in the red and infrared portion of the electromagnetic spectrum, respectively [40].

3. Results

3.1. Identification of Potential Landslides

The surface phase change and deformation information of the section from Baishugong to Shangjiangxiang along the Jinsha River was obtained using SBAS-InSAR technology. The phase change of surface displacement can be represented by a continuous color band. When the color band changes rapidly, it indicates that the displacement changes greatly, and it is regarded as a potential landslide area. Therefore, the surface phase map of the study area is interpreted to obtain the potential landslide area. Then, combined with the landform, deformation rate, and Google Earth images of the study area, the basic elements of the landslide are identified, and the landslide boundary is preliminarily delineated and confirmed as the identification of a potential landslide.

By interpreting the phase change information of surface displacement obtained from ALOS-2 and Sentinel-1A data, a total of 13 potential landslides and two deformation areas were identified, and their distribution is shown in Figure 3. There are more landslides distributed on the right bank, and potential landslides are named from north to south according to the location. We can find that 12 of them were identified by ALOS-2 data and 2 potential landslides were identified by Sentinel-1A data. The KZG landslide was identified in both datasets. The NDVI of the study area obtained by using ArcGIS software (Figure 4) shows that the NDVI value in the north is small, which indicates that the vegetation is sparse and the coverage is low. The vegetation in the south is dense and the coverage is high. The NDVI value is large, which can be up to 0.609. Therefore, the number of landslides in the north is higher than that in the south. The NDVI and slope values of potential landslides are shown in Table 2. The slope of the landslide is obtained through the slope extraction function of DEM by ArcGIS software. We can find that the slope of landslides is between 20° and 40°, with a maximum slope of 39.4° (Xiaohekou landslide), and a minimum slope of 20.7° (Wulucun landslide). The vegetation coverage of the southern landslides is obviously higher than that of the northern landslides, and the vegetation coverage of the Tacheng landslide is the most intensive. The KZG landslide is easily monitored by the two datasets due to low vegetation coverage and slope.

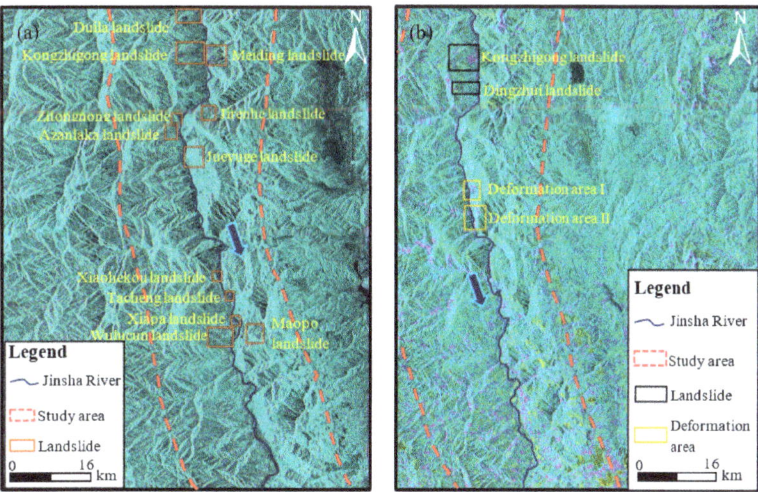

Figure 3. (**a**) Distribution of potential landslides interpreted from ALOS-2 data and (**b**) distribution of potential landslides interpreted from Sentinel-1A data.

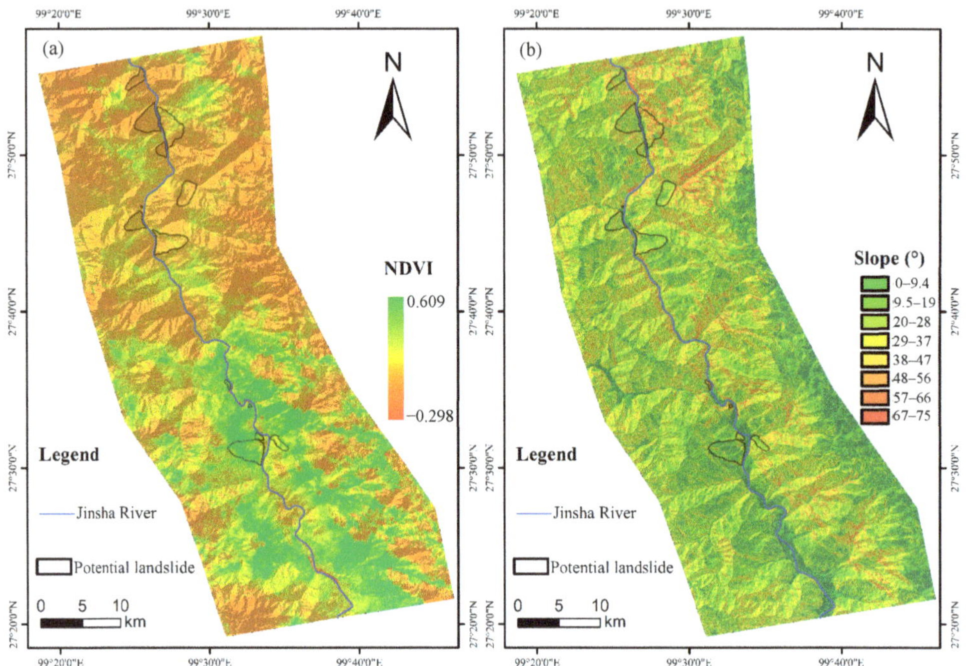

Figure 4. The NDVI value (**a**) and slope (**b**) of the study area.

Table 2. Vegetation coverage and slope of potential landslides.

Landslide	NDVI	Slope (°)	Landslide	NDVI	Slope (°)
Duila	0.058	31	Jueyuge	0.090	24.5
Kongzhigong	0.060	28.8	Xiaohekou	0.252	39.4
Meiding	0.099	37.4	Tacheng	0.413	37.9
Dingzhui	0.068	31.4	Xiapa	0.223	31.9
Jirenhe	0.074	26.9	Maopo	0.214	23.8
Zitongnong	0.083	36	Wulucun	0.304	20.7
Azanlaka	0.089	27.9	/	/	/

Field investigation is the most effective way of verifying the accuracy of potential landslide identification results. Therefore, we went to the study area on 25 April 2021 to carry out a field geological survey for the identified landslide for 17 days. Through the field investigation, we found that the identified macro damage characteristics such as cracks at the trailing edge of the landslide, landslide accumulation, and gullies are obvious, and the failure position is roughly the same as that with a large deformation rate. According to the investigation results, we finally delineated the landslide boundary. These landslides are deformed and have the possibility of occurrence, and some of them are the deformation of old landslide deposits, which proves that the identification results are more accurate. Then, typical landslides were selected for detailed analysis.

3.2. On-Site Investigation and Mechanism Analysis of Typical Landslides

Six typical landslides including Duila (DL), Kongzhigong (KZG), Jirenhe (JRH), Maopo (MP), Zitongnong (ZTN), and Xiaohekou (XHK) were selected for detailed analysis. According to the interpretation of remote sensing images and SBAS-InSAR deformation rate, DL, KZG, JRH, and MP landsides are local deformation (Figure 5). Combined with the field investigation, the deformation mechanism was further analyzed. The ZTN and XHK landslides are deforming landslides. Due to the influence of dense vegetation, it is difficult to conduct detailed geological investigation. Therefore, the deformation mechanism was preliminarily analyzed based only on the two aspects of remote sensing images and deformation rate.

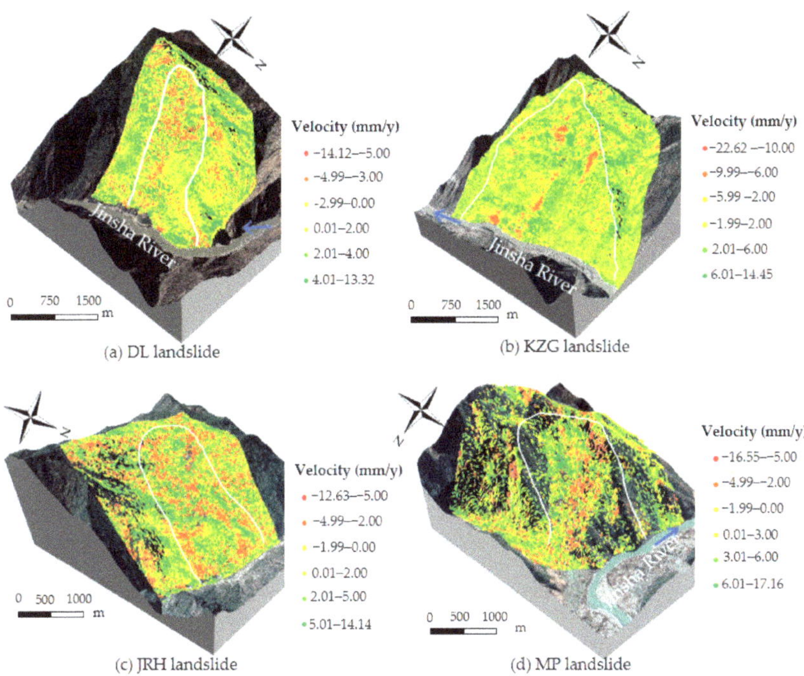

Figure 5. Deformation features of old landslide deposits with ALOS-2 data.

3.2.1. Old Landslide Deformation Features

The DL and KZG landslides are distributed on the right bank of the Jinsha River under study, and the front edge of the landslide extends to the Jinsha River terrace. The front edge of the landslide has typical river terrace landform characteristics. The JRH and MP landslides are distributed on the left side of the Jinsha River, and the front edge of the JRH landslide extends to the bottom of the valley, while the front edge of the MP landslide is a gentle and open accumulation fan. Four landslides are located near the junction of the Zhongdian-Dagu strike–slip fault, the Tuoding-Kaiwen fault, and the Jinsha River-Honghe thrust fault zone. The KZG, JRH, and MP landslides are crossed by the faults, so the underlying bedrock in this area has poor integrity and is prone to landslides. The main strata in the study area are gray-green micromorphites, sericite micromorphites intercalated with phyllite in the Lower Cambrian System; dolomite, marble intercalated with marl in the Middle and Upper Silurian System; Lower and Middle phyllites, slates, etc.; gray-green phyllites intercalated with micromorphites and dacites in the Upper Permian; and purplish-red conglomerates intercalated with mudstones and limestones in the Upper Triassic. The field investigation found that there was almost no exposed bedrock, and the main material composition was gravel soil. Among them, the overlying strata under DL, JRH, and MP were thin-layered phyllite with low strength, showing moderate to strong weathering, with pores. The medium-thin layer of slate or dolomite in the overlying strata under the KZG landslide is moderately weathered, while the weathered phyllite and dolomite have a higher degree of weathering and lower strength. The slope direction of KZG landslide is 78°, the slope is about 28°, and the rock formation is 102°–135°∠39°–60°. It was found through on-site investigation that in the middle of the intersection of dip angles on the slope outside the steep slope, all rock layers are present. It can be inferred that the deformation mechanism of the DL and KZG landslides is bedding slip with traction failure caused by bedding phyllite slip and bending. The bedrock layer in the upper part of the JRH landslide is $130^\circ\angle 20^\circ$, and the bedrock layer is almost orthogonal to the bank slope. The large angle intersects, and the deformation of the rock stratum caused by thin phyllite is mainly interlayer slip and bending. The long axis direction of the MP landslide is 317°, and the occurrence of the phyllite bedrock layer beneath the landslide body is $195^\circ\angle 65^\circ$, the bedrock layer intersects with the landslide surface at a large angle and slightly inclines to the inside of the slope. Based on this, it is speculated that in the early stage of the JRH landslide and the MP landslide, the fragmentation slopes were cut by multiple groups of structural planes, meaning that the deformation of the accumulation was mainly through the deformation of the structure. The phyllite rock mass was fully disintegrated during the sliding deformation process of the landslide, which eventually leads to instability failure, and the failure mode of it was the creep–pull failure.

The middle and lower reaches of the study area are densely covered with vegetation. Figure 5 shows the landslide deformation rate obtained from ALOS-2 data. ALOS-2 data has better penetrability through vegetation and can monitor the surface deformation in the densely vegetated areas. As can be seen from Figure 5, the local deformation of the four landslides is large and obvious. According to the field investigation results, the boundary of the landslide was finally determined, as shown in the white curve in Figure 5. There are gullies developed on the surface of the DL and KZG landslides, and the deformation is distributed near the gullies. Figure 5a shows that the deformation is distributed in the middle and lower parts of DL. According to the field investigation of the DL landslide (Figure 6a), in the houses seriously damaged by the internal and external sliding of the accumulation near the gully, there are four cracks on the side wall of the house, all of which penetrate through the wall, and they have obvious characteristics of a wide bottom and narrow top. Gullies also appear in the upper part of the JRH landslide. The deformation rate of the JRH landslide is between −12.63 and 14.14 mm/y, and the deformation is mostly located in the middle of the slope (Figure 5a). The upper part of the gullies is 3–6 m wide, 2 m deep, and has a slope of 20°; the lower part is 2–4 m wide and 2 m deep and has a slope of 25.5° (Figure 6e). The material in the gully section is mainly silty sand, and the rest is phyllite crushed stone, with a crushed stone content of about 40%. After on-site verification, part of the surface deformation was found to be caused by the local sliding of the surface rock and soil mass. The sliding mostly formed surface features such as tension cracks, small accumulations, or small steps, as shown in Figure 6c for the KZG landslide. Regarding the small landslide in the direction of the highway, the width of the road is 4.5 m, and the width of the arc-shaped tensile crack at the trailing edge of the landslide is 4 cm. This belongs to the landslide formed by the local instability of the superficial soil mass, which was induced by human engineering activities. Figure 6f is the tension crack of the MP landslide, which is the trailing crack in Figure 6g, with a dislocation height difference of 30–45 cm, a length of about 40 m, a strike of roughly 335°, and an irregular arc shape. Figure 6b shows the deposit of the DL landslide. The exposed material of the deposit here is gravel soil, with crushed stone accounting for about 60%. Poor, angular–sub-angular, and low shrubs can be seen on the surface of the accumulation. Figure 6d shows an artificial excavation slope profile of the KZG landslide. The outcrop height of it is 6 m, and the material composition mainly consists of blocks of stone soil, of which the content of boulders is about 35%. The lithology of the block rock is mainly slate and limestone, the size is mixed, there is no sorting, the particle size is generally 2–10 cm, the larger ones can reach 45 cm, the rounding is poor, they are angular or sub-angular, and the vegetation on the accumulation is flourishing. Figure 6g shows a small soil landslide on MP; according to the field investigation, it is an active landslide, which slides every rainy season and forms a multi-level ledge. The height difference between the trailing edge of the landslide and the point is 12 m, the horizontal distance is 30 m, and the slope is 21.5°. There are six (I–VI) platforms from the upper part to the lower part. The staggered height difference of the I platform is 40 cm, the width of the platform between the I–II sills is 4.3 m, and the length is 9 m. The height difference of the grade III platform sill is 55cm, the width of the platform between grades III and IV is 4.9 m, and the length is 10.7 m. The platform between them is 0.75 m wide and 10.5 m long; the height difference between the V platform sills is 47 cm and the platform between the V–VI platform sills is 8.4 m wide and 25 m long; and the VI platform sills are staggered and the height difference is 2.3 m.

Figure 6. Field investigation of old landslide deposits. (**a**) Cracks on the surface of the house; (**b**) the downstream boundary of the deposit; (**c**) a small landslide in the free direction of the highway; (**d**) a typical section at the top of the downstream side of the KZG Village; (**e**) the gully between platform IV and platform V; (**f**) a tension crack in MP landslide; (**g**) a small soil landslide in MP landslide.

3.2.2. Integral Deforming Landslides

ZTN and XHK landslides (Figure 7) are distributed on the right bank of the Jinsha River. Combining NDVI and Google Earth images, it can be found that the surface vegetation of ZTN and XHK landslides is dense, and the landform of the front edge of landslides is river terrace landform. Due to the influence of vegetation, a detailed field investigation on landslides was not carried out, so the landslide boundary was comprehensively determined by Google image, landform, and SBAS-InSAR (white line in Figure 7). Because ALOS-2 data has strong penetrability through vegetation, the two landslides were identified after data processing by SBAS-InSAR technology. According to the interpretation results, it was found that both the ZTN and XHK landslides had large deformation as a whole and the deformation rate of the two landslides can reach −13 mm/y. The larger deformation area is mainly distributed in the upper and lower parts. According to the distribution of the deformation area, it is inferred that the deformation of the ZTN is caused by the lower rock and soil mass loosening and sliding downward due to the influence of precipitation or human activities, so the landslide is a retrogressive landslide; the deformation of XHK may be due to the fact that the slope is steep and the slope is affected by precipitation, causing the surface rock and soil to loosen and slide downward.

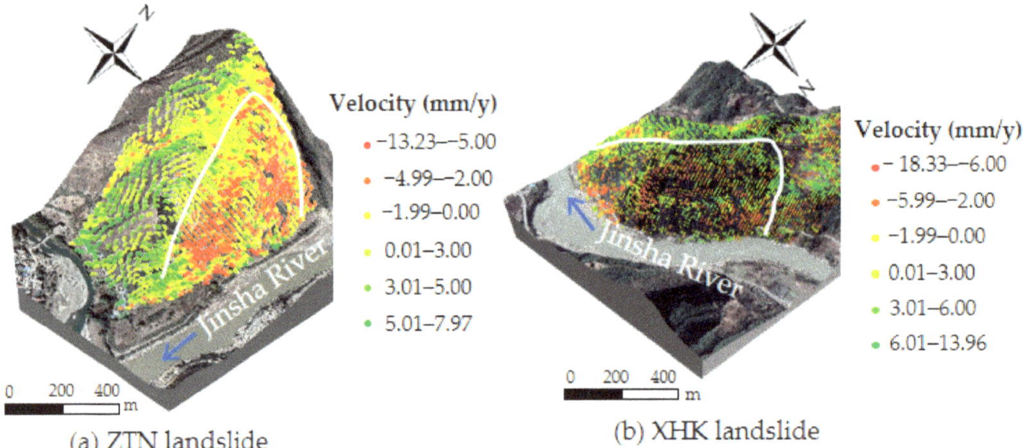

Figure 7. Deformation features of potential landslides with ALOS-2 data.

3.3. Deformation Results of KZG Landslide Based on SAR Data

The surface deformation rate of the KZG landslide was obtained by processing Sentinel-1A and ALOS-2 satellite data as shown in Figure 8. From the deformation rate map, it can be seen that the surface deformation rate obtained by Sentinel-1A data is between −27.99 and 14.28 mm/y, and the rate obtained by the ALOS-2 data is between −22.62 and 14.45 mm/y, indicating that the results obtained from the two data are roughly the same in value. In terms of the density of deformation points, the deformation points obtained by Sentinel-1A are relatively scattered and seriously lacking, especially in areas with dense vegetation coverage, with a total of 7715 deformation points, while the results of ALOS-2 data are relatively dense, with a total of 35,961 deformation points. The density of points is significantly greater than that of the Sentinel-1A data, at 4.5 times that of the Sentinel-1A data. The deformation points are counted and divided into five intervals. The statistical results are shown in Figure 8c. Except for the interval with the rate of <−5 mm/y, the number of deformation points obtained by ALOS-2 data is greater than Sentinel-1A. The deformation rate obtained by Sentinel-1A is mainly concentrated at −15–−5 mm/y, with 5546 deformation points, accounting for 71.8% of the total. The deformation rate obtained by ALOS-2 data is mainly concentrated at -5–5mm/y, with 30,278 deformation points, accounting for 84.1% of the total. The number of points is much larger than that of Sentinel-1A in this range. From the distribution of landslide deformation, the result of the ALOS-2 data shows that the deformation is mainly concentrated in four regions of the slope, numbered I to IV (in Figure 8). The deformation results obtained by Sentinel-1A data show that the landslide deformation is mainly concentrated in the middle and lower parts of the slope, and the deformation range is large, making it difficult to obtain the specific deformation area. Comparing Figure 8a,b, except for the III region, there are obvious differences in the deformation of the areas of I, II, and IV.

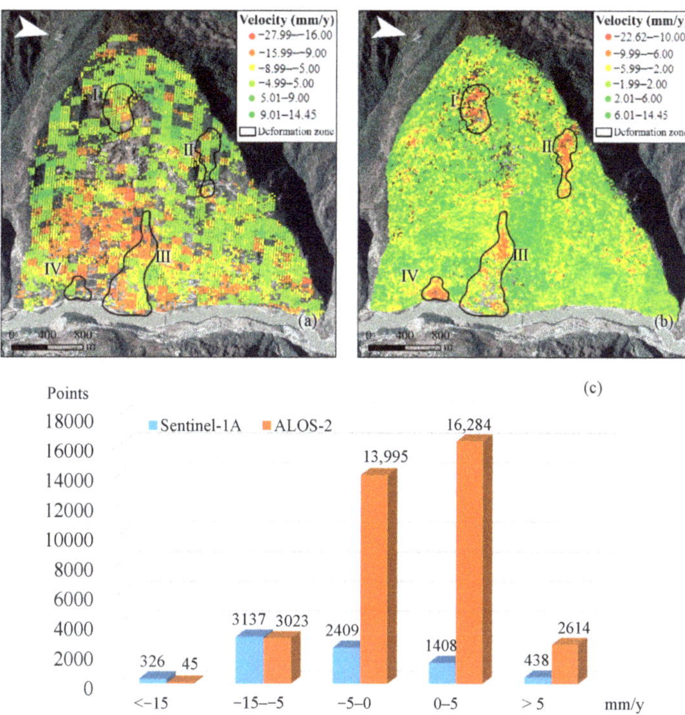

Figure 8. Deformation details and features of the KZG landslide with different data: (**a**) Sentinel-1A data, (**b**) ALOS-2 data, and (**c**) statistical diagram of deformation points.

4. Discussion

InSAR technology is one of the most effective methods for the early identification of potential landslides. In most studies, both Sentinel-1 and ALOS-2 satellite data are used to obtain surface deformation information. In this study, we also selected these two satellite datasets to identify potential landslides in the study area.

In terms of the number of the potential landslides, more landslides can be identified in the study area using ALOS-2 data, while Sentinel-1A data were relatively less applicable, with fewer identifiable landslides. Both Sentinel-1 and ALOS-2 satellites have the advantages of penetrating clouds and all-weather observations. However, there are some different drawbacks and advantages in the identification of potential landslides in large areas. The Sentinel-1 satellite in the C-band has a wavelength of 5.6 cm, with a width of 250 km and a large coverage area. The satellite revisit time is 12 days for a single satellite [12], which ensures the coherence of the data. ALOS-2 is the only operational L-band satellite that can provide high-resolution SAR images, with a width of 70 km and a wavelength of 23.5 cm. The microwave signals of the satellite will be interfered by vegetation, resulting in the decoherence of data and the loss of deformation data [28]. Compared with other SAR sensors, the penetration ability of L-band to vegetation cover is more effective than C and X bands, and it can more accurately detect the change in the ground information [32,41]. Therefore, for the watershed with relatively sparse vegetation, it is more appropriate to use Sentinel-1A data for geological hazard identification with relatively low cost and less data. However, for the upper reaches of Jinsha River with dense vegetation (such as the section from Baishugong to Shangjiangxiang, Figure 4), the ALOS-2 data is more suitable, and the results of the deformation monitoring are more accurate. Additionally, the HH polarization used in ALOS-2 data is more sensitive to surface deformation than the VV polarization used in Sentinel-1A data [42].

Through early identification and field investigation, it was found that the ALOS-2 data, which has a strong penetrating ability through vegetation, has a strong advantage in identifying landslides in alpine and canyon areas covered with dense vegetation.

After on-site investigation and verification and deformation mechanism analysis of the landslides identified in the study area, the deformation mechanisms of the landslides identified in the upstream of the study area are all bedding sliding, and most of the landslides identified in the middle and lower reaches are slip–bending–creep–crack-type landslides, which can be seen in the downstream. Additionally, traction landslides can be found in the lower reach. It is easy to explain this in combination with the distribution of strata and lithology: most of the landslides identified in the upper reaches of the study area are distributed on the right bank of the Jinsha River. On the lateral slope outside the slope, the bedding plane directly affects the stability of the rock slope, and a shear slip plane is easily formed along the bedding plane [43]. The bedding slope is affected by factors such as precipitation and human activities. It is very easy to slip in the lower reaches of the study area; the landslides in the middle and lower reaches of the study area are mainly distributed on the left bank of the Jinsha River, and the landslide rock layers on the left bank tend to be inward, which is conducive to the stability of the slope, but because the underlying bedrock is broken, and the rocks are mostly thin layers with low-strength phyllite, soft rock (phyllite) interlayers may eventually lead to the formation of landslides, and precipitation concentration in the study area plays a key role in the formation of landslides [44], easily expanding and contracting under the action of precipitation. The existence of the phyllite interlayer leads to the reduction in its mechanical strength [45–47] and coupled with the steepness of the slope and the influence of human activities, it is very easy to make the surface rock mass of the slope slip. Slip bending occurs, and then develops into a creep–crack-type landslide.

The field investigation found that the areas with large deformation identified by InSAR were mostly tensile cracks, gullies, and accumulations. The small retaining wall in the MP landslide is used to surround a small landslide behind. According to the cracks distributed on the surface of the retaining wall and anti-slide pile, it can be seen that the retaining wall and anti-slide pile play a certain protective role. At the same time, the small landslide is also observed in the continuous deformation; in the large deformation area of KZG, a house with deformation damage was also found. There are four cracks on the side of the house, which are wide at the bottom and narrow at the top. The cracks are obviously caused by the deformation of the ground caused by the deformation of the lower rock and soil. A large deformation occurred in this area, which verifies the reliability of the InSAR identification results.

For the ZTN and XHK landslides, the integral deformation occurs. Due to the dense vegetation on the surface of the landslide, it is impossible to conduct on-site field investigations, and it is impossible to align the deformation mechanism for accurate analysis. According to the distribution of deformation areas obtained by InSAR, it can be seen that most areas with large deformation are concentrated in the slope body. In the lower part, combined with the steep gradient and lithological distribution, it is inferred that most of the landslides in this area are traction landslides. According to historical case studies, some large-scale landslides are the result of the evolution of degenerative landslides. Combined with remote sensing images and field surveys, it is found that the identified landslides (such as the KZG and JRH landslides, etc.) are populated by residents, and most of the identified landslide fronts are built across with carrying traffic. Since the landslides in the study area have already undergone great deformation, they will continue to deform under the influence of rainfall and human activities in the future. Therefore, local government departments need to do a good job in disaster prevention and mitigation work to avoid unnecessary losses.

Twelve deformation points were selected in the areas of I, II, III, and IV to plot the time series curves using Sentinel-1A and ALOS-2 deformation data. As shown in Figure 9b, it can be seen that the deformation of regions I and II is small and mostly maintains a stable state. The deformation occurs mostly in regions III and IV, showing a downward trend, and the deformation variable can reach −95 mm. The deformation was in a state of continuous deformation during the study period and intensified during the rainy season, which can be divided into three stages: accelerated deformation, slow deformation, and accelerated deformation again. The time series deformation curve obtained by ALOS-2 data in October 2014 to May 2020 is shown in Figure 9c; the deformation shows a certain periodicity—that is, from October 2014 to February 2015, the landslide was in a stable state. Then, affected by the rainy season, the rainfall gradually increased, and the landslide was affected by erosion, which made the slope of the curve become larger, and then it was in a state of accelerated deformation. When the rainy season ends, the landslide gradually returns to a stable state, which makes the landslide change periodically during the study period—that is, a stable state, accelerated deformation, and a stable state again. Based on the above analysis, it can be seen that the deformation of the landslide is greatly affected by the erosion of rainwater. Under the action of rainfall, the loose material source on the landslide surface slides down, which causes the slope body to show a downward trend. The results obtained by ALOS-2 data have obvious regularity, showing that the slope deformation is greatly affected by rainfall, and the greater the rainfall is, the more obvious the landslide deformation is. The results of Sentinel-1A data also show the characteristics of rainfall, but the regularity is not obvious. Therefore, ALOS-2 data can better highlight the deformation characteristics of landslides.

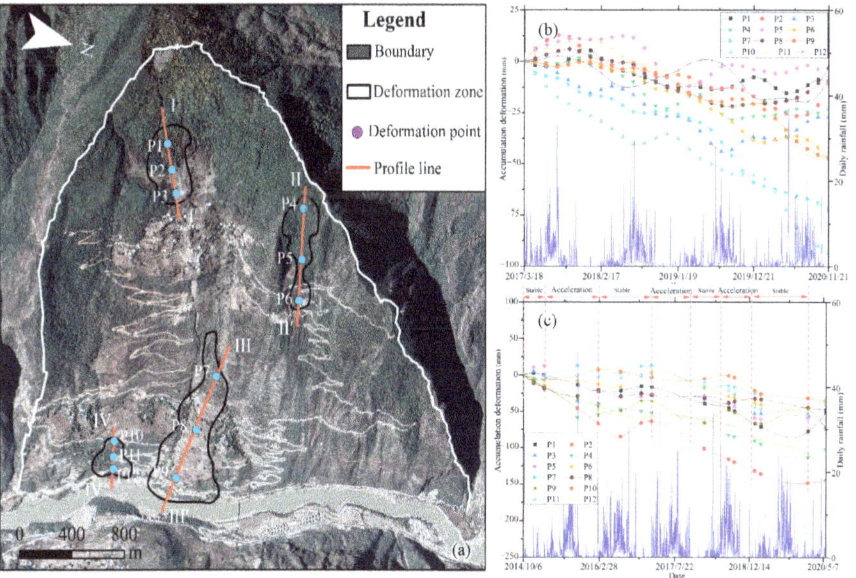

Figure 9. Time series deformation of KZG landslide, (**a**) the distribution map of the selected deformation points, (**b**) Sentinel-1A data, and (**c**) ALOS-2 data.

As observed during the field investigation, area I is located at the top of the KZG landslide. It can be found that it was a small landslide, which occurred about 5 or 6 years ago. It has an elliptical area, and many fragments are visible, with the length between 3 and 10 cm. A chair-shaped groove can be seen in the lower part of the landslide. It is speculated that this may be the tensile crack at the trailing edge of the landslide (Figure 10a), which is tens of meters long and about 20 cm wide. Affected by surface sedimentation, the crack is basically closed and difficult to identify. With the results of a rainstorm, the crack may develop again and evolve into a new trailing edge tensile fracture. When it is connected to the potential sliding surface inside the accumulation, the landslide may occur. Area II is located on the right edge of the landslide. The deformation of the area is affected by human engineering activities. There is an artificial excavation section with a height of about 8 m. There is a gray-white collapse deposit in the upper part, mostly composed of rock and soil. The bottom is slate bedrock with an occurrence of 110°∠53° and a spacing of 15cm. The bedrock is visible at about 80 m downstream of the slope, and the surface rock mass is seriously broken. Area III is a small accumulation in the middle of the front edge of the landslide. The artificial excavation section is visible inside, with a height of about 10 m. The exposed material is gray-white soil, and occasionally blocks of stones with a particle size of 10–20 cm. In addition, multiple yellow bands can be seen in the profile, which is a staggered zone. Area IV is located in the leading edge of the KZG landslide, where the bedrock lithology is medium-thick dolomitic limestone. The occurrence of the bedding plane is 47°∠58° near the upstream, and 50°∠38°, 70°∠38°, 53°∠60° near the downstream. The spacing of the bedding plane is about 30 cm. The bedding plane is rough, and its integrity is good as a whole. The local fragmentation structure is moderate-weak weathering, and the bank slope structure is a consequent slope outside the steep slope. In addition, two groups of structural planes were developed. One group was outside the steep slope, and the other group was a near-vertical structural plane in the inclined slope, both of which played a cutting role. The upper part of the bedrock is the accumulation body, and its thickness can reach 1.5 m. According to the field investigation results, the deformation in regions I to IV is serious. Therefore, from the deformation rate of slope in Figure 9, it can be seen that the landslide deformation results obtained from ALOS-2 data are more in line with the actual change in slope, and the results are more reliable.

Figure 10. Field photos of areas I to IV. (**a**) The tensile crack at the trailing edge of the small landslide in area I; (**b**) the artificial excavation section in area II; (**c**) the artificial excavation section in area III; (**d**) bedrock outcrop at the bottom of the KZG landslide.

Therefore, although Sentinel-1A data has many advantages, ALOS-2 data is more suitable for identifying potential landslides in the lower reaches of the Jinsha River with dense vegetation and complex geological conditions.

5. Conclusions

In this paper, the typical reach of the Jinsha River from Baishugong to Shangjiangxiang was selected as the study area. Landslides occurred frequently in the region, which can easily lead to river blockages. Therefore, it is necessary to identify potential landslides in the study area.

The SBAS-InSAR technology was used to obtain the surface deformation information of the study area. Then, potential landslides were identified, and the identification results were verified by field investigation. Thirteen potential landslides were identified using C-band Sentinel-1A data and L-band ALOS-2 satellite data. Most of them were distributed on the right bank of the Jinsha River, and the slope was between 20° and 40°. Twelve potential landslides were identified by ALOS-2 data, two landslides were identified by Sentinel-1A data, and KZG landslide was identified by both data. The vegetation in the north of the study area is sparse and the vegetation in the south is dense, so there are more landslides in the north, and the landslides identified by Sentinel-1A data are all in the north. Therefore, in terms of quantity and vegetation penetration ability, ALOS-2 data can identify more landslides, having good applicability.

The field investigation showed that the potential landslides were deformed, which confirmed the reliability of the results. Combined with remote sensing data and field investigation, the deformation mechanisms of the DL landslide and KZG landslide are bedding slip, and those of the JRH landslide and MP landslide are creep–pull failure. Through the analysis of the deformation results of the two datasets for the KZG landslide, it was found that the ALOS-2 data results are relatively dense, and there are four deformation zones in the landslide, while the Sentienl-1A data results are relatively dispersed, and the deformation zone is different from the ALOS-2 results, where only region III has deformation. Therefore, the field investigation and time series analysis of these areas showed that the deformation of KZG landslide is serious and still in a state of deformation, and its deformation is affected by rainfall and human activities. Areas I to IV are deformed and seriously damaged, indicating that ALOS-2 data are more reliable and more suitable for identifying potential landslides in areas with complex terrain and dense vegetation cover.

Comparing the advantages of ALOS-2 and Sentinel-1A data and the identification results of potential landslides, ALOS-2 data have higher reliability, applicability, and advantages in the identification of potential landslides in areas with dense vegetation and complex geological conditions similar to the study area. The research can provide a basis for the use and selection of satellite data for potential landslide identification in alpine and canyon areas, so as to carry out effective disaster prevention and reduction, which has important practical significance.

Author Contributions: Conceptualization, C.C. and K.Z.; methodology, T.S. and J.B.; software, K.Z.; validation, C.C., K.Z. and J.B.; formal analysis, W.Z.; investigation, J.C.; resources, C.C.; data curation, C.C.; writing—original draft preparation, K.Z., T.S. and J.B.; writing—review and editing, C.C.; visualization, C.C., K.Z., T.S. and B.J; supervision, W.Z.; project administration, C.C.; funding acquisition, C.C., W.Z., J.C. and S.S. All authors have read and agreed to the published version of the manuscript.

Funding: This research was funded by the 2nd Research Announcement on the Earth Observations (EO-RA2) (NO. ER2A2N110, 2019.04–2022.04), the Interdisciplinary Integration and Innovation Project of Jilin University (JLUXKJC2021ZZ17), and the National Nature Science Foundation of China (No. 42022053, 41877220, 42007261).

Acknowledgments: Thanks to the ALOS-2 and Sentinel-1A data provided by the Japan Aerospace Exploration Agency (JAXA) and the European Space Agency (ESA) respectively, which ensured the success of the study.

Conflicts of Interest: The authors declare no conflict of interest.

References

1. Zhang, L.; Dai, K.; Deng, J.; Ge, D.; Liang, R.; Li, W.; Xu, Q. Identifying Potential Landslides by Stacking-InSAR in Southwestern China and Its Performance Comparison with SBAS-InSAR. *Remote Sens.* **2021**, *13*, 3662. [CrossRef]
2. Wang, Y.; Zhao, B.; Li, J. Mechanism of the catastrophic June 2017 landslide at Xinmo Village, Songping River, Sichuan Province, China. *Landslides* **2018**, *15*, 333–345. [CrossRef]
3. Chen, J.; Peng, W.; Sun, X.; Wang, Q.; Han, X. Comparisons of several methods for landslide susceptibility mapping: Case of the Benzilan and Waka Towns, Southwest China. *Arab. J. Geosci.* **2021**, *14*, 1622. [CrossRef]
4. Sun, X.; Han, X.; Chen, J.; Bao, Y.; Peng, W. Numerical simulation of the Qulong Paleolandslide Dam event in the late pleistocene using the finite volume type shallow water model. *Nat. Hazards* **2022**, *111*, 439–464. [CrossRef]
5. Larsen, I.J.; Montgomery, D.R. Landslide erosion coupled to tectonics and river incision. *Nat. Geosci.* **2012**, *5*, 468–473. [CrossRef]
6. Li, Y.; Chen, J.; Zhou, F.; Li, Z.; Mehmood, Q. Stability evaluation and potential damage of a giant paleo-landslide deposit at the East Himalayan Tectonic Junction on the Southeastern margin of the Qinghai-Tibet Plateau. *Nat. Hazards* **2022**, *111*, 2117–2140. [CrossRef]
7. Tian, S.; Chen, N.; Wu, H.; Yang, C.; Zhong, Z.; Rahman, M. New insights into the occurrence of the Baige landslide along the Jinsha River in Tibet. *Landslides* **2020**, *17*, 1207–1216. [CrossRef]
8. Yao, J.M.; Lan, H.X.; Li, L.P.; Cao, Y.M.; Wu, Y.M.; Zhang, Y.X.; Zhou, C.D. Characteristics of a rapid landsliding area along Jinsha River revealed by multi-temporal remote sensing and its risks to Sichuan-Tibet railway. *Landslides* **2022**, *19*, 703–718. [CrossRef]
9. Chen, J.; Dai, F.; Lv, T.; Cui, Z. Holocene landslide-dammed lake deposits in the Upper Jinsha River, SE Tibetan Plateau and their ages. *Quat. Int.* **2013**, *298*, 107–113. [CrossRef]
10. Yan, J.; Chen, J.; Zhou, F.; Li, Y.; Zhang, Y.; Gu, F.; Zhang, Y.; Li, Y.; Li, Z.; Bao, Y.; et al. Numerical simulation of the Rongcharong paleolandslide river-blocking event: Implication for the longevity of the landslide dam. *Landslides* **2022**. [CrossRef]
11. Adnan, M.S.G.; Rahman, M.S.; Ahmed, N.; Ahmed, B.; Rabbi, M.F.; Rahman, R.M. Improving Spatial Agreement in Machine Learning-Based Landslide Susceptibility Mapping. *Remote Sens.* **2020**, *12*, 3347. [CrossRef]
12. Zhu, K.; Xu, P.; Cao, C.; Zheng, L.; Liu, Y.; Dong, X. Preliminary Identification of Geological Hazards from Songpinggou to Feihong in Mao County along the Minjiang River Using SBAS-InSAR Technique Integrated Multiple Spatial Analysis Methods. *Sustainability* **2021**, *13*, 1017. [CrossRef]
13. Chen, D.; Chen, H.; Zhang, W.; Cao, C.; Zhu, K.; Yuan, X.; Du, Y. Characteristics of the Residual Surface Deformation of Multiple Abandoned Mined-Out Areas Based on a Field Investigation and SBAS-InSAR: A Case Study in Jilin, China. *Remote Sens.* **2020**, *12*, 3752. [CrossRef]
14. Sterlacchini, S.; Ballabio, C.; Blahut, J.; Masetti, M.; Sorichetta, A. Spatial agreement of predicted patterns in landslide susceptibility maps. *Geomorphology* **2011**, *125*, 51–61. [CrossRef]
15. Guzzetti, F.; Reichenbach, P.; Ardizzone, F.; Cardinali, M.; Galli, M. Estimating the quality of landslide susceptibility models. *Geomorphology* **2006**, *81*, 166–184. [CrossRef]
16. Rossi, M.; Guzzetti, F.; Reichenbach, P.; Mondini, A.C.; Peruccacci, S. Optimal landslide susceptibility zonation based on multiple forecasts. *Geomorphology* **2010**, *114*, 129–142. [CrossRef]
17. Zhao, C.; Lu, Z.; Zhang, Q.; de la Fuente, J. Large-area landslide detection and monitoring with ALOS/PALSAR imagery data over Northern California and Southern Oregon, USA. *Remote Sens. Environ.* **2012**, *124*, 348–359. [CrossRef]
18. Ren, T.; Gong, W.; Gao, L.; Zhao, F.; Cheng, Z. An Interpretation Approach of Ascending-Descending SAR Data for Landslide Identification. *Remote Sens.* **2022**, *14*, 1299. [CrossRef]
19. Guo, L.; Gong, H.; Li, J.; Zhu, L.; Xue, A.; Liao, L.; Sun, Y.; Li, Y.; Zhang, Z.; Hu, L.; et al. Understanding Uneven Land Subsidence in Beijing, China, Using a Novel Combination of Geophysical Prospecting and InSAR. *Geophys. Res. Lett.* **2020**, *47*, e2020GL088676. [CrossRef]
20. Liu, F.; Elliott, J.R.; Craig, T.J.; Hooper, A.; Wright, T.J. Improving the Resolving Power of InSAR for Earthquakes Using Time Series: A Case Study in Iran. *Geophys. Res. Lett.* **2021**, *48*, e2021GL093043. [CrossRef]
21. Zhou, J.; Li, Z.; Li, X.; Liu, S.; Chen, Q.; Xie, C.; Tian, B. Movement estimate of the Dongkemadi Glacier on the Qinghai-Tibetan Plateau using L-band and C-band spaceborne SAR data. *Int. J. Remote Sens.* **2011**, *32*, 6911–6928. [CrossRef]
22. Motagh, M.; Wetzel, H.-U.; Roessner, S.; Kaufmann, H. A TerraSAR-X InSAR study of landslides in southern Kyrgyzstan, Central Asia. *Remote Sens. Lett.* **2013**, *4*, 657–666. [CrossRef]
23. Catani, F.; Farina, P.; Moretti, S.; Nico, G.; Strozzi, T. On the application of SAR interferometry to geomorphological studies: Estimation of landform attributes and mass movements. *Geomorphology* **2005**, *66*, 119–131. [CrossRef]
24. Ng, A.H.-M.; Ge, L.; Du, Z.; Wang, S.; Ma, C. Satellite radar interferometry for monitoring subsidence induced by longwall mining activity using Radarsat-2, Sentinel-1 and ALOS-2 data. *Int. J. Appl. Earth Obs. Geoinf.* **2017**, *61*, 92–103. [CrossRef]
25. Minh, D.H.T.; Tran, Q.C.; Pham, Q.N.; Dang, T.T.; Nguyen, D.A.; El-Moussawi, I.; Toan, T.L. Measuring Ground Subsidence in Ha Noi Through the Radar Interferometry Technique Using TerraSAR-X and Cosmos SkyMed Data. *IEEE J. Sel. Top. Appl. Earth Obs. Remote Sens.* **2019**, *12*, 3874–3884. [CrossRef]
26. Hanssen, R.F.; Weckwerth, T.M.; Zebker, H.A.; Klees, R. High-resolution water vapor mapping from interferometric radar measurements. *Science* **1999**, *283*, 1297–1299. [CrossRef]

27. Chen, Z.; White, L.; Banks, S.; Behnamian, A.; Montpetit, B.; Pasher, J.; Duffe, J.; Bernard, D. Characterizing marsh wetlands in the Great Lakes Basin with C-band InSAR observations. *Remote Sens. Environ.* **2020**, *242*, 111750. [CrossRef]
28. Cao, C.; Zhang, W.; Chen, J.; Shan, B.; Song, S.; Zhan, J. Quantitative estimation of debris flow source materials by integrating multi-source data: A case study. *Eng. Geol.* **2021**, *291*, 106222. [CrossRef]
29. Arikawa, Y.; Kankaku, Y.; Saruwatari, H.; Hatooka, Y.; Suzuki, S. ALOS-2 launch and initial checkout status. In Proceedings of the Conference on Sensors, Systems, and Next-Generation Satellites XVIII, Amsterdam, The Netherlands, 22–25 September 2014.
30. Darwish, N.; Kaiser, M.; Koch, M.; Gaber, A. Assessing the Accuracy of ALOS/PALSAR-2 and Sentinel-1 Radar Images in Estimating the Land Subsidence of Coastal Areas: A Case Study in Alexandria City, Egypt. *Remote Sens.* **2021**, *13*, 1838. [CrossRef]
31. Cui, H.; Jiang, L.; Paloscia, S.; Santi, E.; Pettinato, S.; Wang, J.; Fang, X.; Liao, W. The Potential of ALOS-2 and Sentinel-1 Radar Data for Soil Moisture Retrieval With High Spatial Resolution Over Agroforestry Areas, China. *IEEE Trans. Geosci. Remote Sens.* **2022**, *60*, 1–17. [CrossRef]
32. Sekertekin, A.; Marangoz, A.M.; Abdikan, S. ALOS-2 and Sentinel-1 SAR data sensitivity analysis to surface soil moisture over bare and vegetated agricultural fields. *Comput. Electron. Agric.* **2020**, *171*, 105303. [CrossRef]
33. Xiong, Z.; Feng, G.; Feng, Z.; Miao, L.; Wang, Y.; Yang, D.; Luo, S. Pre- and post-failure spatial-temporal deformation pattern of the Baige landslide retrieved from multiple radar and optical satellite images. *Eng. Geol.* **2020**, *279*, 105880. [CrossRef]
34. Dong, J.; Zhang, L.; Li, M.; Yu, Y.; Liao, M.; Gong, J.; Luo, H. Measuring precursory movements of the recent Xinmo landslide in Mao County, China with Sentinel-1-and ALOS-2 PALSAR-2 datasets. *Landslides* **2018**, *15*, 135–144. [CrossRef]
35. Bayik, C.; Abdikan, S.; Ozdemir, A.; Arlkan, M.; Sanli, F.B.; Dogan, U. Investigation of the landslides in Beylikduzu-Esenyurt Districts of Istanbul from InSAR and GNSS observations. *Nat. Hazards* **2021**, *109*, 1201–1220. [CrossRef]
36. Chen, Y.; Sun, Q.; Hu, J. Quantitatively Estimating of InSAR Decorrelation Based on Landsat-Derived NDVI. *Remote Sens.* **2021**, *13*, 2440. [CrossRef]
37. Vafaei, S.; Soosani, J.; Adeli, K.; Fadaei, H.; Naghavi, H.; Pham, T.D.; Bui, D.T. Improving Accuracy Estimation of Forest Aboveground Biomass Based on Incorporation of ALOS-2 PALSAR-2 and Sentinel-2A Imagery and Machine Learning: A Case Study of the Hyrcanian Forest Area (Iran). *Remote Sens.* **2018**, *10*, 172. [CrossRef]
38. Berardino, P.; Fornaro, G.; Lanari, R.; Sansosti, E. A new algorithm for surface deformation monitoring based on small baseline differential SAR interferograms. *IEEE Trans. Geosci. Remote Sens.* **2002**, *40*, 2375–2383. [CrossRef]
39. Lanari, R.; Mora, O.; Manunta, M.; Mallorquí, J.J.; Berardino, P.; Sansosti, E. A Small-Baseline Approach for Investigating Deformations on Full-Resolution Differential SAR Interferograms. *IEEE Trans. Geosci. Remote Sens.* **2004**, *42*, 1377–1386. [CrossRef]
40. Sezer, E.A.; Pradhan, B.; Gokceoglu, C. Manifestation of an adaptive neuro-fuzzy model on landslide susceptibility mapping: Klang valley, Malaysia. *Expert Syst. Appl.* **2011**, *38*, 8208–8219. [CrossRef]
41. Zhang, T.; Xie, S.; Fan, J.; Huang, B.; Wang, Q.; Yuan, W.; Zhao, H.; Chen, J.; Li, H.; Liu, G.; et al. Detection of Active Landslides in Southwest China using Sentinel-1A and ALOS-2 Data. *Procedia Comput. Sci.* **2021**, *181*, 1138–1145. [CrossRef]
42. Holah, N.; Baghdadi, N.; Zribi, M.; Bruand, A.; King, C. Potential of ASAR/ENVISAT for the characterization of soil surface parameters over bare agricultural fields. *Remote Sens. Environ.* **2005**, *96*, 78–86. [CrossRef]
43. Luo, G.; Hu, X.; Gu, C.; Wang, Y. Numerical simulations of kinetic formation mechanism of Tangjiashan landslide. *J. Rock Mech. Geotech. Eng.* **2012**, *4*, 149–159. [CrossRef]
44. Zhao, J.; Yu, J.; ML, X.; HJ, C.; Tao, L.; Fan, B.; Bing, L. Physical model studies on fill embankment slope deformation mechanism under rainfall condition. *Rock Soil Mech.* **2018**, *39*, 2933–2940. [CrossRef]
45. Skempton, A.W. Long-Term Stability of Clay Slopes. *Géotechnique* **1964**, *14*, 77–102. [CrossRef]
46. Shuzui, H. Process of slip-surface development and formation of slip-surface clay in landslides in Tertiary volcanic rocks, Japan. *Eng. Geol.* **2001**, *61*, 199–219. [CrossRef]
47. Sun, S.; Huang, R.; CJ, Z.; XJ, P. Types and formation law of structural planes of Emeishan basalt in Sichuan, China. *Chengdu Univ. Technol.* **2015**, *42*, 463–475. [CrossRef]

Article

Interferometric SAR Observation of Permafrost Status in the Northern Qinghai-Tibet Plateau by ALOS, ALOS-2 and Sentinel-1 between 2007 and 2021

Lichuan Zou [1,2,3], Chao Wang [1,2,3,*], Yixian Tang [1,2,3], Bo Zhang [1,2,3], Hong Zhang [1,2,3] and Longkai Dong [1,2,3]

[1] Key Laboratory of Digital Earth Science, Aerospace Information Research Institute, Chinese Academy of Sciences, Beijing 100094, China; zoulichuan20@mails.ucas.ac.cn (L.Z.); tangyx@aircas.ac.cn (Y.T.); zhangbo@radi.ac.cn (B.Z.); zhanghong@radi.ac.cn (H.Z.); donglk@aircas.ac.cn (L.D.)

[2] International Research Center of Big Data for Sustainable Development Goals, Beijing 100094, China

[3] College of Resources and Environment, University of Chinese Academy of Sciences, Beijing 100049, China

[*] Correspondence: wangchao@radi.ac.cn; Tel.: +86-10-821-781-61

Abstract: With global warming, permafrost is undergoing degradation, which may cause thawing subsidence, collapse, and emission of greenhouse gases preserved in previously frozen permafrost, change the local hydrology and ecology system, and threaten infrastructure and indigenous communities. The Qinghai-Tibet Plateau (QTP) is the world's largest permafrost region in the middle and low latitudes. Permafrost status monitoring in the QTP is of great significance to global change and local economic development. In this study, we used 66 scenes of ALOS data (2007–2009), 73 scenes of ALOS-2 data (2015–2020) and 284 scenes of Sentinel-1 data (2017–2021) to evaluate the spatial and temporal permafrost deformation over the 83,000 km^2 in the northern QTP, passing through the Tuotuohe, Beiluhe, Wudaoliang and Xidatan regions. We use the SBAS-InSAR method and present a coherence weighted least squares estimator without any hypothetical model to calculate long-term deformation velocity (LTDV) and maximum seasonal deformation (MSD) without any prior knowledge. Analysis of the ALOS results shows that the LTDV ranged from −20 to +20 mm/year during 2007–2009. For the ALOS-2 and Sentinel-1 results, the LTDV ranged from −30 to 30 mm/year during 2015–2021. Further study shows that the expansion areas of permafrost subsidence are concentrated on braided stream plains and thermokarst lakes. In these areas, due to glacial erosion, surface runoff and river alluvium, the contents of water and ground ice are sufficient, which could accelerate permafrost subsidence. In addition, by analyzing LTDV and MSD for the different periods, we found that the L-band ALOS-2 is more sensitive to the thermal collapse of permafrost than the C-band sensor and the detected collapse areas (LTDV < −10 mm/year) are consistent with the GF-1/2 thermal collapse dataset. This research indicates that the InSAR technique could be crucial for monitoring the evolution of permafrost and freeze-thaw disasters.

Keywords: permafrost; InSAR; Qinghai-Tibet Plateau; ALOS; ALOS-2; Sentinel-1; thermal melting collapse

Citation: Zou, L.; Wang, C.; Tang, Y.; Zhang, B.; Zhang, H.; Dong, L. Interferometric SAR Observation of Permafrost Status in the Northern Qinghai-Tibet Plateau by ALOS, ALOS-2 and Sentinel-1 between 2007 and 2021. *Remote Sens.* **2022**, *14*, 1870. https://doi.org/10.3390/rs14081870

Academic Editors: Takeo Tadono and Masato Ohki

Received: 28 February 2022
Accepted: 9 April 2022
Published: 13 April 2022

1. Introduction

The Qinghai-Tibet Plateau (QTP) is known as the Asian water tower, with an average altitude of more than 4000 m [1]. It is bounded by the Pamir Plateau in the west, Hengduan Mountain in the east, the southern end of the Himalayas in the south and Kunlun Altun Mountain, Qilian Mountain in the north [2]. The QTP is a high terrain and thus receives more solar radiation energy than lower elevation areas [3]. The Chinese mainland climate is affected by the South Asian and East Asian monsoons, resulting in a diversity of climates in different regions, such as the rainy climate in China's southern part of the Yangtze River and drought in Northwest China [4]. In addition, the QTP has many glaciers, lakes, groundwater and surface rivers, making the QTP a super water tower in the plateau area, which affects the water system layout of all of Asia [5]. The QTP is a region with a large amount of permafrost at high latitudes [6]. As a key component of the Earth's cryosphere, permafrost plays an important role in the surface energy balance, carbon and water cycles, terrestrial ecosystem, and hydrological system [7]. In recent years, with global warming, permafrost degradation has accelerated [8], and degradation has had an impact on the environment and the energy and material balance. Therefore, it is very important to monitor the permafrost status on a large scale for a long time series [7].

Traditional measurement methods of permafrost deformation include GPS [9], leveling surveys [10], and drilling [11]. However, due to the harsh environment of the QTP, these methods cannot monitor permafrost on a large scale [8]. The multitemporal interferometric synthetic aperture radar (MT-InSAR) technique is a useful tool to map ground deformation [12]. MT-InSAR has been used to monitor the freeze-thaw cycle of permafrost [13–30], and to retrieve the thickness of the active layer [31–36] and permafrost degradation [37–40]. In these studies, some researchers have been committed to monitoring permafrost for a long time. Zhang [8] used Sentinel-1, ENVISAT and ERS-1 data to evaluate the ground deformation of permafrost and the risk along the Qinghai-Tibet Railway (QTR) from 1997 to 2018. The results show that the estimated deformation rate ranged from −20 to +10 mm/year and most of the QTR appeared to be stable. Daout [41] used ENVISAT and Sentinel-1 data to construct the spatial and temporal dynamics of permafrost deformation in the northeastern QTP from 2003 to 2019. The results show the pervasive subsidence of the permafrost of up to ~2 cm/year, increasing by a factor of 2 to 5 from 2003 to 2019. However, because the C-Band SAR data are easily affected by the region's vegetation and the atmosphere, the results may be affected by spatial and temporal decorrelation. The ALOS Phased Array type L-band Synthetic Aperture Rada (PALSAR) is preferred for ground subsidence monitoring in areas covered by vegetation and where there is a high rate of ground deformation [42]. Therefore, in order to improve the coherence of targets, we used L-band datasets to monitor the ground deformation of permafrost from 2007 to 2021.

The ground deformation process of permafrost is complex. With tectonic activity, erosion, and sedimentation all interacting in the QTP [43], it is difficult to accurately describe the freezing and thawing cycle of permafrost. Therefore, research has attempted to understand the deformation characteristics of permafrost. The sinusoidal model [44,45] and degree-day model [8,46] were used to describe the seasonal variation in the ground surface due to up-down deformation cycles of permafrost. However, it remains controversial which type of model is better at describing seasonal deformation [47]. To extract the temporal characteristics of permafrost directly from the SAR data, Wang [47] directly converted the network of interferograms into a deformation time series without a preset deformation model. Then, the long-term deformation velocity and seasonal deformation were extracted. However, for seasonal deformation, Wang assumed that the highest terrain elevation occurred from January to February, and the lowest elevation occurred from August to October. Wang also averaged the intra-annual deformation value. The average intra-annual deformation may smooth the features of the permafrost deformation. In addition, using prior knowledge may not be suitable for application to the QTP with spatial heterogeneity. In this study, we proposed a long-term deformation velocity and maximum seasonal deformation model without any prior knowledge to directly extract the deformation features of permafrost.

To reveal the status of the permafrost, we extracted time series deformation directly. First, we used 66 scenes of ALOS data (2007–2009), 73 scenes of ALOS-2 data (2015–2020) and 284 scenes of Sentinel-1 data (2017–2021) to reveal the spatial and temporal permafrost deformation in the northern QTP. Second, thermal collapse of permafrost was detected. Finally, we revealed the relationship between the maximum seasonal deformation and the long-term deformation velocity.

2. Study Area and Dataset

2.1. Study Area

The study area is located in the Hoh Xil region of the QTP. Hoh Xil is located in Qinghai Province, China, between Kunlun Mountain and Hoh Xil Mountain [48]. The Hoh Xil area, located at a high altitude, covers an area of approximately 2.35×10^5 square kilometers, with an average altitude of 5000 m [49]. The cold climate conditions make Hoh Xil a natural permafrost field. More than 90% of the land is covered by permafrost, with a thickness of 80~120 m [50]. There are a large number of glaciers and lakes. Glaciers have created a large-scale glacial erosion of mountains and produced a large number of deposits [51]. These deposits freeze and thaw repeatedly in the extremely cold environment, and are decomposed into sand of different particle sizes, to form different landscapes [52]. In addition, the ice and snow on the high mountains continue to flow into the Hoh Xil basin and finally converge in the lake [53]. Eventually, the Hoh Xil region evolved many types of landforms. The study area (Figure 1) is located in the northern part of the QTP, passing through the Tuouohe, Beiluhe, Wudaoliang and Xidatan (90.713~95.828°E, 33.981~36.197°N). The mean annual ground temperature of the study area is close to 0 °C, with elevations from 2600 to 7000 m. The annual precipitation is between 50 mm and 400 mm [54]. The typical geomorphic types include glaciers, lakes, hot melt lakes, alpine meadows, and alpine grasslands. There is a large amount permafrost and seasonally frozen soil which provides abundant opportunities for us to study the permafrost deformation.

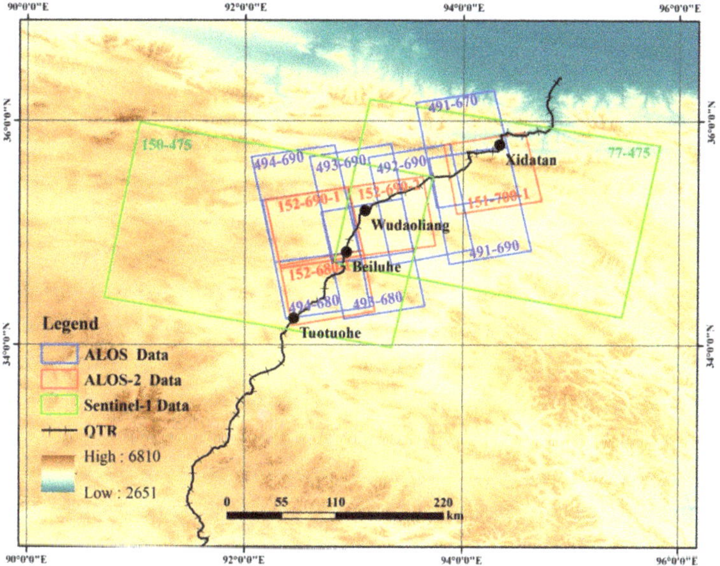

Figure 1. The study area with topography and surface traces of four overlapping ALOS ascending tracks, three ALOS-2 ascending tracks and two Sentinel-1 descending tracks in the northern QTP.

2.2. ALOS and ALOS-2 Data

The first SAR dataset was acquired by the Japanese Advanced Land Observing Satellite (L-band Synthetic Aperture Radar ALOS PALSAR). The wavelength of ALOS PALSAR is 23.62 cm with a pixel resolution of 12.5 m and a swath of 70 km. The incidence angles range between 36° and 38°. The images correspond to the ascending orbit that is northeast-looking [55]. In this study, a total of 66 scenes of ascending stripmap mode ALOS PALSAR data along four adjacent tracks covering seven frames from 2007 to 2009, were collected in the northern QTP. The acquired ALOS data are shown in Table 1. The second SAR dataset was acquired by the Advanced Land Observing Satellite-2 (L-band ALOS-2 PALSAR-2) launched in May 2014 by the Japan Aerospace Exploration Agency (JAXA). The revisit cycle of the ALOS-2 PALSAR-2 is 14 days [56]. In this study, a total of 73 scenes of ascending ALOS-2 PALSAR-2 data along three tracks covering four frames from 2015 to 2020 were collected in the northern QTP. The acquired ALOS-2 data are shown in Table 2.

Table 1. ALOS/PALSAR Data.

ALOS/PALSAR Path-Frame	Number of SAR Images	Acquisition Time	Resolution (Azimuth)	Ground Coverage
491-690	10	16 June 2007~ 12 December 2009	10 m	70 km × 100 km
491-700	10	16 June 2007~ 12 December 2009	10 m	70 km × 100 km
492-690	17	15 February 2007~ 23 February 2010	10 m	70 km × 100 km
493-680	12	17 January 2007~ 09 March 2009	10 m	70 km × 100 km
493-690	12	17 January 2007~ 09 March 2009	10 m	70 km × 100 km
494-680	11	21 June 2007~ 08 February 2009	10 m	70 km × 100 km
494-690	11	21 June 2007~ 08 February 2009	10 m	70 km × 100 km

Table 2. ALOS-2/PALSAR-2 Data.

ALOS-2/PALSAR-2 Path-Row-Id	Number of SAR Images	Acquisition Time	Resolution (Azimuth)	Ground Coverage
152-680-1	20	24 February 2015~ 09 November 2021	10 m	70 km × 70 km
152-690-1	15	15 December 2015~ 03 September 2019	10 m	70 km × 70 km
152-690-2	17	30 June 2015~ 21 December 2021	10 m	70 km × 70 km
151-700-1	20	23 July 2015~ 16 January 2020	10 m	70 km × 70 km

2.3. Sentinel-1 Data

Sentinel-1 is a two satellites constellation that carries C-band SAR sensors (~5.6 cm wavelength) under the Copernicus Programme coordinated and managed by the European Space Agency (ESA). Sentinel-1 data are provided to users free of charge. The revisit cycle of Sentinel-1 single satellite A or B is 12 days in all weather conditions, day and night [57]. In this study, a total of 284 scenes of descending Sentinel-1 TOPS data along two tracks covering two frames from 2017 to 2021, were collected in the northern QTP. The Sentinel-1 datasets were acquired in interferometric wide swath (IW) mode. The acquired Sentinel-1 data are shown in Table 3.

Table 3. Sentinel-1 Data.

Sentinel-1 Path-Frame	Number of SAR Images	Acquisition Time	Resolution (Azimuth)	Ground Coverage
77-475	161	16 March 2017~ 12 December 2021	20 m	250 km × 167 km
150-475	123	11 October 2017~ 01 December 2021	20 m	250 km × 167 km

3. Methodology

A coherence weighted least square estimator without any hypothetical model was used to calculate the long-term deformation velocity (LTDV) and maximum seasonal deformation (MSD) without any prior knowledge. The main steps are shown in Figure 2. The main steps are: (1) importing ALOS, ALOS-2 and Sentinel-1 images and images coregistration; (2) Optimal interference pair selection and interferogram generation; (3) Multilook filtering, phase unwrapping and coherent target selection; (4) Least squares inversion; (5) Atmospheric phase correction and topographic residual correction; (6) Geocoding the average deformation map and extracting the time series deformation; (7) Extracting the long-term deformation velocity and maximum seasonal deformation; and (8) Analyzing the deformation of permafrost with auxiliary data.

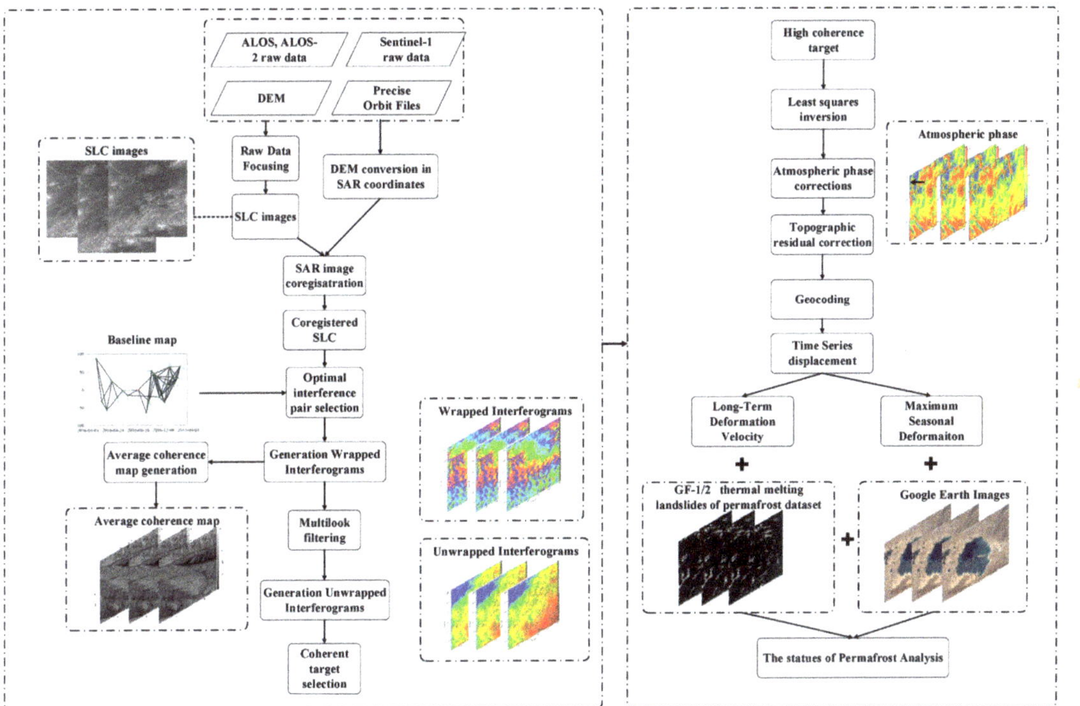

Figure 2. Flowchart of the processing approach.

3.1. Interferogram and Phase Unwrapping

First, we converted raw SAR data (ALOS, ALOS-2, Sentinel-1) to SLC format and used DEM data to register the slave image with the master image. Then we selected the optimal interferometric pair by calculating the temporal and spatial baselines of the primary and secondary images. After optimal interference pair selection, the next step is to generate interferograms. In the interferogram, each point has a phase value, which contains much important information. Let us assume point P in the master and slave image. Each SAR image contains phase and amplitude information. For point P, the phase can be modeled as:

$$\phi(P) = \varphi + \frac{4\pi}{\lambda}\mathrm{r} + \alpha + \mathrm{n} \tag{1}$$

where φ is the original phase controlled by target scatter attributes, such as roughness and moisture, λ is the wavelength of the SAR sensor, r is the distance between the sensor and the target, α is the atmospheric phase, and n represents the noise phase. In the interferogram, for point P, the phase can be modeled as:

$$\Delta\phi(P) = \Delta\varphi + \frac{4\pi}{\lambda}\Delta r + \Delta\alpha + \Delta\nu \tag{2}$$

where $\Delta\varphi$ is the phase difference generated by the change in the target's properties, $\frac{4\pi}{\lambda}\Delta r$ is the distance difference between the two observations to point P, $\Delta\alpha$ is the different atmospheric phase, and $\Delta\nu$ is the noise phase.

Through Equation (2), the interferogram can be generated from two registered SAR images. The range of the interferogram is $-\pi$ to π. After generating the interferogram, the next step is phase unwrapping. Phase unwrapping is the process of estimating the whole phase difference between a reference pixel and other pixels. The phase unwrapping algorithm calculates the phase difference between adjacent pixels and then integrates these phase differences, which is equivalent to calculating the number of color fringes starting from the reference pixel. For most unwrapping algorithms, minimizing the residual weighting function between the estimated unwrapping phase difference and the original phase difference is used. In this study, we used a minimum cost flow (MCF) unwrapping algorithm. The phase unwrapping algorithm target is to minimize Equation (3):

$$\sum_{i,j} w_{i,j}^{(x)}\left|\Delta\phi_{i,j}^{(x)} - \Delta\psi_{i,j}^{(x)}\right| + \sum_{i,j} w_{i,j}^{(y)}\left|\Delta\phi_{i,j}^{(y)} - \Delta\psi_{i,j}^{(y)}\right| \tag{3}$$

$\Delta\psi^{(y)}$ is the unwrapping phase difference in the y direction and $\Delta\psi^{(x)}$ is the wrapping phase difference in the x direction.

We used ISCE software (https://github.com/isce-framework/isce2, accessed on 20 March 2022) for this p;rocessing. To increase the signal-to-noise ratio, during processing, we set 14:4, 16:8 and 3:9 multilook in the azimuth and range direction for ALOS, ALOS-2 and Sentinel-1 interferograms, respectively. After the multilook processing, the spatial resolution (azimuth pixel resolution × range pixel resolution) of ALOS, ALOS-2 and Sentinel-1 was 50 m × 10 m, 5 m × 35 m and 23 m × 3 m, respectively. To accelerate the processing rate, we set 15 threads based on an Intel(R) Xeon(R) Gold 6129 CPU with 16 CPU cores for parallel processing.

3.2. Time Series Deformation Processing

After the phase unwrapping, we selected highly coherent targets (coherence > 0.7) to extract the time series deformation. Let us assume that there are N + 1 scene SAR images coregistered to the master image, and the time order is $(t_0 \ldots t_N)$ which can generate M interferograms. The original phase of each SAR image is $\phi = \left[0, \phi^1 \ldots \phi^N\right]$, where we assume $\phi^0 = 0$. ϕ contains the ground deformation phase, atmospheric phase, topographic phase and noise phase. For each pixel, the model is described as:

$$\Delta\phi = A\phi \tag{4}$$

where $\Delta\phi$ represents the interferometric phase and A is the $M \times N - 1$ coefficient matrix. In matrix A, 1 represents the master image, and -1 represents the slave image. Then, the coherence weighted least square method is used to obtain raw phase ϕ [58,59]:

$$\hat{\phi} = argmin\left|\left|W^{\frac{1}{2}}(\Delta\phi - A\phi)\right|\right|_2 = \left(A^T W A\right)^{-1} A^T W \Delta\phi \tag{5}$$

The estimated original phase $\hat{\phi}$ includes the following items:

$$\hat{\phi} = \hat{\phi}_{dis} + \hat{\phi}_{tropo} + \hat{\phi}_{geom} + \hat{\phi}_{resid} \tag{6}$$

where $\hat{\phi}_{dis}$ represents the ground deformation phase, $\hat{\phi}_{tropo}$ represents the atmospheric phase, $\hat{\phi}_{geom}$ represents the topographic phase, and $\hat{\phi}_{resid}$ represents the residual phase. W is a M × M diagonal weight matrix:

$$\mathrm{W} = \mathrm{diag}\left\{\frac{2L\gamma^{1^2}}{1-\gamma^{1^2}}, \ldots, \frac{2L\gamma^{M^2}}{1-\gamma^{M^2}}\right\} \tag{7}$$

$$\gamma^{j^2} = \frac{1}{\sigma^2_{\Delta\phi^j}} \tag{8}$$

where $\sigma^2_{\Delta\phi^j}$ is the phase variance of the j_{th} interferogram calculated through the integration of the phase probability distribution function and L is the number of looks. After the raw phase is calculated by the least squares method, the displacement time series can be obtained:

$$\hat{\phi}_{dis} = \hat{\phi} - \hat{\phi}_{tropo} - \hat{\phi}_{geom} - \hat{\phi}_{resid} \tag{9}$$

For atmospheric phase removal, global atmospheric models (GAMs), particularly ERA5, have great potential in atmospheric phase screen (APS) correction for InSAR applications on the Tibetan plateau [60]. The ERA5 model has been successfully applied in the APS correction in the QTP [41,47,54]. Therefore, in this study, we used ERA-5 reanalysis data to estimate the tropospheric delay [61]. For the topographic residual correction, the topographic phase residual caused by a DEM error was estimated based on the proportionality with the perpendicular baseline time series [62]:

$$\hat{\phi}^i - \hat{\phi}^i_{tropo} = \frac{-4\pi}{\lambda}\left(\frac{B^i_\perp}{r sin(\theta)} z_\varepsilon + \sum_{k=0}^{2} \frac{c_k (t_i - t_1)^k}{k!} + \sum_{l \in I_s} s_l H(t_i - t_1)\right) + \phi^i_{resid} \tag{10}$$

where $B^i_\perp$ is the perpendicular baseline, θ is the incidence angle, $H(t_i - t_1)$ is a Heaviside step function and z_ε, c_k and s_l are the unknown parameters, which can be estimated by minimizing the L^2-norm of the residual phase. The time series deformation processing was implemented by MintPy (https://github.com/insarlab/MintPy, accessed on 20 March 2022) [63].

3.3. Long-Term Deformation Velocity and Maximum Seasonal Deformation

We can extract two import indices from the time series deformation: long-term deformation trend and maximum seasonal deformation. The long-term deformation may be used to describe the water storage in the active layer. The maximum seasonal deformation describes the active layer thickness. These two indices were obtained by direct calculation without any assumed model. Wang [47] used the intra-annual highest-lowest terrain elevation difference to represent the intra-annual seasonal deformation after extracting and separating the linear trend from the deformation time series. Wang used prior knowledge and assumed that the highest terrain elevation occurred from January–February, and the lowest elevation occurred from August to October. In addition, the averaged value was used for the final intra-annual deformation value. The average intra-annual deformation may smooth the features of the permafrost deformation. In addition, using prior knowledge may not be suitable for application to the QTP with spatial heterogeneity. The long-term linear trend velocity can be modeled as:

$$D'(t) = D(t) - v \cdot t \tag{11}$$

where $D(t)$ is the time series deformation, t is the temporal span of image acquisition dates away from the first acquisition date, v is the long-term linear trend velocity, and $D'(t)$

is the deformation time series minus the long-term linear trend. The maximum seasonal deformation can be modeled as:

$$msd = max_{i \epsilon T}\left(D'_{max} - D'_{min}\right) \quad (12)$$

where D'_{max} is the highest terrain elevation in year i and D'_{min} is the lowest terrain elevation in year i. T is the number of years spanned by the observation period.

4. Results and Analysis

4.1. Long-Term Deformation and Time Series Deformation in the Northern QTP

The coherence weighted least squares method with 66 scenes of ALOS data (2007–2009), 73 scenes of ALOS-2 data (2015–2020) and 284 scenes of Sentinel-1 data (2017–2021) was used to retrieve the permafrost deformation in the northern QTP. The results illuminate on a large scale the permafrost deformation in the northern QTP from 2007 to 2021. The ground deformation results are shown in Figure 3.

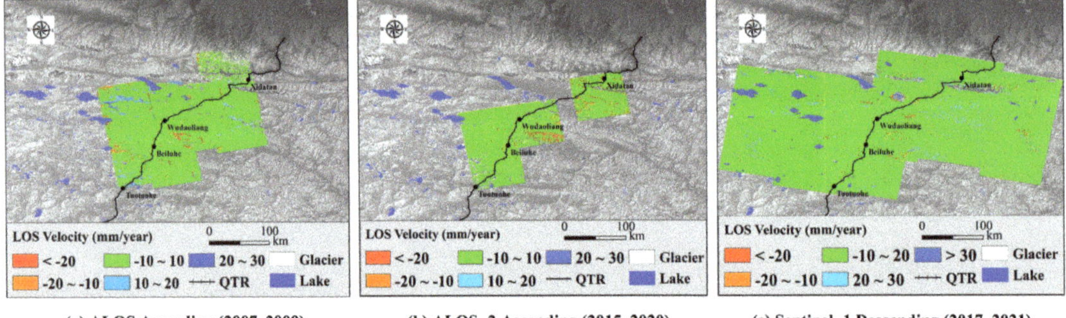

Figure 3. Long-term deformation results of the study area. Background is the DEM. (**a**) ALOS result (2007–2009). (**b**) ALOS-2 result (2015–2020). (**c**) Sentinel-1 result (2017–2021).

In Figure 3, the red color indicates the subsidence in the LOS direction, and the blue color is the uplift toward the satellite. There are a large number of lakes (deep blue color) in the western part of the study area. From the three different deformation maps obtained by using three different SAR sensors, we find that the long-term deformation velocity in the northern QTP ranges from −20 to +20 mm/year (ALOS, 2007–2009), −30 to 15 mm/year (ALOS-2, 2015–2020), and −25 to 30 mm/year (Sentinel-1, 2017–2021). Further study shows that the expansion areas of permafrost subsidence are concentrated on braided stream plains and thermokarst lakes. In these areas, due to glacial erosion, surface runoff and river alluvium, the contents of water and ground ice are sufficient, which could accelerate permafrost subsidence.

In Figure 4a, we selected two typical areas, Salt Lake and a braided stream plain. Salt Lake (35.532°N, 93.409°E) is located in the northeastern section of Hoh Xil (Kekexili) in the QTP. The water of Salt Lake mainly comes from seasonal rivers. On 14 September 2011, Zonag Lake burst its banks, which caused the area of Salt Lake to expand significantly. The outburst of Zonag Lake could affect the freeze-thaw behavior of the surrounding permafrost. We used Google Earth images to show the historical expansion of Salt Lake (Figure 4b,d). From the different period images, we find that the Salt Lake experienced significant expansion and that the distance from Salt Lake to the QTR is decreasing. Figure 4c shows the result processed from stacks of ascending ALOS datasets from 2007 to 2009. The red color indicates ground deformation away from the satellite. We find that the degradation of the permafrost mainly occurred in the northeast direction. This may be due to the existence of glaciers and braided stream plains in northeast of Salt Lake. Glaciers have created a large-scale glacial erosion of mountains, resulting in a large number of deposits.

In addition, with sufficient moisture, the effect of thawing processes is stronger than that of freezing processes. This could increase the thickness of the active layer. Figure 4d shows the result processed from stacks of descending Sentinel-1 datasets from 2017 to 2021. Similarly, the red color represents the subsidence along the LOS direction. Compared with the 2007–2009 result and 2017–2021 result, the area of permafrost degradation is expanding. We used ArcMap software to analyze the temporal and spatial changes in the permafrost within a 30-km rectangle around Salt Lake. The calculation result shows an increase of 51.7 km^2 areas (deformation rate > 10 mm/year) within the analysis rectangle after the outburst of Zonag Lake in 2011. The second landscape example is a braided stream plain (34.8791°N, 93.1845°E). We used Google Earth image to show the spatial properties of the braided stream plain (Figure 4f). This braided stream plain is surrounded by glaciers in the north, south and east. The terrain in the west is relatively low. There are many lake tributaries in this area. The surrounding glaciers provide many water resources for the braided stream plain. With sufficient moisture, the permafrost may undergo degradation. To illuminate the freeze-thaw cycle of permafrost, we used the ALOS, ALOS-2 and Sentinel-1 datasets to map the ground deformation of permafrost around the braided stream plain from 2007 to 2021. The results are shown in Figure 4g–i.

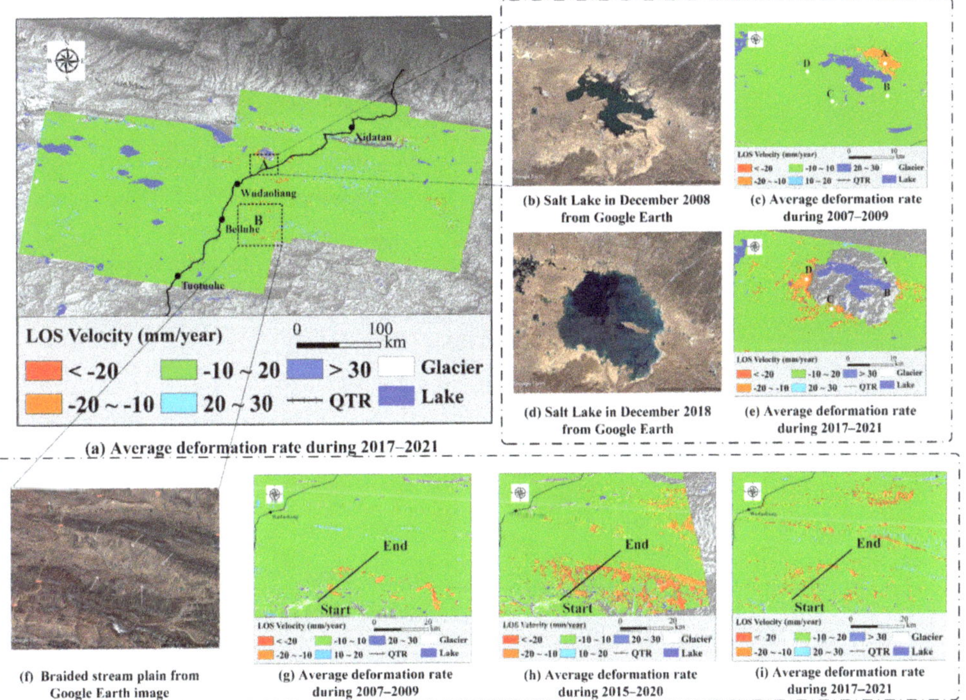

Figure 4. Permafrost evolution in two typical areas, Salt Lake and braided stream plain. (**a**) Average deformation rate during 2017–2021. (**b**) Salt Lake in December 2008 from Google Earth. (**c**) Average deformation rate during 2007–2009. (**d**) Salt Lake in December 2018 from Google Earth. (**e**) Average deformation rate during 2017–2021. (**f**) Braided stream plain from Google Earth image. (**g**) Average deformation rate during 2007–2009. (**h**) Average deformation rate during 2015–2020. (**i**) Average deformation rate during 2017–2021.

To further study the time series deformation of permafrost around Salt Lake, we selected four points (Figure 4c). The details for these points are provided in Table 4. A

25 km profile (Figure 4g) was selected to reveal the relationship between the permafrost deformation and the elevation.

Table 4. The details of the four selection points around Salt Lake.

Point	Position	Long-Term Deformation Velocity during 2007–2009	Long-Term Deformation Velocity during 2017–2021
Point A	93.467°E, 35.547°N	−12.5 mm/year	-
Point B	93.473°E, 35.485°N	−4.3 mm/year	-
Point C	93.363°E, 35.476°N	8.7 mm/year	−15.6 mm/year
Point D	93.312°E, 35.532°N	−2.3 mm/year	−20.3 mm/year

Figure 5a,b shows the long-term deformation trend deformation of selected points around Salt Lake. From March 2007 to April 2008, the four points could balance the freeze-thaw process. However, from July 2009 to March 2010, point A breaks the cycle, and the ability to thaw is stronger than the ability to freeze, which may increase the thickness of the active layer. From Figure 5b and Table 3, at point C, the average uplift deformation from 2007 to 2008 changed into the average subsidence deformation from 2017 to 2021. At point D, the results reveal that the subsidence of the ground increased 10 times from 2007 to 2021. From Figure 5c, we find that there is a negative correlation between the subsidence and the elevation. This may be due to the melting of glaciers around the braided stream plain, resulting in a large amount of water, accumulating on the slope. Finally, the thickness of the permafrost active layer is increased, which causes permafrost degradation.

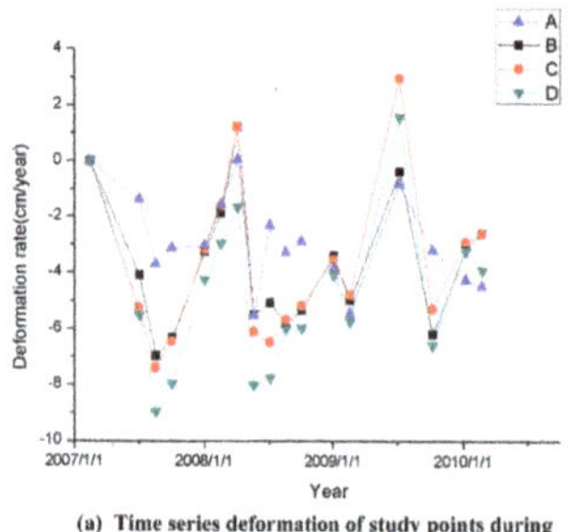

(a) Time series deformation of study points during 2007–2010

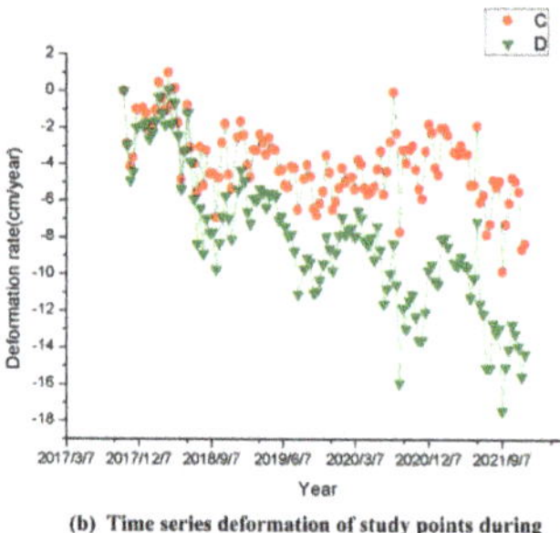

(b) Time series deformation of study points during 2017–2021

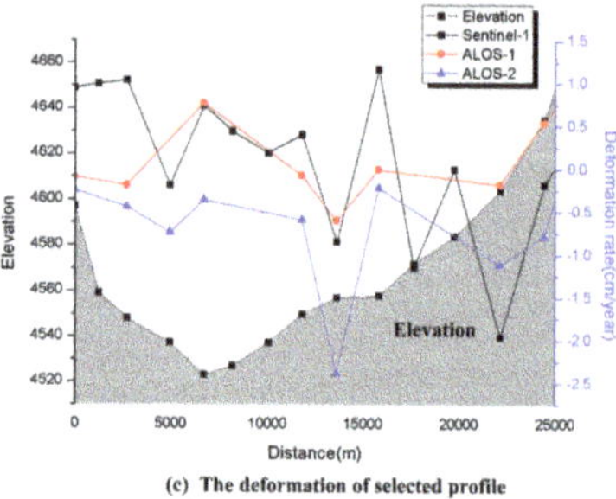

(c) The deformation of selected profile

Figure 5. The time series deformation of the study points and the deformation of the selected profile. (**a**) Time series deformation of study points during 2007–2009. (**b**) Time series deformation of study points during 2017 to 2021. (**c**) The deformation of selected profile.

In the freezing and thawing process of permafrost, the moisture in permafrost can move and change, which may cause rock damage, ground deformation and thawing disasters. The sudden thawing of permafrost may lead to environmental disasters, such as ground collapse and rapid erosion. The thawing mud flow of the permafrost occurs on the slope of the permafrost. In summer, the thawing of the active layer could not flow downslope due to the existence of permafrost, resulting in abundant water in the active layer. Under the condition of excessive moisture, the permafrost creeps downward along the slope, resulting in the thermal melting collapse. We combined optical data (GF-1, GF-2) [64] to reveal the thermal melting collapse of permafrost.

In Figure 6a, the red color indicates the positions of thermal melting collapse, which were generated from 2019–2020 GF-1/2 data. In Figure 6bc, the red color indicates the subsidence in the LOS direction. In Figure 6b, we used the ALOS-2 dataset to reveal the thermal melting collapses of permafrost from 2015 to 2020. The result shows that L-band ALOS-2 can effectively detect the location of thermal melting collapses. The result shows spatial consistency with the GF-1/2 datasets. Compared with the ALOS-2 results, C-band Sentinel-1 rarely detected the location of thermal melting collapses. The results show that the ALOS-2 data can detect displacement that cannot be detected by C-band sensors. To reveal the relationship between the time series deformation and the thermal melting collapses. We selected four points and plotted the time series deformation. The details of the four points are provided in Table 5.

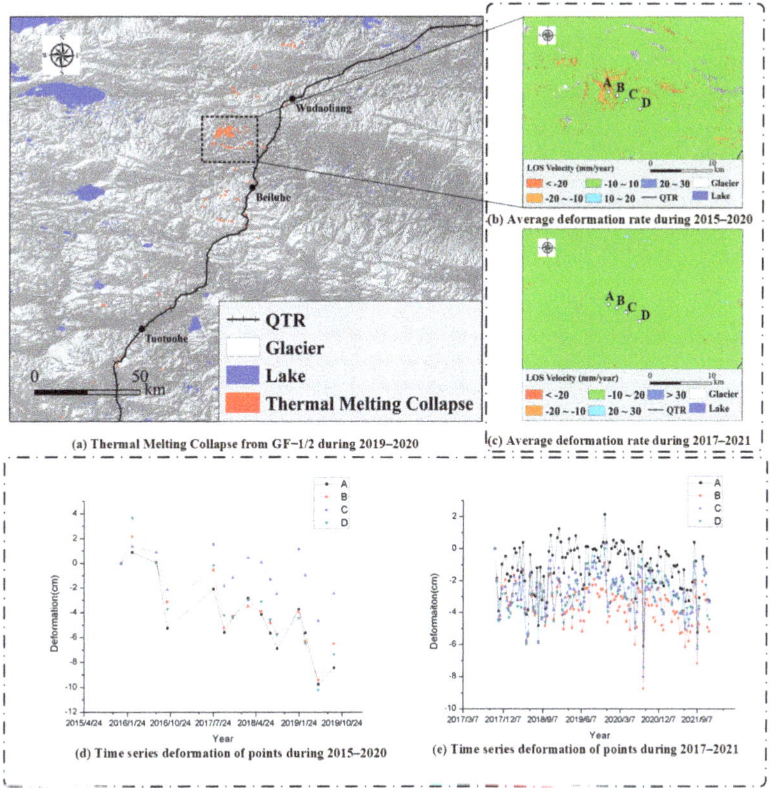

Figure 6. The thermal melting collapse, average deformation and time series deformation of the study points. (**a**) Thermal melting collapse from GF-1/2 during 2019–2020. (**b**) Average deformation rate during 2015–2020. (**c**) Average deformation rate during 2017–2021. (**d**) Time series deformation of points during 2015–2020. (**e**) Time series deformation of points during 2017–2021.

Table 5. The details of the four selection points.

Point	Position	Long-Term Deformation Velocity during 2015–2020	Long-Term Deformation Velocity during 2017–2021
Point A	92.771° E, 35.059° N	−24.9 mm/year	3.4mm/year
Point B	92.790° E, 35.051° N	−8.1 mm/year	9.1 mm/year
Point C	92.803° E, 35.044° N	7.1 mm/year	13.0 mm/year
Point D	92.823° E, 35.032° N	−11.1 mm/year	14.1 mm/year

From Table 5, from 2015 to 2020, the L-band ALOS-2 result is different from the C-band Sentinel-1 result. For the ALOS-2 result, the permafrost under points A, B and C is undergoing subsidence. The Sentinel-1 result shows that the permafrost is still stable. Due to the different observation periods of the two sensors, ALOS-2 has a longer observation period. In addition, results are also affected by the internal characteristics of the sensor, such as wavelength and incident angle.

4.2. Long-Term Deformation Velocity and Maximum Seasonal Deformation in the Northern QTP

After determining the time series deformation derived from the raw phase, the long-term deformation trend and maximum seasonal deformation magnitude can be calculated. The spatial distributions of the maximum seasonal deformation magnitude are shown in Figure 7.

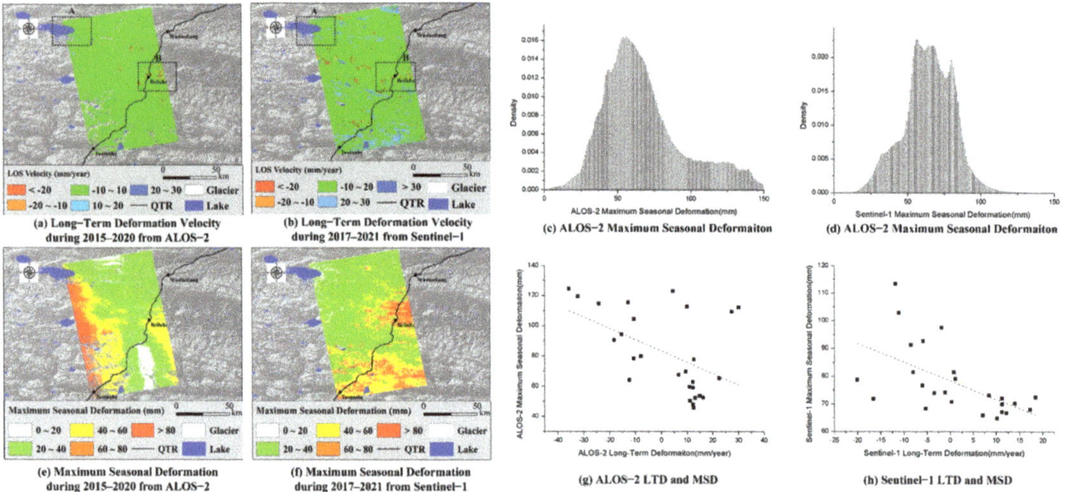

Figure 7. The long-term deformation velocity and maximum seasonal deformation. (**a**) Long-term deformation velocity during 2015–2020 from ALOS-2. (**b**) Long-term deformation velocity during 2017–2021 from Sentinel-1. (**c**) ALOS-2 maximum seasonal deformation. (**d**) Sentinel-1 maximum seasonal deformation. (**e**) Maximum seasonal deformation during 2015–2020 from ALOS-2. (**f**) Maximum seasonal deformation during 2017–2021 from Sentinel-1. (**g**) ALOS-2 LTD and MSD. (**h**) Sentinel-1 LTD and MSD.

Due to the limitation of the revisit period of ALOS data, we calculated the long-term deformation velocity (Figure 7a,b) and the maximum seasonal deformation (Figure 7e,f) of ALOS-2 and Sentinel-1 data. The experimental results show that the average maximum seasonal deformation of the two sensors is 66 mm. As shown in Figure 7a,b,e,f, the area with large maximum seasonal deformation may tend to undergo strong subsidence and uplift. From the ALOS-2 result, the average maximum seasonal deformation during 2015–2020 is 68.74 mm. According to the Sentinel-1 result, the average MSD during 2017–2021 is 64.85 mm. The areas with high MSD are mainly concentrated in glacier outwash plains and braided stream plains. We selected two typical areas, A and B. The upper left of study area A is located in Duoergai Co Lake, surrounded by glaciers to the north. The MSD is large, which may be due to glacial erosion. A large amount of ice deposits is brought to low-lying areas, affecting the water storage capacity and ice melting capacity of permafrost. In addition, a large MSD occurred in thermal karst landforms, such as area B, which is characterized by the formation of hot melt lakes and proglacial lakes. Due to the thermal influence of underground ice in the permafrost, the land shrinks and settles, and finally melted water flows along the groove of the slope. In addition, we reveal the relationship between the maximum seasonal deformation and the long-term deformation velocity. From Figure 7g,h, there is a weak correlation between MSD and LTD, with coefficients of determination (R^2) [65] of 0.208 for ALOS-2 and 0.301 for Sentinel-1. This shows that the relation between MSD and LTD is not simply linear and is related to geomorphic types, moisture, ice and other factors.

5. Discussion

To evaluate the permafrost deformation in the northern QTP, we compared other studies that are shown in Table 6.

Table 6. Previous studies on permafrost deformation in the QTP.

Study Area	InSAR Method	SAR Dataset	Observation Time	Average Ground Deformation Rate	Authors
Beiluhe	PSI	Envisat	2003–2007	−20~3 mm/year	Xie et al. (2010)
Beiluhe	IPTA and SBAS	ALOS and Envisat	2004–2010	−20~20 mm/year	Chen et al. (2012)
Beiluhe	SBAS	ALOS	2007–2010	−25~10 mm/year	Chen et al. (2013)
Naqu-Lhasa	SBAS	Envisat and TerraSAR-X	2003–2012	−20~20 mm/year	Zhang et al. (2018)
Yangbaijin	MTInSAR	TerraSAR-X	2014–2015	−30–10 mm/year	Li et al. (2017)
Wudaoliang-Tuotuohe	MT-InSAR	Sentinel-1, Envisat and ERS-1	1997~2018	−20~10 mm/year	Zhang et al. (2019)
Wudaoliang	StaMPS-InSAR	Sentinel-1 and TerraSAR-X	2017–2018	−12~7 mm/year	Han et al. (2020)
Golmud-Lhasa	NSBAS	Sentinel-1	2017~2020	−20~20 mm/year	Wang et al. (2021)

Xie et al. [66] used the PSI technique with Envisat datasets to reveal the ground deformation in Beiluhe. The result shows that the average deformation from 2003 to 2007 was −20~3 mm/year. Chen [67] used multisource datasets (ALOS and Envisat) based on the IPTA and SABS techniques to illuminate the displacement of permafrost in Beiluhe. The result shows that the average displacement is −20~20 mm/year. Chen et al. [28] found that the average deformation of permafrost in Beiluhe is −25~10 mm/year by using the SBAS method with ALOS data. From 2014 to 2020, Li et al. and Wang et al. [68] used the MT-InSAR method to research the deformation of permafrost in the QTP. Li et al. [69] found that the average deformation in Yangbaijin ranged from −30 to 10 mm/year from 2014 to 2015. Wang et al. [54] used Sentinel-1 data based on NSBAS to reveal the deformation of the whole QTP. The result shows that the average deformation rate is −20~20 mm/year. Zhang et al. [8] used three different sensor datasets (ERS-1, Envisat and Sentinel-1) to reveal the freeze-thaw cycle of permafrost in Wudaoliang and Tuouohe from 1997 to 2008. The result shows that the average deformation rate ranges from −20 to 10 mm/year. From the above related research, our results are consistent with the previous results.

In addition, to evaluate the accuracy of the SBAS-InSAR results in this paper, we used leveling data from other researchers in the QTR [70]. The locations of the three leveling points are shown in Figure 8. The elevations of the three leveling points were collected on 9 January 2018, and 11 February 2018. The deformation results of 3 January 2018 and 8 February 2018, were extracted from 150-475 (Path-Frame) Sentinel-1 InSAR results. In order to ensure the accuracy, we convert the leveling results to the displacement of the LOS direction based on the radar incident angle. The comparison between the InSAR experiment results and the leveling results of A, B and C is shown in Table 7. Table 7 shows that the absolute errors of the leveling data and InSAR measurements at points A, B and C were 3.1, 5.7 and 6.8 mm, respectively, which were relatively small.

Through a comparison with previous studies and leveling data, this paper provides a theoretical basis and data support for the study of permafrost evolution. However, there are some limitations in our study. First, there are some errors in the InSAR processing, such as orbit errors, unwrapping phase errors, residual topographic errors, and ERA5 APS correction errors. These errors may lead to incorrect permafrost deformation. Second, from 2010 to 2015, due to the lack of data, we have no ability to monitor the deformation of permafrost in the QTP during this period. In the future, we will combine more SAR data and auxiliary data to help us better understand the freeze-thaw cycle and degradation of permafrost in the QTP.

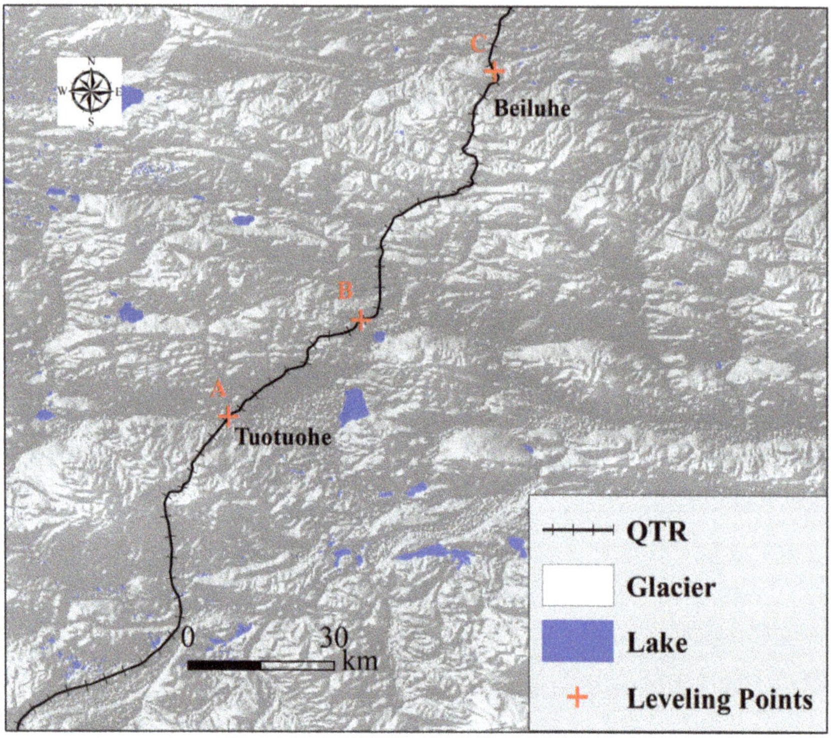

Figure 8. Geographic location map of leveling points along the QTR.

Table 7. The absolute error between the measured values of leveling points A, B, and C and the InSAR observed values.

Benchmark	A	B	C
Leveling (unit: mm)	−2.2	8.8	16.6
InSAR results (unit: mm)	−5.3	14.5	9.8
Absolute error (unit: mm)	3.1	5.7	6.8

6. Conclusions

In this study, we used 66 scenes of ALOS data (2007–2009), 73 scenes of ALOS-2 data (2015–2020) and 284 scenes of Sentinel-1 data (2017–2021) based on SBAS-InSAR to reveal the spatial and temporal deformation of permafrost in the northern QTP, passing through the Tuotuohe, Beiluhe, Wudaoliang and Xidatan regions between 2007 and 2021. In addition, a coherence weighted least squares estimator without any hypothetical model was used to calculate the long-term deformation velocity (LTDV) and maximum seasonal deformation (MSD) without any prior knowledge. The conclusions are summarized as follows:

1. Analysis of the ALOS results shows that the LTDV ranged from −20 to +20 mm/year during 2007–2009. For the ALOS-2 and Sentinel-1 results, the LTDV ranged from −30 to 30 mm/year during 2015–2021. Over a 15-year observation period, permafrost degradation is accelerating. The expansion areas of permafrost degradation are concentrated on braided stream plains, thermokarst lakes and glacier outwash plains. In addition, the high MSDs are mainly concentrated in glacier outwash plains and

braided stream plains, which may be due to glacial erosion and a large amount of ice deposits brought to low-lying areas, affecting the water storage capacity and ice melting capacity of permafrost.

2. For the geological disaster of thermal melting collapses, we compared the permafrost monitored by the L-band and the C-band sensors. The results show that the L-band ALOS-2 is more sensitive to the thermal collapses of permafrost than C-band sensor, and the detected collapse areas (LTDV < -10 mm/year) are consistent with the GF-1/2 thermal collapses dataset.
3. We extracted the time series deformation directly from the interferograms without a preset deformation model. The long-term deformation velocity (LTDV) and maximum seasonal deformation (MSD) were calculated from time series deformation without any prior knowledge, demonstrating that our methods can be effectively used to extract the deformation features of permafrost.

In the future, more SAR datasets and auxiliary datasets will be used to illuminate the freeze-thaw cycle and retrieve the active layer thickness on the QTP on a large scale.

Author Contributions: Conceptualization, L.Z., C.W. and Y.T.; formal analysis, L.Z.; methodology, L.Z. and C.W.; software, L.Z., H.Z., B.Z. and L.D.; validation, L.Z., C.W. and Y.T.; writing—original draft, L.Z.; writing—review and editing, C.W.; Y.T. and H.Z.; visualization, L.Z.; supervision, C.W.; project administration, H.Z.; funding acquisition, C.W. All authors have read and agreed to the published version of the manuscript.

Funding: This paper was funded by the National Natural Science Foundation of China, Grant No. 41930110.

Data Availability Statement: ALOS is provide by JAXA and available from the Alaska Satellite Facility (ASF) (https://vertex.daac.asf.alaska.edu, accessed on 14 February 2022). ALOS-2 data is provided and available from JAXA EORA2 program. Sentinel-1 data is provided by the European Space Agency (ESA) and available from the Alaska Satellite Facility (ASF) (https://vertex.daac.asf.alaska.edu, accessed on 14 February 2022). GF-1/2 is provided by the National Tibetan Plateau Data Center and available from National Tibetan Plateau Data Center (http://data.tpdc.ac.cn, accessed on 14 February 2022). SRTM DEM data is available at https://srtm.csi.cgiar.org/srtmdata/, accessed on 14 February 2022.

Acknowledgments: The authors would like to thank JAXA EORA2 program (project id: ER2A2N030) and its ALOS, ALOS-2 data, NASA, CGIAR-CSI for SRTM DEM data, ESA for Sentinel-1 data and National Tibetan Plateau Data Center for GF-1/2 data. Profs. Qingbai Wu, Ji Chen, Jing Luo and Beiluhe Permafrost Station of the Northwest Institute of Eco-Environment and Resources of CAS, and Drs. Fan Wu and Lu Xu of our team are acknowledged for helpful discussion and field support.

Conflicts of Interest: The authors declare no conflict of interest.

References

1. Ma, Y.; Wang, Y.; Wu, R.; Hu, Z.; Yang, K.; Li, M.; Ma, W.; Zhong, L.; Sun, F.; Chen, X. Recent Advances on the Study of Atmosphere-Land Interaction Observations on the Tibetan Plateau. *Hydrol. Earth Syst. Sci.* **2009**, *13*, 1103–1111. [CrossRef]
2. Liu, X.; Sun, H.; Miao, Y.; Dong, B.; Yin, Z.-Y. Impacts of Uplift of Northern Tibetan Plateau and Formation of Asian Inland Deserts on Regional Climate and Environment. *Quat. Sci. Rev.* **2015**, *116*, 1–14. [CrossRef]
3. Hu, G.-R.; Li, X.-Y.; Yang, X.-F. The Impact of Micro-Topography on the Interplay of Critical Zone Architecture and Hydrological Processes at the Hillslope Scale: Integrated Geophysical and Hydrological Experiments on the Qinghai-Tibet Plateau. *J. Hydrol.* **2020**, *583*, 124618. [CrossRef]
4. Zhai, P.; Yu, R.; Guo, Y.; Li, Q.; Ren, X.; Wang, Y.; Xu, W.; Liu, Y.; Ding, Y. The Strong El Niño of 2015/16 and Its Dominant Impacts on Global and China's Climate. *J. Meteorol. Res.* **2016**, *30*, 283–297. [CrossRef]
5. Şerban, R.-D.; Jin, H.; Şerban, M.; Luo, D.; Wang, Q.; Jin, X.; Ma, Q. Mapping Thermokarst Lakes and Ponds across Permafrost Landscapes in the Headwater Area of Yellow River on Northeastern Qinghai-Tibet Plateau. *Int. J. Remote Sens.* **2020**, *41*, 7042–7067. [CrossRef]
6. Chen, B.; Luo, S.; Lü, S.; Zhang, Y.; Ma, D. Effects of the Soil Freeze-Thaw Process on the Regional Climate of the Qinghai-Tibet Plateau. *Clim. Res.* **2014**, *59*, 243–257.
7. Vonk, J.E.; Tank, S.E.; Bowden, W.B.; Laurion, I.; Vincent, W.F.; Alekseychik, P.; Amyot, M.; Billet, M.F.; Canário, J.; Cory, R.M. Reviews and Syntheses: Effects of Permafrost Thaw on Arctic Aquatic Ecosystems. *Biogeosciences* **2015**, *12*, 7129–7167. [CrossRef]

8. Zhang, Z.; Wang, M.; Wu, Z.; Liu, X. Permafrost Deformation Monitoring Along the Qinghai-Tibet Plateau Engineering Corridor Using InSAR Observations with Multi-Sensor SAR Datasets from 1997–2018. *Sensors* **2019**, *19*, 5306. [CrossRef]
9. Chen, D.; Fu, Y.-S.; Cai, B.; Yuan, Y.-X. Modeling and Algorithms of GPS Data Reduction for the Qinghai–Tibet Railway. *IEEE Trans. Intell. Transp. Syst.* **2010**, *11*, 753–758. [CrossRef]
10. Qin, Y.; Zhang, J.; Li, G.; Qu, G. Settlement Characteristics of Unprotected Embankment along the Qinghai–Tibet Railway. *Cold Reg. Sci. Technol.* **2010**, *60*, 84–91. [CrossRef]
11. Niu, F.; Luo, J.; Lin, Z.; Liu, M.; Yin, G. Thaw-induced slope failures and susceptibility mapping in permafrost regions of the Qinghai–Tibet Engineering Corridor, China. *Nat. Hazards* **2014**, *74*, 1667–1682. [CrossRef]
12. Antonova, S.; Sudhaus, H.; Strozzi, T.; Zwieback, S.; Kääb, A.; Heim, B.; Langer, M.; Bornemann, N.; Boike, J. Thaw Subsidence of a Yedoma Landscape in Northern Siberia, Measured in Situ and Estimated from TerraSAR-X Interferometry. *Remote Sens.* **2018**, *10*, 494. [CrossRef]
13. Short, N.; Brisco, B.; Couture, N.; Pollard, W.; Murnaghan, K.; Budkewitsch, P. A Comparison of TerraSAR-X, RADARSAT-2 and ALOS-PALSAR Interferometry for Monitoring Permafrost Environments, Case Study from Herschel Island, Canada. *Remote Sens. Environ.* **2011**, *115*, 3491–3506. [CrossRef]
14. Chen, J.; Wu, Y.; O'Connor, M.; Cardenas, M.B.; Schaefer, K.; Michaelides, R.; Kling, G. Active Layer Freeze-Thaw and Water Storage Dynamics in Permafrost Environments Inferred from InSAR. *Remote Sens. Environ.* **2020**, *248*, 112007. [CrossRef]
15. Meng, Y.; Lan, H.; Li, L.; Wu, Y.; Li, Q. Characteristics of Surface Deformation Detected by X-Band SAR Interferometry over Sichuan-Tibet Grid Connection Project Area, China. *Remote Sens.* **2015**, *7*, 12265–12281. [CrossRef]
16. Wang, L.; Marzahn, P.; Bernier, M.; Jacome, A.; Poulin, J.; Ludwig, R. Comparison of TerraSAR-X and ALOS PALSAR Differential Interferometry With Multisource DEMs for Monitoring Ground Displacement in a Discontinuous Permafrost Region. *IEEE J. Sel. Top. Appl. Earth Obs. Remote Sens.* **2017**, *10*, 4074–4093. [CrossRef]
17. Zhang, Z.; Wang, M.; Liu, X.; Chen, Q.; Wang, C.; Zhang, H. Deformation Monitoring of Qinghai-Tibet Railway from 1997–2018 Using SAR Interferometry with Multi-Sensors SAR Datasets. In Proceedings of the 2019 SAR in Big Data Era (BIGSARDATA), Beijing, China, 5–6 August 2019; IEEE: New York, NY, USA, 2019; pp. 1–4.
18. Hu, J.; Wang, Q.; Li, Z.; Zhao, R.; Sun, Q. Investigating the Ground Deformation and Source Model of the Yangbajing Geothermal Field in Tibet, China with the WLS InSAR Technique. *Remote Sens.* **2016**, *8*, 191. [CrossRef]
19. Chen, J.; Wu, T.; Zou, D.; Liu, L.; Wu, X.; Gong, W.; Zhu, X.; Li, R.; Hao, J.; Hu, G.; et al. Magnitudes and Patterns of Large-Scale Permafrost Ground Deformation Revealed by Sentinel-1 InSAR on the Central Qinghai-Tibet Plateau. *Remote Sens. Environ.* **2022**, *268*, 112778. [CrossRef]
20. Zhao, R.; Li, Z.; Feng, G.; Wang, Q.; Hu, J. Monitoring Surface Deformation over Permafrost with an Improved SBAS-InSAR Algorithm: With Emphasis on Climatic Factors Modeling. *Remote Sens. Environ.* **2016**, *184*, 276–287. [CrossRef]
21. Zhang, X.; Tang, Y.; Zhang, H.; Wang, C.; Zhang, B.; Wu, F. Permafrost Subsidence Monitoring of Qinghai-Tibet Plateau Using Sentinel-1 Data. In Proceedings of the 2019 SAR in Big Data Era (BIGSARDATA), Beijing, China, 5–6 August 2019; IEEE: New York, NY, USA, 2019; pp. 1–4.
22. Short, N.; LeBlanc, A.-M.; Sladen, W.; Oldenborger, G.; Mathon-Dufour, V.; Brisco, B. RADARSAT-2 D-InSAR for Ground Displacement in Permafrost Terrain, Validation from Iqaluit Airport, Baffin Island, Canada. *Remote Sens. Environ.* **2014**, *141*, 40–51. [CrossRef]
23. Rouyet, L.; Lauknes, T.R.; Christiansen, H.H.; Strand, S.M.; Larsen, Y. Seasonal Dynamics of a Permafrost Landscape, Adventdalen, Svalbard, Investigated by InSAR. *Remote Sens. Environ.* **2019**, *231*, 111236. [CrossRef]
24. Rouyet, L.; Liu, L.; Strand, S.M.; Christiansen, H.H.; Lauknes, T.R.; Larsen, Y. Seasonal InSAR Displacements Documenting the Active Layer Freeze and Thaw Progression in Central-Western Spitsbergen, Svalbard. *Remote Sens.* **2021**, *13*, 2977. [CrossRef]
25. Bartsch, A.; Leibman, M.; Strozzi, T.; Khomutov, A.; Widhalm, B.; Babkina, E.; Mullanurov, D.; Ermokhina, K.; Kroisleitner, C.; Bergstedt, H. Seasonal Progression of Ground Displacement Identified with Satellite Radar Interferometry and the Impact of Unusually Warm Conditions on Permafrost at the Yamal Peninsula in 2016. *Remote Sens.* **2019**, *11*, 1865. [CrossRef]
26. Robson, G.; Treitz, P.; Lamoureux, S.F.; Murnaghan, K.; Brisco, B. Seasonal Surface Subsidence and Frost Heave Detected by C-Band DInSAR in a High Arctic Environment, Cape Bounty, Melville Island, Nunavut, Canada. *Remote Sens.* **2021**, *13*, 2505. [CrossRef]
27. Wang, L.; Marzahn, P.; Bernier, M.; Ludwig, R. Sentinel-1 InSAR Measurements of Deformation over Discontinuous Permafrost Terrain, Northern Quebec, Canada. *Remote Sens. Environ.* **2020**, *248*, 111965. [CrossRef]
28. Chen, F.; Lin, H. Tibetan Plateau Permafrost Evolution Monitoring Using C- and L-Band Spaceborne SAR Interferometry. In Proceedings of the 2012 IEEE International Geoscience and Remote Sensing Symposium, Munich, Germany, 22–27 July 2012; IEEE: New York, NY, USA, 2012; pp. 1884–1887.
29. Zhang, X.; Zhang, H.; Wang, C.; Tang, Y.; Zhang, B.; Wu, F.; Wang, J.; Zhang, Z. Time-Series InSAR Monitoring of Permafrost Freeze-Thaw Seasonal Displacement over Qinghai–Tibetan Plateau Using Sentinel-1 Data. *Remote Sens.* **2019**, *11*, 1000. [CrossRef]
30. Wang, C.; Zhang, H.; Zhang, B.; Tang, Y.; Zhang, Z.; Liu, M.; Zhao, L. New Mode TerraSAR-X Interferometry for Railway Monitoring in the Permafrost Region of the Tibet Plateau. In Proceedings of the 2015 IEEE International Geoscience and Remote Sensing Symposium (IGARSS), Milan, Italy, 26–31 July 2015; IEEE: New York, NY, USA, 2015; pp. 1634–1637.
31. Wang, C.; Zhang, Z.; Zhang, H.; Zhang, B.; Tang, Y.; Wu, Q. Active Layer Thickness Retrieval of Qinghai–Tibet Permafrost Using the TerraSAR-X InSAR Technique. *IEEE J. Sel. Top. Appl. Earth Obs. Remote Sens.* **2018**, *11*, 4403–4413. [CrossRef]

32. Zhang, Z.; Wang, C.; Zhang, H.; Tang, Y.; Liu, X. Analysis of Permafrost Region Coherence Variation in the Qinghai–Tibet Plateau with a High-Resolution TerraSAR-X Image. *Remote Sens.* **2018**, *10*, 298. [CrossRef]
33. Chen, R.H.; Michaelides, R.J.; Sullivan, T.D.; Parsekian, A.D.; Zebker, H.A.; Moghaddam, M.; Schaefer, K.M. Joint Retrieval of Soil Moisture and Permafrost Active Layer Thickness Using L-Band Insar and P-Band Polsar. In Proceedings of the IGARSS 2020-2020 IEEE International Geoscience and Remote Sensing Symposium, Waikoloa, HI, USA, 26 September–2 October 2020; IEEE: New York, NY, USA, 2020; pp. 4606–4609.
34. Schaefer, K.; Liu, L.; Parsekian, A.; Jafarov, E.; Chen, A.; Zhang, T.; Gusmeroli, A.; Panda, S.; Zebker, H.; Schaefer, T. Remotely Sensed Active Layer Thickness (ReSALT) at Barrow, Alaska Using Interferometric Synthetic Aperture Radar. *Remote Sens.* **2015**, *7*, 3735–3759. [CrossRef]
35. Chen, F.; Lin, H.; Zhou, W.; Hong, T.; Wang, G. Surface Deformation Detected by ALOS PALSAR Small Baseline SAR Interferometry over Permafrost Environment of Beiluhe Section, Tibet Plateau, China. *Remote Sens. Environ.* **2013**, *138*, 10–18. [CrossRef]
36. Parsekian, A.D.; Chen, R.H.; Michaelides, R.J.; Sullivan, T.D.; Clayton, L.K.; Huang, L.; Zhao, Y.; Wig, E.; Moghaddam, M.; Zebker, H.; et al. Validation of Permafrost Active Layer Estimates from Airborne SAR Observations. *Remote Sens.* **2021**, *13*, 2876. [CrossRef]
37. Rudy, A.C.A.; Lamoureux, S.F.; Treitz, P.; Short, N.; Brisco, B. Seasonal and Multi-Year Surface Displacements Measured by DInSAR in a High Arctic Permafrost Environment. *Int. J. Appl. Earth Obs. Geoinf.* **2018**, *64*, 51–61. [CrossRef]
38. Lu, P.; Han, J.; Li, Z.; Xu, R.; Li, R.; Hao, T.; Qiao, G. Lake Outburst Accelerated Permafrost Degradation on Qinghai-Tibet Plateau. *Remote Sens. Environ.* **2020**, *249*, 112011. [CrossRef]
39. Eshqi Molan, Y.; Kim, J.-W.; Lu, Z.; Wylie, B.; Zhu, Z. Modeling Wildfire-Induced Permafrost Deformation in an Alaskan Boreal Forest Using InSAR Observations. *Remote Sens.* **2018**, *10*, 405. [CrossRef]
40. Li, R.; Li, Z.; Han, J.; Lu, P.; Qiao, G.; Meng, X.; Hao, T.; Zhou, F. Monitoring Surface Deformation of Permafrost in Wudaoliang Region, Qinghai–Tibet Plateau with ENVISAT ASAR Data. *Int. J. Appl. Earth Obs. Geoinf.* **2021**, *104*, 102527. [CrossRef]
41. Daout, S.; Dini, B.; Haeberli, W.; Doin, M.-P.; Parsons, B. Ice Loss in the Northeastern Tibetan Plateau Permafrost as Seen by 16 Yr of ESA SAR Missions. *Earth Planet. Sci. Lett.* **2020**, *545*, 116404. [CrossRef]
42. Ng, A.H.-M.; Chang, H.-C.; Ge, L.; Rizos, C.; Omura, M. Assessment of Radar Interferometry Performance for Ground Subsidence Monitoring Due to Underground Mining. *Earth Planets Space* **2009**, *61*, 733–745. [CrossRef]
43. Fort, F.; van Vliet-Lanoe, B. Permafrost and Periglacial Environment of Western Tibet. *Landf. Anal.* **2007**, *5*, 25–29.
44. Li, Z.; Zhao, R.; Hu, J.; Wen, L.; Feng, G.; Zhang, Z.; Wang, Q. InSAR Analysis of Surface Deformation over Permafrost to Estimate Active Layer Thickness Based on One-Dimensional Heat Transfer Model of Soils. *Sci. Rep.* **2015**, *5*, 15542. [CrossRef]
45. Jia, Y.; Kim, J.-W.; Shum, C.; Lu, Z.; Ding, X.; Zhang, L.; Erkan, K.; Kuo, C.-Y.; Shang, K.; Tseng, K.-H.; et al. Characterization of Active Layer Thickening Rate over the Northern Qinghai-Tibetan Plateau Permafrost Region Using ALOS Interferometric Synthetic Aperture Radar Data, 2007–2009. *Remote Sens.* **2017**, *9*, 84. [CrossRef]
46. Liu, L.; Zhang, T.; Wahr, J. InSAR Measurements of Surface Deformation over Permafrost on the North Slope of Alaska. *J. Geophys. Res. Earth Surf.* **2010**, *115*. [CrossRef]
47. Wang, L.; Zhao, L.; Zhou, H.; Liu, S.; Du, E.; Zou, D.; Liu, G.; Wang, C.; Li, Y. Permafrost Ground Ice Melting and Deformation Time Series Revealed by Sentinel-1 InSAR in the Tanggula Mountain Region on the Tibetan Plateau. *Remote Sens.* **2022**, *14*, 811. [CrossRef]
48. Xia, L.; Yang, Q.; Li, Z.; Wu, Y.; Feng, Z. The Effect of the Qinghai-Tibet Railway on the Migration of Tibetan Antelope Pantholops Hodgsonii in Hoh-Xil National Nature Reserve, China. *Oryx* **2007**, *41*, 352–357. [CrossRef]
49. Zhou, H.; Li, D.; Zhang, Y.; Yang, T.; Liu, Y. 'Genetic Diversity of Microsatellite DNA Loci of Tibetan Antelope (Chiru, Pantholops Hodgsonii) in Hoh Xil National Nature Reserve, Qinghai, China. *J. Genet. Genom.* **2007**, *34*, 600–607. [CrossRef]
50. Niu, F.; Lin, Z.; Liu, H.; Lu, J. Characteristics of Thermokarst Lakes and Their Influence on Permafrost in Qinghai–Tibet Plateau. *Geomorphology* **2011**, *132*, 222–233. [CrossRef]
51. Zhou, S.; Li, J. The Sequence of Quaternary Glaciation in the Bayan Har Mountains. *Quat. Int.* **1998**, *45*, 135–142. [CrossRef]
52. Ma, Q.; Zhang, K.; Jabro, J.D.; Ren, L.; Liu, H. Freeze–Thaw Cycles Effects on Soil Physical Properties under Different Degraded Conditions in Northeast China. *Environ. Earth Sci.* **2019**, *78*, 321. [CrossRef]
53. Yu, J.; Gao, C.; Cheng, A.; Liu, Y.; Zhang, L.; He, X. Geomorphic, Hydroclimatic and Hydrothermal Controls on the Formation of Lithium Brine Deposits in the Qaidam Basin, Northern Tibetan Plateau, China. *Ore Geol. Rev.* **2013**, *50*, 171–183. [CrossRef]
54. Wang, J.; Wang, C.; Zhang, H.; Tang, Y.; Duan, W.; Dong, L. Freeze-Thaw Deformation Cycles and Temporal-Spatial Distribution of Permafrost along the Qinghai-Tibet Railway Using Multitrack InSAR Processing. *Remote Sens.* **2021**, *13*, 4744. [CrossRef]
55. LaRocque, A.; Phiri, C.; Leblon, B.; Pirotti, F.; Connor, K.; Hanson, A. Wetland Mapping with Landsat 8 OLI, Sentinel-1, ALOS-1 PALSAR, and LiDAR Data in Southern New Brunswick, Canada. *Remote Sens.* **2020**, *12*, 2095. [CrossRef]
56. Adriano, B.; Yokoya, N.; Miura, H.; Matsuoka, M.; Koshimura, S. A Semiautomatic Pixel-Object Method for Detecting Landslides Using Multitemporal ALOS-2 Intensity Images. *Remote Sens.* **2020**, *12*, 561. [CrossRef]
57. Imamoglu, M.; Kahraman, F.; Cakir, Z.; Sanli, F.B. Ground Deformation Analysis of Bolvadin (W. Turkey) by Means of Multi-Temporal InSAR Techniques and Sentinel-1 Data. *Remote Sens.* **2019**, *11*, 1069. [CrossRef]
58. Berardino, P.; Fornaro, G.; Lanari, R.; Sansosti, E. A New Algorithm for Surface Deformation Monitoring Based on Small Baseline Differential SAR Interferograms. *IEEE Trans. Geosci. Remote Sens.* **2002**, *40*, 2375–2383. [CrossRef]

59. Shen, Z.; Liu, Z. Integration of GPS and InSAR Data for Resolving 3-Dimensional Crustal Deformation. *Earth Space Sci.* **2020**, *7*, e2019EA001036. [CrossRef]
60. Wang, Y.; Chang, L.; Feng, W.; Samsonov, S.; Zheng, W. Topography-Correlated Atmospheric Signal Mitigation for InSAR Applications in the Tibetan Plateau Based on Global Atmospheric Models. *Int. J. Remote Sens.* **2021**, *42*, 4361–4379. [CrossRef]
61. Jolivet, R.; Agram, P.S.; Lin, N.Y.; Simons, M.; Doin, M.-P.; Peltzer, G.; Li, Z. Improving InSAR Geodesy Using Global Atmospheric Models. *J. Geophys. Res. Solid Earth* **2014**, *119*, 2324–2341. [CrossRef]
62. Fattahi, H.; Amelung, F. DEM Error Correction in InSAR Time Series. *IEEE Trans. Geosci. Remote Sens.* **2013**, *51*, 4249–4259. [CrossRef]
63. Yunjun, Z.; Fattahi, H.; Amelung, F. Small Baseline InSAR Time Series Analysis: Unwrapping Error Correction and Noise Reduction. *Comput. Geosci.* **2019**, *133*, 104331. [CrossRef]
64. Niu, F.; Luo, J. Distribution Data of Freezing and Thawing Disasters in Qinghai Tibet Project Corridor (2019–2020). National Tibetan Plateau Data Center. 2021. Available online: https://doi.org/10.11888/RemoteSen.tpdc.271883 (accessed on 22 February 2022).
65. Li, H.; Kato, T.; Hayashi, M.; Wu, L. Estimation of Forest Aboveground Biomass of Two Major Conifers in Ibaraki Prefecture, Japan, from PALSAR-2 and Sentinel-2 Data. *Remote Sens.* **2022**, *14*, 468. [CrossRef]
66. Xie, C.; Li, Z.; Xu, J.; Li, X. Analysis of Deformation over Permafrost Regions of Qinghai-Tibet Plateau Based on Permanent Scatterers. *Int. J. Remote Sens.* **2010**, *31*, 1995–2008. [CrossRef]
67. Chen, F.; Lin, H.; Li, Z.; Chen, Q.; Zhou, J. Interaction between Permafrost and Infrastructure along the Qinghai–Tibet Railway Detected via Jointly Analysis of C-and L-Band Small Baseline SAR Interferometry. *Remote Sens. Environ.* **2012**, *123*, 532–540. [CrossRef]
68. Wang, C.; Zhang, Z.; Zhang, H.; Wu, Q.; Zhang, B.; Tang, Y. Seasonal Deformation Features on Qinghai-Tibet Railway Observed Using Time-Series InSAR Technique with High-Resolution TerraSAR-X Images. *Remote Sens. Lett.* **2017**, *8*, 1–10. [CrossRef]
69. Li, Y.; Zhang, J.; Li, Z.; Luo, Y.; Jiang, W.; Tian, Y. Measurement of Subsidence in the Yangbajing Geothermal Fields, Tibet, from TerraSAR-X InSAR Time Series Analysis. *Int. J. Digit. Earth* **2016**, *9*, 697–709. [CrossRef]
70. Hong, Z.; Jin, S. Permafrost deformation in Qinghai-Tibet Plateau time-series PS-InSAR. *Bull. Surv. Mapp.* **2021**, *1*, 35–40. (In Chinese)

Article

Estimation of Forest Aboveground Biomass of Two Major Conifers in Ibaraki Prefecture, Japan, from PALSAR-2 and Sentinel-2 Data

Hantao Li [1], Tomomichi Kato [2,3,*], Masato Hayashi [4] and Lan Wu [5]

1 Graduate School of Agriculture, Hokkaido University, Sapporo 060-8589, Hokkaido, Japan; lihantao@env.agr.hokudai.ac.jp
2 Research Faculty of Agriculture, Hokkaido University, Sapporo 060-8589, Hokkaido, Japan
3 Global Center for Food, Land and Water Resources, Research Faculty of Agriculture, Hokkaido University, Sapporo 060-8589, Hokkaido, Japan
4 Earth Observation Research Center, Japan Aerospace Exploration Agency (JAXA), Tsukuba 305-8505, Ibaraki, Japan; hayashi.masato2@jaxa.jp
5 College of Ecology and Environment, Hainan University, Haiko 570228, China; wulan@hainanu.edu.cn
* Correspondence: tkato@cen.agr.hokudai.ac.jp; Tel.: +81-11-706-4942

Abstract: Forest biomass is a crucial component of the global carbon budget in climate change studies. Therefore, it is essential to develop a credible way to estimate forest biomass as carbon stock. Our study used PALSAR-2 (ALOS-2) and Sentinel-2 images to drive the Random Forest regression model, which we trained with airborne lidar data. We used the model to estimate forest aboveground biomass (AGB) of two significant coniferous trees, Japanese cedar and Japanese cypress, in Ibaraki Prefecture, Japan. We used 48 variables derived from the two remote sensing datasets to predict forest AGB under the Random Forest algorithm, and found that the model that combined the two datasets performed better than models based on only one dataset, with R^2 = 0.31, root-mean-square error (*RMSE*) = 54.38 Mg ha^{-1}, mean absolute error (*MAE*) = 40.98 Mg ha^{-1}, and relative *RMSE* (*rRMSE*) of 0.35 for Japanese cedar, and R^2 = 0.37, *RMSE* = 98.63 Mg ha^{-1}, *MAE* = 76.97 Mg ha^{-1}, and *rRMSE* of 0.33 for Japanese cypress, over the whole AGB range. In the satellite AGB map, the total AGB of Japanese cedar in 17 targeted cities in Ibaraki Prefecture was 5.27 Pg, with a mean of 146.50 Mg ha^{-1} and a standard deviation of 44.37 Mg ha^{-1}. The total AGB of Japanese cypress was 3.56 Pg, with a mean of 293.12 Mg ha^{-1} and a standard deviation of 78.48 Mg ha^{-1}. We also found a strong linear relationship with between the model estimates and Japanese government data, with R^2 = 0.99 for both species and found the government information underestimates the AGB for cypress but overestimates it for cedar. Our results reveal that combining information from multiple sensors can predict forest AGB with increased accuracy and robustness.

Keywords: ecosystem carbon cycle; L-band SAR; vegetation index; random forest regression; plantation

Citation: Li, H.; Kato, T.; Hayashi, M.; Wu, L. Estimation of Forest Aboveground Biomass of Two Major Conifers in Ibaraki Prefecture, Japan, from PALSAR-2 and Sentinel-2 Data. *Remote Sens.* **2022**, *14*, 468. https://doi.org/10.3390/rs14030468

Academic Editor: Janne Heiskanen

Received: 13 December 2021
Accepted: 14 January 2022
Published: 19 January 2022

1. Introduction

Forests play a significant role in the global carbon budget, as they store a large share of terrestrial carbon in their biomass [1]. About 90% of the total carbon in the world's vegetation stock comprises forests, which cover 65% of the land area [2]. The forest aboveground biomass (AGB) is therefore considered one of the most important factors in evaluating forest carbon pools [3]. To better understand the amount of stored carbon in forest, spatially explicit and temporally consistent estimates of AGB are urgently needed [4]. Field biometric studies to quantify AGB, usually using the diameter at breast height (DBH) and tree height as inputs for allometries based on destructive sampling, have provided simple and useful models, but constructing reliable allometric relationships over large areas is difficult, time-consuming, and expensive [5].

Remote sensing techniques can be scaled up to cover large areas, thereby allowing efficient collection of forest biophysical information and repeated analysis to reveal changes over time [6]. Among the available techniques, lidar (light detection and ranging) is one of the most accurate remote-sensing technologies for assessing forest canopy characteristics [7]. Lidar data are particularly useful for mapping vertical structural attributes of ecosystems such as carbon storage, biomass, and stand volume [8]. These advantages let lidar-based approaches provide high-quality assessment of AGB, even in forests with high biomass per unit area, and can retrieve numerous forest parameters in a single survey [9–11].

Despite the ability of airborne lidar to provide highly accurate assessments of parameters such as tree density at an urban scale, the high cost of airborne lidar data can prevent its use in larger areas [12]. In addition, the sparse coverage of land areas by space-borne lidar (e.g., ICESat2, GEDI) reduces the availability of these data for large-area estimation of forest AGB. This suggests the need to develop a robust and consistent large-area model for AGB estimation that takes advantage of airborne lidar datasets, but combines them with other data sources to perform estimation over long time periods and large areas [13]. Studies often use two main sources of AGB training data based on ground-truthing (e.g., forest inventory data) and airborne lidar [14].

Synthetic aperture radar (SAR) provides information on the dielectric (essentially, moisture content) and structural properties of the targeted objects, which include soil surfaces and plants in wetlands, agricultural land, and forests [15–17]. However, when estimating forest AGB, this kind of dataset depends on the degree of saturation, which refers to the AGB level at which the signal's sensitivity (e.g., backscatter, reflectance) becomes too small to be measurable or where the signal fails to penetrate the forest canopy [18]. These phenomena lead to drastic deterioration of accuracy at high levels of AGB. Thus, saturation levels limit the role that SAR sensors can play in direct measurement of forest biomass for global inventories [19].

One strategy that can be used to overcome this problem is to combine SAR images with optical images [20]. Multispectral optical imagery contains information on the photosynthetic parts of the vegetation, which are rich in chlorophyll, and optical satellite images have a long history of being used for estimation of forest parameters and assessment of different wood quality results. Unfortunately, optical satellite signals are strongly affected by weather and other atmospheric conditions; under unsuitable conditions, optical images are prone to significant errors [21]. Another drawback is that optical satellite signals cannot measure vegetation structure directly and suffer from spectral saturation in densely vegetated environments, which limits their ability to map AGB in some cases, as is the case for SAR data [22]. However, because the two kinds of satellite data have different limitations, it may be possible to combine them to estimate forest AGB, with the advantages of one method offsetting the disadvantages of the other. Selection of appropriate regression models for modeling AGB is crucial because optical remote sensing data and SAR data have different relationships with AGB. For example, some studies have shown a strong linear relationship between SAR and AGB [23,24]. Other studies found that non-linear regression models provided a better fit for this relationship [25,26]. Because of these contradictory results, it is worth considering alternatives to simple regression.

Here, we selected the Random Forest regression model because it has worked well with both SAR data [27] and optical satellite data [28]. Random Forest is an efficient machine learning method proposed by Breiman [29]. It is a type of ensemble machine learning algorithm based on bootstrap aggregation also called bagging [29]. The model lets researchers combine different sources of satellite images in a single model [30]. It is well suited to analyzing complex non-linear and possibly hierarchical interactions in large datasets [31]. Moreover, it can determine the importance of the variables to provide a plausible strategy for combining variables from different datasets. This approach has been used successfully in many cases with different combinations of satellite datasets [32–34].

Although Random Forest estimates AGB well, it tends to overestimate AGB at low values of AGB and underestimate it at high values [35]. Despite these limitations, many researchers have estimated forest AGB in different countries by using SAR and optical satellite datasets, including Mexico, China, Russia, the USA, and Cameroon [11,36–39]. However, few areas have been studied using Random Forest in Japan [40]. To provide more data on this approach, we selected two species of forest tree as our targets. Japanese cedar (*Cryptomeria japonica*) and Japanese cypress (*Chamaecyparis obtusa*) play important roles in Japanese forest ecosystems, and cover 28% of the forested area in Japan, which is equivalent to 19% of Japan's land surface [41].

Our objectives here were to:

(1) assess the potential of combining two types of satellite data (SAR and optical sensors) to improve AGB estimation performance;
(2) estimate the spatial extent of forest AGB for two major forest types in northern Ibaraki Prefecture, Japan; and
(3) benchmark the AGB estimates using forest register data collected by the Ibaraki Prefecture government.

2. Materials and Methods

2.1. Study Area

We focused on plantations of two forest tree species, both in the cypress family (Cupressaceae), growing in Japan's Ibaraki Prefecture, central Japan. The prefecture has an active forest industry, supported by *C. japonica* and *C. obtusa*. These are major plantation species throughout Japan, occupying 4.44×10^6 and 2.60×10^6 ha of forest (equivalent to 18% and 10% of the forest area in Japan), respectively [41]. Both are important timber resources, and are also associated with public functions such as conservation of natural land, prevention of global warming, and recharge of water sources.

Our study area was located in northern Ibaraki Prefecture, which is the prefecture's main forest area. The southern part of the prefecture is dominated by agricultural land with almost no forests (Figure 1). The regional average elevation is 22 m above sea level and the highest point is 1021 m, with rugged terrain; the target forests are distributed mainly in mountainous topography.

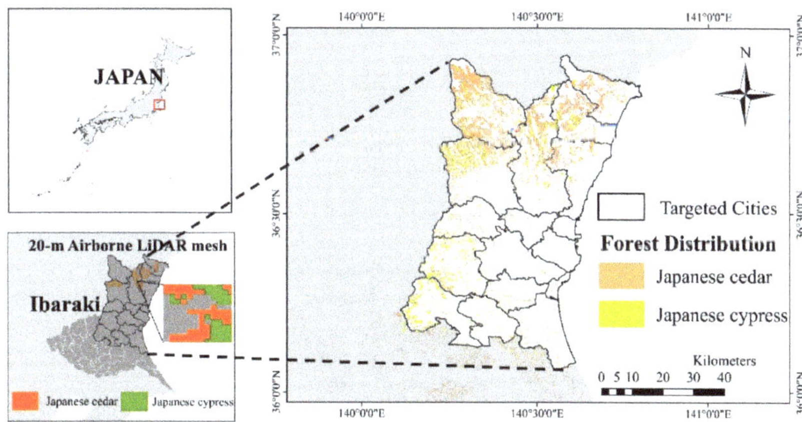

Figure 1. Location of the study area in Japan's Ibaraki prefecture.

2.2. Data Analysis Process

Figure 2 illustrates the work flow we used for the AGB estimation, which comprises: (1) satellite data collection and preprocessing (resampling, application of a unified coordinate system, filtering, and image clipping); (2) extraction of landscape textures from PALSAR-2 data; (3) computation of indices derived from the satellite images (e.g., the HV and HH polarization ratios; vegetation indices such as *NDVI*, *EVI*); (4) model development (selection of the optimal variables, tuning of the hyperparameters); and (5) mapping and estimation of the AGB of the two species in the targeted cities in Ibaraki Prefecture.

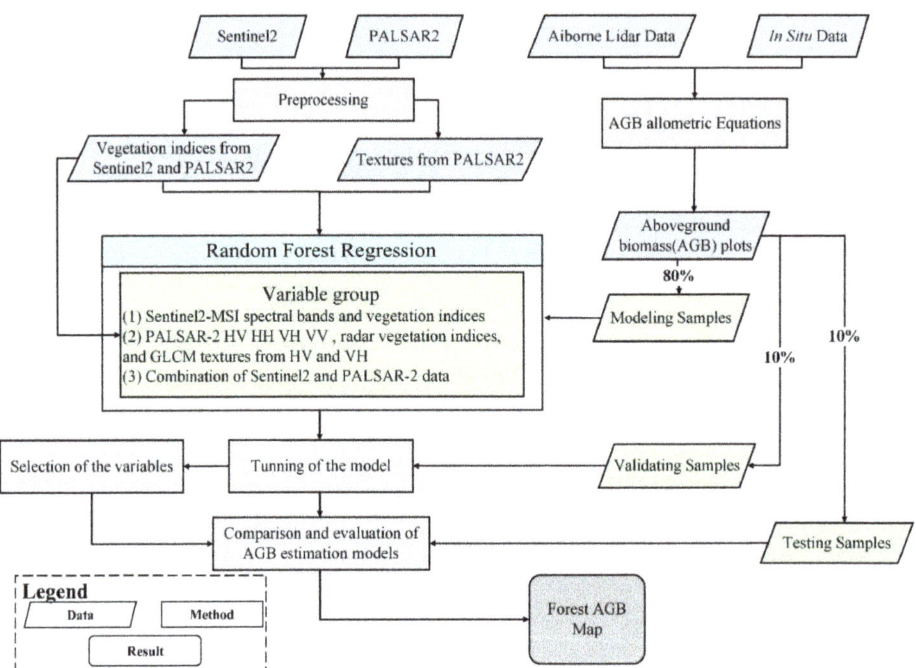

Figure 2. Research flow for calculating aboveground biomass (AGB). Abbreviations: GLCM, gray-level co-occurrence matrix; HH, horizontal transmit–horizontal channel; HV, horizontal transmit–vertical channel; MSI, multispectral instrument; VH, vertical transmit–horizontal channel; VV, vertical transmit–vertical channel.

2.3. Forest AGB Observed by Airborne Lidar

Airborne lidar data obtained from the Ibaraki Prefecture government was utilized for training the Random Forest model. Ibaraki Prefectural Government preprocessed the airborne lidar product and generated the stem volume through the following procedures:

(1) collecting and using 40 human measured points (20 for each forest species), each point covering 0.04 ha, for ground-based calibration to evaluate the accuracy of the airborne lidar data related to stem volume calculation,
(2) collecting airborne lidar data in northern Ibaraki Prefecture on 31 July 2020,
(3) determining the values of parameters related to the stem biomass calculation calibrated by ground measured plots (i.e., tree species, tree height, and diameter at breast height [DBH]; Table 1 and
(4) calculating the stem volume from the tree height and DBH using the conventional allometric equations for these species in Japan [42].

Table 1. Descriptive statistics of the forest parameters derived from airborne lidar data of Ibaraki Government.

Species	Stand Variable	Mean	Standard Deviation	Min.	Max.	Sample Size
Japanese cedar	Tree height (m)	24.1	5.2	2.1	46.7	201,854
	Diameter at breast height (cm)	24.1	5.3	9.9	78.0	
	Stem volume (m^3 ha^{-1})	403.6	170.7	0.3	1516.3	
	Biomass (Mg ha^{-1})	155.9	65.9	0.1	585.6	
Japanese cypress	Tree height (m)	19.2	4.3	2.3	39.6	69,374
	Diameter at breast height (cm)	27.0	5.8	10.1	72.0	
	Stem volume (m^3 ha^{-1})	585.9	246.2	0.3	1800.0	
	Biomass (Mg ha^{-1})	295.7	124.3	0.1	908.4	

The accuracy of the forest parameters used in stem volume calculation are evaluated by the root mean square error (*RMSE*) with the 40 fields measured data mentioned above; the accuracy of the parameters is shown in Table 2.

$$RMSE = \sqrt{\frac{1}{N}(Y_i - Y_i')^2} \tag{1}$$

where N is the number of validation plots collected in the field. Y_i refers to the field measured parameters. Y_i' refers to lidar-based predicted parameters in the corresponding position i. The maps of volume for each of the two forest species were generated in the lidar covered area and then converted into AGB values at a 20-m mesh size using a biomass–volume equation with a biomass expansion factor and the tree volume and density by Equation (2) [43]. AGB values are presented as Mg ha^{-1}. It is important to note that only cedar and cypress were considered in the AGB calculation; this is acceptable because we focused on plantations, which are essentially single-species forests. We obtained 201,854 airborne lidar AGB samples for cedar and 69,374 for cypress and used these data as the modeling samples in our subsequent analysis.

$$AGB = V \times WD \times BEF \tag{2}$$

where V is the volume, WD is wood density and BEF is biomass expansion factor [43].

Table 2. Accuracy of the forest parameters derived from airborne lidar data of Ibaraki Government.

Stem Variables		*RMSE*	Sample Size
Japanese cedar	Tree height (m)	1.1	20
	Diameter at breast height (cm)	3.7	
Japanese cypress	Tree height (m)	1.1	20
	Diameter at breast height (cm)	2.8	

2.4. Remote Sensing Data

2.4.1. Processing of PALSAR-2 Data

The Advanced Land Observing Satellite-2 (ALOS-2) is a follow-on mission from ALOS. ALOS-2 has the Phased Array L-band Synthetic Aperture Radar-2 (PALSAR-2), a microwave sensor that can observe, day and night, under any weather conditions. Here, we obtained the 25-m PALSAR2 L-band global mosaic data from May 2019 from the Japan Aerospace Exploration Agency (JAXA; https://www.eorc.jaxa.jp/ALOS/en/palsar_fnf/data/index.html, accessed on 11 June 2021). JAXA preprocessed the PALSAR-2 data, including geometrical calibration against the AW3D30 digital elevation model after 2019. The original SAR data for Japan used the highly sensitive Beam Quad mode, which provides full polarizations, including HH, HV, VV, and VH. The PALSAR-2 signal can be converted into gamma naught backscattering coefficients by using the following equation:

$$\gamma^0 = 10 \log_{10}(DN^2) - CF \quad (3)$$

where γ_0 is the backscattering coefficient (gamma naught), *DN* is the digital number value of each pixel, and *CF* is the calibration factor, −83 [44]. Moreover, we applied a LEE speckle filter with a kernel window size of 3 × 3 to smooth the images [45]. Before LEE filtering, the radar images were also averaged using a 3 × 3 pixel mean filter to reduce the effect of speckle and spatial heterogeneity of the forest stands and to alleviate the problem of noise from dark spots [24]. Because the plot boundaries of airborne lidar samples may overlie several pixels, using a 3 × 3 window improved performance compared with the single-pixel extraction method [46].

In addition to correcting for backscatter, we calculated the radar vegetation indices for different polarizations and calculated the texture information for HV and VH using a gray-level co-occurrence matrix (GLCM) with a 3 × 3 window size and with a relative displacement vector (d = 1, θ = 45°). The displacement vector explains the spatial distribution of the level pairs separated by d with direction θ [47]. In AGB estimation, the GLCM-derived texture is considered as a kind of predictor that can improve the accuracy of estimation [48]. The texture information can also enlarge the saturation range between AGB and satellite images [49]. We adopted eight popular texture parameters for the VH and HV polarization: mean, variance, homogeneity, contrast, dissimilarity, entropy, second moment, and correlation. Table 3 summarizes the SAR-derived variables used for modeling.

Table 3. List of variables from the PALSAR-2 data. In the texture calculations, h represents high of row number, k represents column number of image window and m_{hk} refers the value in the cell h, k of the image window.

Variables (Abbreviation)		Definition
Polarization	HV	Horizontal transmit-vertical channel
	HH	Horizontal transmit-horizontal channel
	VV	Vertical transmit-vertical channel
	VH	Vertical transmit-horizontal channel
Radar Indices	I1 [50]	HH − HV
	I2 [51]	HV + HH
	I3 [52]	(HH − HV)/(HV + HH)
	I4 [53]	HV/HH
	I5 [50]	VH − VV
	I6 [51]	VH + VV
	I7 [52]	(VH − VV)/(VH + VV)
	I8 [53]	VH/VV
	I9 [54]	8 × HV/(HH + VV + 2 × HV)
Texture (HV, VH)	Mean (ME)	$\sum_h\sum_k h * m_{hk}$
	Variance (VA)	$\sum_h\sum_k h * m_{hk} * (h - Mean)^2$
	Homogeneity (HO)	$\sum_h\sum_k \frac{m_{hk}}{1+(h-k)^2}$
	Contrast (CON)	$\sum_h\sum_k (h-k)^2 m_{hk}$
	Dissimilarity (DIS)	$\sum_h\sum_k \lvert h-k \rvert m_{hk}$
	Entropy (ENT)	$-\sum_h\sum_k m_{hk}\, lg(m_{hk})$
	Second Moment (SM)	$\sum_h\sum_k (m_{hk})^2$
	Correlation (COR)	$\frac{\sum_h\sum_k h*k*m_{hk} - \mu_X\mu_y}{\sigma_X\sigma_y}$

2.4.2. Processing of Sentinel2-MSI Data

Sentinel2-MSI data collected from the European Space Agency (ESA: https://scihub.copernicus.eu/dhus/, accessed on 11 June 2021) were used as optical satellite variables to drive the Random Forest model. Because the optical satellite sensor is easily affected by clouds owing to the wavelengths it uses, we selected data from days with low cloud cover and as close as possible to the time of the AGB data sample collection. From the remaining data, we acquired L2-A level data that had been preprocessed by the ESA, including atmospheric correction and scene classification to L1-B data. The image was acquired on 8 May 2019. The Sentinel-2 MSI sensor provides multispectral data with a spatial resolution ranging from 10 to 60 m. We excluded the 60-m data in this study. We also averaged the Sentinel2 images using a 3 × 3 pixel mean filter to extract the values. We then computed the vegetation indices from the Sentinel-2 data (Table 4) and used those data for modeling.

Table 4. List of variables from the Sentinel2-MSI data. Vegetation indices: DVI, difference vegetation index; EVI, enhanced vegetation index; GARI, green atmospherically resistant vegetation index; GDVI, generalized difference vegetation index; GNDVI, green normalized-difference vegetation index; GRVI, green/red vegetation index; NDVI, normalized-difference vegetation index; SAVI, soil-adjusted vegetation index; SR, simple ratio vegetation index.

Variables Bands, Indices (Abbreviation)		Definition (Central Wavelength)
Multispectral Bands	Band2 (B2)	Blue, 490 nm
	Band3 (B3)	Green, 560 nm
	Band4 (B4)	Red, 665 nm
	Band5 (B5)	Red edge, 705 nm
	Band6 (B6)	Red edge, 749 nm
	Band7 (B7)	Red edge, 783 nm
	Band8 (B8)	Near Infrared (NIR), 842 nm
	Band8A (B8a)	Near Infrared (NIR), 865 nm
	Band11 (B11)	SWIR-1, 1610 nm
	Band12 (B12)	SWIR-2, 2190 nm
Vegetation Indices	NDVI [55]	$\frac{(NIR-Red)}{(NIR+Red)}$
	EVI [56]	$\frac{2.5*(NIR-Red)}{(NIR+6*Red-7.5*Blue+1)}$
	DVI [57]	$NIR - Red$
	GARI [58]	$\frac{NIR-[Green-1.7*(Blue-Red)]}{NIR+[Green-1.7*(Blue-Red)]}$
	SAVI [59]	$\frac{1.5*(NIR-Red)}{(NIR+Red+0.5)}$
	GNDVI [60]	$\frac{(NIR-Green)}{(NIR+Green)}$
	GDVI [61]	$NIR - Green$
	SR [62]	$\frac{NIR}{Red}$
	GRVI [63]	$\frac{NIR}{Green}$

2.4.3. Extraction of Satellite Images Values from Forest AGB Plots

The AGB plots and satellite images were first unified into the Universal Transverse Mercator (UTM) coordinate system (zone 54 N), with datum of WGS84. Then all of the satellite images were resampled in 20 m resolution using bilinear convolution to meet the resolution of airborne Lidar metric. The geometric center of every airborne Lidar plot was represented as the position of the AGB and extracted the corresponding values of all predictors from the satellite images. Finally, a total of 48 predictors were utilized in a regression model for our analysis: 10 Sentinel-2 MSI spectral bands, 9 Sentinel-2 MSI–derived vegetation indices, 4 ALOS-PALSAR-2 radar backscatter coefficient bands, 16 texture information variables (8 textures each for VH and HV respectively), and 9 radar backscatter coefficient-derived indices.

2.5. Random Forest Regression

Modeling datasets were randomly split into 80%, 10%, and 10% bins for training, validation, and testing samples, respectively, using the train_test_split function in the sklearn package for the Python language.

Random Forest predicts AGB from the remote-sensing predictors by growing many decision trees and averaging every result for each tree. We not only performed inversion modeling for each type of remote sensing data, but also identified (through filtering) the best performing variables for each tree species in this model according to the impurity-based feature importance of the variables, and then we combined the selected variables and used them to process the Random Forest model again. This filtering is necessary because the presence of many redundant variables contributes little to the model, it results in the inclusion of repetitive information, and increases the complexity of the model. For each experiment, we assessed the accuracy of the predictions using the testing data, and then tuned the hyperparameters using the validation data.

Four error statistics were selected to evaluate the model's performance: the root-mean-square error (*RMSE*) in Equation (2), the coefficient of determination (R^2) in Equation (3), the mean absolute error (*MAE*) in Equation (4), and the relative *RMSE* (*rRMSE*) in Equation (5) as *RMSE* divided by the mean of the observed AGB values. In the comparison between *RMSE* and *MAE*, *RMSE* is harder to interpret and is more sensitive to outliers than *MAE*. However, a detailed interpretation is not critical, because variations of the same model will have similar error distributions. Therefore, *RMSE* is more appropriate as a loss function to tune the hyperparameters for the model as in our case [64]. However, it is still necessary to use *MAE* together with *RMSE* to evaluate the variation of model errors [65]. Overall, lower values of *RMSE*, *rRMSE*, and *MAE* and higher R^2 indicate better performance of a model. In addition, the smaller the difference between *RMSE* and *MAE*, the smaller the variance between errors will be.

$$RMSE = \sqrt{\frac{1}{N}(Y_i - Y_i')^2} \tag{4}$$

$$R^2 = 1 - \frac{\sum_{i=1}^{N}(Y_i - Y_i')^2}{\sum_{i=1}^{N}(Y_i - \overline{Y})^2} \tag{5}$$

$$MAE = \frac{1}{N}\sum_{i=1}^{N}\sqrt{|Y_i - Y_i'|} \tag{6}$$

$$rRMSE = \frac{RMSE}{\overline{Y}} \tag{7}$$

where N is the number of observed values, Y_i is the observed AGB value for observation i, Y_i' is the predicted AGB value, and $\overline{Y}$ is the mean of the observed AGB values. Even though many variables have potential value for estimating AGB, not all are available to be used in the modeling owing to high inter-variable correlation or weak relationships with AGB [66]. Including such variables provides little improvement of accuracy, although it may increase model flexibility. To eliminate the least useful variables, we used the impurity-based feature importance for each variable: the higher the impurity, the more critical the feature. We computed the importance of a feature as the (normalized) total reduction of the criterion; here, we used the mean squared error (MSE) as the criterion brought by that feature, which is also known as the Gini importance.

2.6. Determination of the Saturation Level

The saturation level for an individual tree species is crucial for evaluating the estimation result. We defined the AGB saturation level as occurring where a clear pattern of AGB leveling was found in the logarithmic regression slope in a plot of the HV backscatter coefficient against AGB since longer wavelength L-bands with HV backscatter are identified as the most sensitive polarizations to AGB [67]. This approach has been used in previous studies to reveal the relationship between AGB and satellite images and the model's performance [68–70]. Since we used a large sample size in our study, it is difficult to accurately determine the location of the saturation point. Therefore, we adopted an interval sampling method in which we counted the average value of the data points in bins equal to 5 Mg ha^{-1} in size as the AGB, and removed points with a value greater than 250 Mg ha^{-1} from the calculation range. Such a kind of approach was accessed in previous research with a huge number of AGB samples [49,71]. Finally, we estimated the saturation levels for each species by examining the slope within every interval (in units of 5 Mg ha^{-1}) with the HV backscatter coefficient. Saturation points in the scatterplot were defined as the points where the slope of each AGB interval was less than 0.1 or where it starts to change in a very disorderly way.

2.7. Evaluation of Forest Resources

After running the models, the best performance model with the highest accuracy was utilized to map the AGB of Japanese cedar and cypress in several target cities with a large area of plantations of the two forest species. The identification of forest area is very crucial for AGB mapping, because mismatched forest distribution maps will cause large estimation errors derived from mismatched estimation models and wrong corresponding forest area. The forest distribution map from the Ibaraki Prefecture government was selected to classify the tree species distribution in our study so that an accurate AGB map could be generated that followed the same standard as the Ibaraki Prefecture AGB map [72]. Finally, we compared the satellite-based AGB map with the forest registered map from Ibaraki Prefecture to evaluate the AGB in the targeted cities.

3. Results

3.1. Determination of the AGB Saturation Level

Figure 3 shows the relationship between the HV backscattering coefficient and AGB. AGB leveled off at a slope of 0.01 dB for the cedar, which represented an AGB of 105 Mg ha^{-1}. However, it was difficult to determine the saturation point for cypress by this method since the slope showed high variation. We defined the saturation point at 175 Mg ha^{-1}, since the HV values leveled off at this point. Nevertheless, when the AGB reached 235 Mg ha^{-1}, the slope continued to increase. The cause of this pattern is unclear; it may have resulted from a relatively small number of samples at high AGB, and the uncertainty of the AGB accuracy rise.

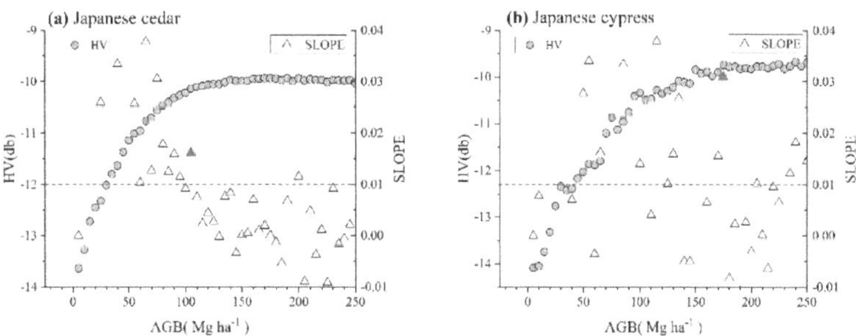

Figure 3. Determination of the aboveground biomass (AGB) saturation level as a function of the horizontal transmit–vertical channel (HV) backscatter coefficient and slope of every pair of neighbor plots. The saturation level is indicated by the red triangles.

3.2. Development of the Random Forest Model

In the Random Forest model, the importance measures for the variables are affected by the number of variable categories and the measurement scale of the predictor variables [73]. Determining the optimal feature space is an important step for model development. Increasing the number of variables might lead to a high time requirement for the calculations despite a low increase of accuracy [74]. Consequently, we divided the variables into two parts (the PALSAR-2 group and the SENTINEL2-MSI group) and selected the optimal variables on the basis of their importance values. We selected the variables whose importance was greater than 0.05 to interact together and assessed the model again. Figure 4 shows the importance results. The most important variables for cypress (i.e., the variables with a Gini importance greater than 0.05) were VH mean, VH variance, HV variance, VH correlation, and HV correlation (Figure 4a), and Band 5, Band 9, Band 8a, Band 11, SR, NDVI, and Band 12 (Figure 4b). The most important variables for cedar were VH mean, VH variance, HV mean, and HV variance (Figure 4c) and Band 12, Band 4, Band 9, Band 11, Band 5, Band 8a, and Band 6 (Figure 4d).

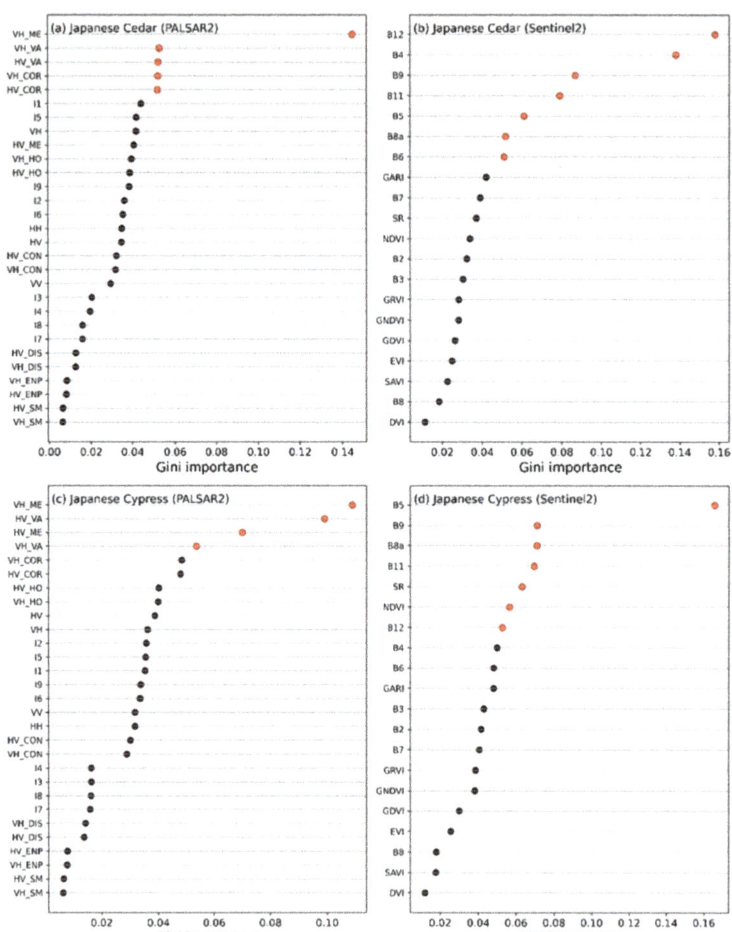

Figure 4. Variables listed in order of importance based on the mean decrease of impurity (i.e., the Gini importance). Variable names are defined in Tables 3 and 4. Red points show the relatively effective variables.

Random Forest algorithm was run repeatedly to obtain the optimal hyperparameters in each PALSAR-2-based model, each Sentinel2-MSI-based model, and the model that combined the two datasets. We chose four input hyperparameters to determine their optimal values in each model: the number of trees in the forest (EST), the maximum depth of the decision tree (MD), the minimum number of samples (MS) required to split in every internal node, and the minimum number of samples required to be at every leaf node (ML). We reserved 10% of the samples for use as the validation samples to determine their values from the *RMSE* score. Owing to the size of our dataset, we didn't perform cross-validation using the validation datasets. We set the number of variables fed to each predictor tree (named max_features in the sklearn package) to the square root of the number of input variables in every model [74], and when we optimized one hyperparameter, we set the others to their default value as MD = 10, ML = 1, MS = 2, and EST = 200. We optimized EST with the determined values of the other three hyperparameters in the last step. The lowest *RMSE* for the EST tuning indicated the best scores with the optimized hyperparameters. Figure 5 shows the results of the hyperparameter tuning. In every case, the combined model had the best performance. Therefore, we used it to estimate AGB in our subsequent analyses.

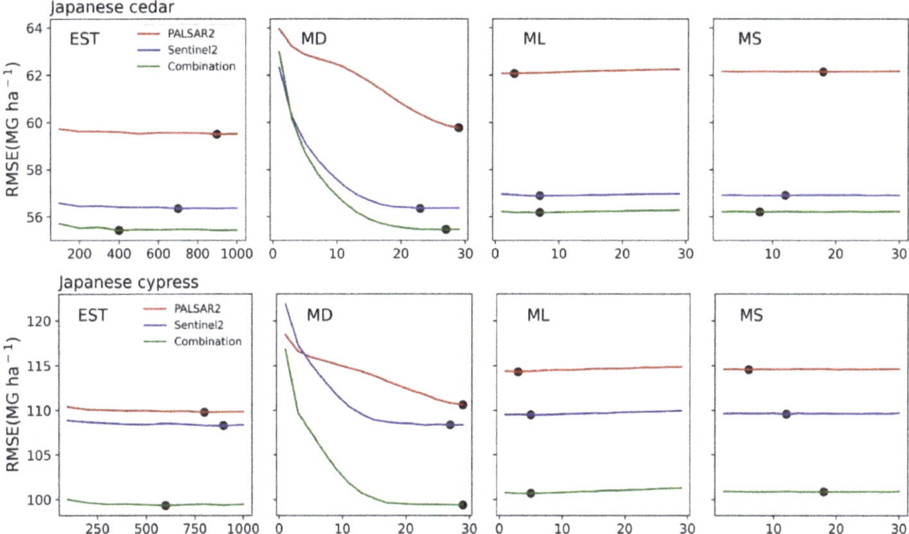

Figure 5. Results of tuning the hyperparameters in the three models: EST, the number of trees in the forest; MD, the maximum decision-tree depth; MS, the minimum number of samples required to split in every internal node; ML, the minimum number of samples required at every leaf node. The value with the minimum root-mean-square error is shown with a black circle.

3.3. Model Accuracy Assessment

Testing data assessed the model's accuracy using our selected error statistical indicators. These testing data represented 10% of the overall sample, and excluded data used in model development to keep robustness. For cedar, the Random Forest algorithm was able to predict the AGB with R^2 = 0.31, *RMSE* = 54.38 Mg ha^{-1}, *MAE* = 40.98 Mg ha^{-1}, and *rRMSE* = 0.35 from the 20,186 test samples (Figure 6). For cypress, it was able to predict the AGB with R^2 = 0.37, *RMSE* = 98.63 Mg ha^{-1}, *MAE* = 76.97 Mg ha^{-1}, and *rRMSE* = 0.33 from the 6938 test samples. For the two tree species, different variables from different remote sensing sensors were important in determining the model's performance.

Thus, it is necessary to retrieve the relationship between AGB and satellite images by separating for each tree species in the forest.

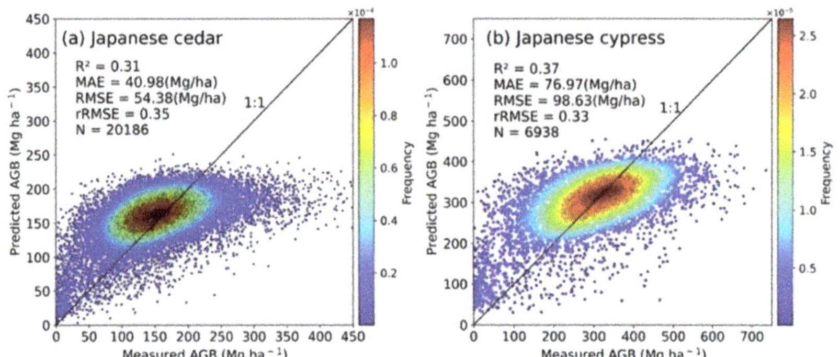

Figure 6. Observed and predicted aboveground biomass (AGB) of the test samples. The color bar on the right indicates the density of points. Note that the color scales differ between the two graphs.

3.4. Mapping AGB

We generated AGB maps for Japanese cedar and Japanese cypress at 20 m resolution using the Random Forest algorithm for the targeted cities in Ibaraki Prefecture. Appendix A compares the data from the Japanese forestry register record for these cities with the predictions of our model. The AGB ranged from 7.49 to 277.02 Mg ha^{-1}, for Japanese cedar and 85.35 to 492.28 Mg ha^{-1} for Japanese cypress in the targeted cities (Figure 7).

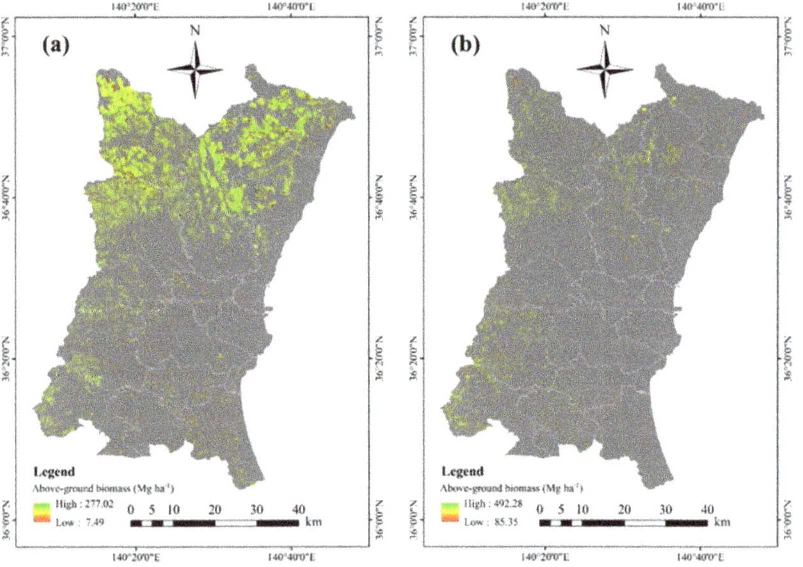

Figure 7. Spatial distribution of aboveground biomass (AGB) of (**a**) Japanese cedar and (**b**) Japanese cypress in the targeted cities in Ibaraki Prefecture.

Japanese cedar and cypress are the main tree species in Japan. Both are evergreen coniferous trees native to Japan. In the overall statistics we analyzed, cypress had a higher AGB than cedar. In terms of the point where AGB saturation occurred according to the HV backscatter coefficient, cypress had a wider range of AGB values than cedar. We think this may be caused by differences in the physical structure of the two species: cypress has a higher biomass range and a larger DBH, which may have led to greater volume scattering, resulting in fluctuations in the relationship between the backscatter coefficient and AGB. Cypress had a much higher mean average AGB (293.12 Mg ha^{-1}) than cedar (146.50 Mg ha^{-1}). However, cedar had a lower standard deviation (44.37 Mg ha^{-1}), with its AGB distributed mainly between 100 and 200 Mg ha^{-1}, whereas Japanese cypress had an AGB distribution with two peaks between 200 and 400 Mg ha^{-1}, with a higher standard deviation (78.48 Mg ha^{-1}). In some previous research, AGB estimation based on the stratification of vegetation types greatly improved the performance [75,76]. We assessed the AGB estimation for two tree species, and found a significant difference in the AGB value distribution (Figure 8). However, the development of an AGB estimation model based on stratification of multiple tree species is more difficult because it requires additional data: (1) vegetation distribution maps for the targeted species, (2) ground-based AGB values classified by species, and (3) a sufficiently large sample size to build a robust model while still leaving data for testing and validation.

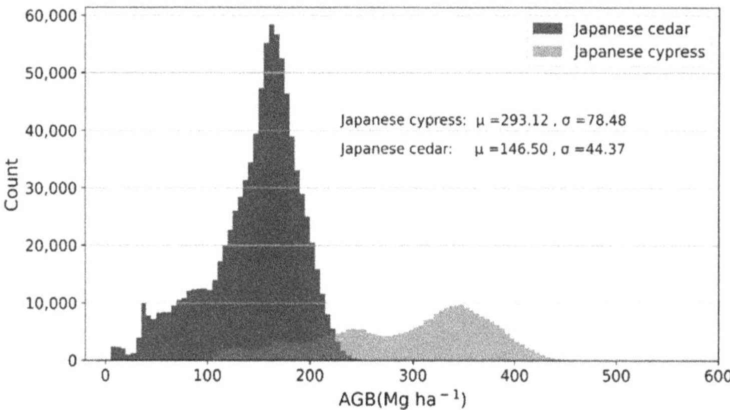

Figure 8. The distributions of aboveground biomass (AGB) in the targeted cities. Data is based on a 20 m resolution and AGB bins with a width of 5 Mg ha^{-1}. Values for each species are means (μ) and standard deviations (σ).

4. Discussion

4.1. Role and Limitation of Satellite-Derived Variables in Accurate Estimation of Japanese Cedar and Japanese Cypress AGB

SAR and optical remote sensing have different drawbacks and advantages for AGB estimation. Either dataset by itself is not enough to accurately estimate forest AGB [77]. SAR is relatively unaffected by weather, since it can penetrate clouds and work all day and night. It can also penetrate through the canopy, soil, and dry snow. However, even L-band SAR becomes saturated at an AGB of 100 Mg ha^{-1} in complex heterogeneous tropical forest structures. In forests with a simple structure and few dominant species, the saturation level could increase to about 250 Mg ha^{-1} [78]. We found that the optical data were more resistant than the SAR data to AGB saturation for Japanese cedar and cypress at high AGB values (Figure 9).

To identify the saturation level for the two tree species, we used the HV backscatter from the SAR data. Cedar became saturated at 105 Mg ha^{-1} and cypress at 175 Mg ha^{-1}. Even though these species are similar in their structure and living conditions, they showed a clear difference in the saturation level with SAR at relatively low values. In contrast, the optical sensors are strongly affected by weather conditions, but also show AGB saturation. Because these different sensor data have different advantages and drawbacks, integration of radar data with optical-sensor data has the potential to improve AGB estimation because it may reduce the number of mixed pixels and data saturation problems [66].

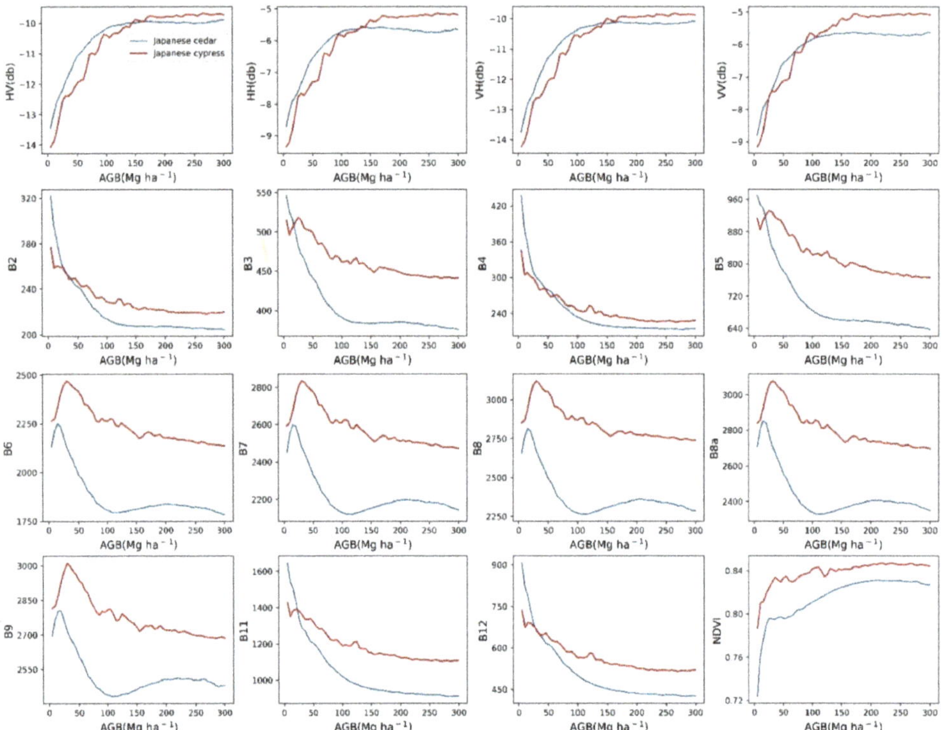

Figure 9. Relationships between aboveground biomass (AGB) and satellite image data of Japanese cedar and Japanese cypress using averages of bins with a width of 5 Mg ha^{-1}. Sensor variables are defined in Tables 3 and 4.

Our aim was to develop models of cedar and cypress for estimating AGB from two types of satellite data: L-band microwave radar data from PALSAR-2 and multispectral optical remote sensing data from Sentinel2-MSI. For our study species, microwave remote sensing was more sensitive to saturation than optical remote sensing. Therefore, the estimation results of the PALSAR-2 model performed worse in both species (Figure 10). In contrast, the model that combined the two datasets performed best in every case. This demonstrates that combining different types of remote sensing data can improve the estimation accuracy and AGB range.

However, underestimation at high AGB values remains large, since satellite information (especially microwave and optical remote sensing data) inevitably became saturated. Although this problem can be alleviated by adding texture information [49] or by combining multi-source remote sensing data, as we did in this study, it is still fundamentally difficult to solve the saturation problem. The airborne lidar data have a high range for estimation of AGB (i.e., high resistance to saturation) owing to their wavelength characteristics, but such data are expensive, which makes it impossible to cover large areas such as a whole country or continent, unlike the space satellite data that are used for large-area studies. Hence, airborne lidar data have mainly been used in small areas [79]. Establishing a model that would extend AGB estimation to large areas by combining data from field plots, airborne lidar, and space satellite data thus has considerable potential to enlarge the area that could be surveyed with airborne lidar data [80]. In such a method, lidar data and satellite remote sensing data could be combined to support large-area AGB predictions, but more tests are still needed, and the analytical framework must be improved to support this use of multisource data. This is because the use of different buffer sizes to combine data from different sources can decrease the accuracy of AGB estimation [80].

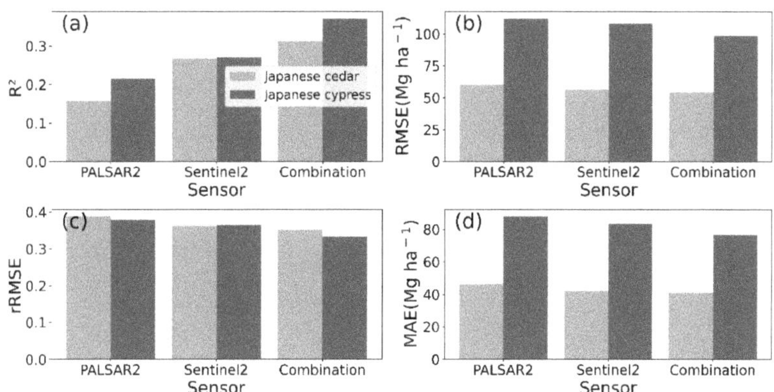

Figure 10. Comparison of the accuracy of the different models (separate models for PALSAR-2 and Sentinel2 and a model that combines both datasets): (**a**) coefficient of determination (R^2), (**b**) root-mean-square error (*RMSE*), (**c**) relative *RMSE* (*rRMSE*), and (**d**) mean absolute error (*MAE*).

The Random Forest model has an excellent predictive ability but has the characteristics of tree-regression. The algorithm operates by constructing many decision trees during training and outputs the predicted mean or mode of the individual decision trees. The AGB prediction averages all variables extracted from the satellite images. However, the algorithm cannot predict the value from the training samples. Using only the satellite-based data, the problem becomes more severe, since saturation occurred in all of the satellite data. Because our approach overlays the underestimation of high AGB values with AGB saturation in the satellite images, it is hard to obtain good performance in forests with high AGB.

4.2. Benchmark AGB Estimated in the Japanese Forest Inventory

The satellite-derived AGB map was compared with government statistics for the total AGB in every targeted city (Figure 11). The forestry statistics in Japan are based on the forest register, which is used for forest management. Japanese forests are managed as land units called sub-compartments. The forest register records forest conditions for every sub-compartment, such as its area, tree species, mean age, and stand volume. Japanese law requires that the forest register be updated every 5 years. The satellite-derived AGB was more significant than the value in the register in Ibaraki, and some previous studies also concluded that the forest register underestimated the forest volume [81]. Japan's Forestry and Forest Products Research Institute compared the register's data with a field measurement at 10,189 sub-compartments throughout Japan, and found that the field-measured forest volume was 1.88 times the value in the forest register for Japan as a whole [82]. These results agree with the present results for cypress, since the estimated AGB was larger than the value in the register. A previous study analyzed airborne lidar data for all of Ehime Prefecture (a prefecture located in southwestern Japan), and found that the total forest volume in Ehime was 2.01 times the value in the register [83]. The authors mentioned that errors in both the lidar estimates and the register contributed to this difference, but suggested that the errors in the register were much larger. These previous studies suggest that our results are reasonable, and the satellite estimates are closer than the register to the actual values. The forest register may underestimate the forest volume for at least two reasons: (1) it records only an estimated value based on the tree species and forest age, not a field-measured value, and (2) the empirical yield tables (used for the estimation that produces the register values) were developed in the 1950s and 1960s and have not been updated since then, so their gap relative to the actual values may have increased [82,84]. One reason for this gap is that there were insufficient measurement data for old-growth forests to support empirical development of yield tables, so the tables underestimate the volume of old-growth forests. Accurate forest resource information is fundamental for forest management, and the forest register, which does not have a monitoring function for actual forests, cannot provide the necessary support. Our approach may solve this problem.

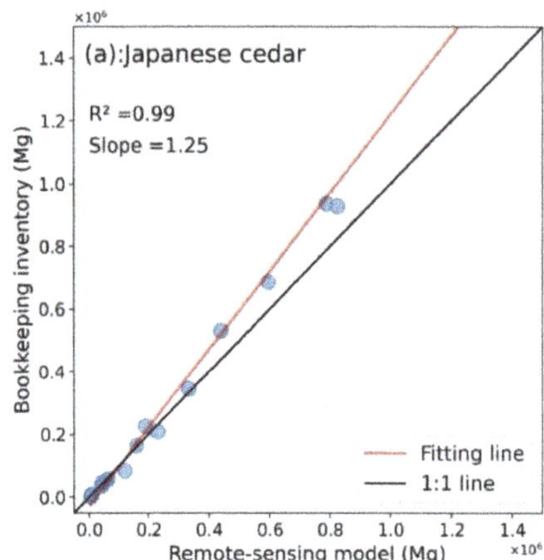

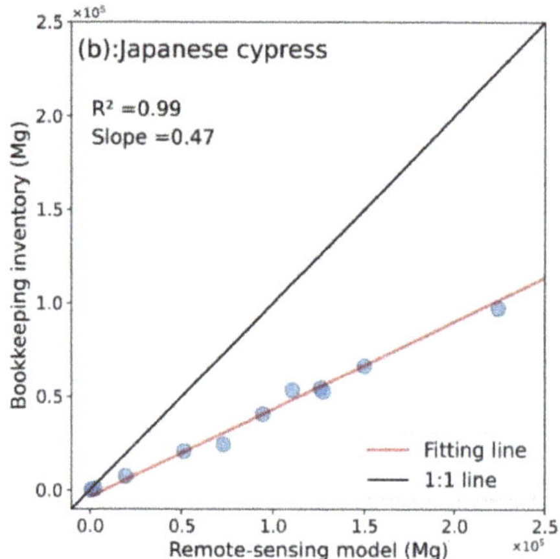

Figure 11. Comparison of total aboveground biomass (AGB) between statistical data in the Japanese forest register and the remote-sensing data. Note that the scales differ greatly between the two species. Each data point represents the cumulative AGB at each of 17 targeted places (villages, towns, and cities) in Ibaraki Prefecture.

4.3. Uncertainty in AGB Estimation

The geographic location between the airborne lidar sample points and satellite pixels will bring significant uncertainty. We use the geometric center of an airborne Lidar sample point with a resolution of 20 m (0.04 hectares) to extract satellite pixel values. Nevertheless, an area of 0.04 hectares cannot cover a plurality of pixels well, leading to missing and biased parts of the data. Although we have alleviated the uncertainty mentioned above through the mean filter, using the mean filter will cause some irrelevant pixels to be calculated into the target pixels, especially when some different tree species are inlaid with each other or in the boundary of the forest area [85]. One method is to select only a cluster of airborne radar sample points gathered by a single tree species as the ground data points and calculate the average value of the pixel points covered by the sample points as the satellite data value corresponding to the airborne lidar plot, but this would significantly increase the complexity of the calculation.

The temporal difference between airborne Lidar and satellite images can also be uncertain. As cloud cover often occurs in the northern part of our research area, in order to avoid the impact of this situation on detection, it is challenging to select satellite data that is the same as or very close to the ground sample point collection time and dozens of days may lead to a difference of forest growth which may lead to temporal variability in satellite images [86]. The change of the period has led to errors in the agreement between the biomass and satellite data. One of the methods focuses on the growth period, ignoring the influence of the year, and selecting ground sample points in other years to collect the satellite data of the corresponding month, but this may lead to forest changes caused by excessive time. Therefore, it is essential to check the forest to see massive changes due to people's affection or natural hazards.

Finally, there is the error caused by the biomass equation because we used the airborne lidar data as the "real" sample data and satellite data for correction. This method provides more sample data than traditional human measurement sample points. A more extensive sample set will avoid the curse of dimensionality caused by too many satellite variables and has better robustness compared to small data set. However, the method of calculating the stock volume by obtaining the parameter values of the trees and converting it through the volume-biomass transferring equation will also have a significant deviation compared with the actual biomass calculation equation of a single tree called recording and grouping errors [79]. However, AGB estimations using only field data over large areas suffer an enormous error rate. Therefore, choosing a large amount of data or a small amount of data with higher accuracy needs to be traded off carefully.

5. Conclusions

We developed robust and effective models to estimate the AGB of Japanese cedar and cypress by a machine learning approach. As far as we know, no other studies have used remote sensing data to retrieve the AGB of those two types of forest at prefecture level. We hope to create a new approach to remedying the lack of forest biomass in Japan. By combining PALSAR-2 and Sentinel2-MSI data and using a large number of validation samples from lidar-based AGB plots, we increased the accuracy of AGB prediction for both species compared with using only one data source. The hyperparameter tuning in Random Forest also improved estimation accuracy, especially for the depth of the tree structure. Because the choice of modeling variables strongly affects the accuracy and simplicity of the models, our approach also helps to select the optimal variables for inclusion in the final model. The texture information for the PALSAR-2 images played an important role in estimating AGB, and this confirms the value of retaining SAR texture information.

Although our method provided a scientific basis for more accurately estimating AGB of the two tree species in our study, more work will be necessary to adapt the method to multi-species forests. The methodology could then be adopted for mapping and estimating forest biomass in Japan and updating the forest register. This use of remote sensing will provide a cost-efficient way to estimate forest conditions and their spatial and temporal variation.

Author Contributions: Conceptualization, T.K. and M.H.; methodology, T.K. and M.H.; software, H.L.; validation, T.K. and M.H.; formal analysis, H.L., T.K. and L.W.; investigation, H.L.; resources, T.K. and M.H.; data curation, H.L. and M.H.; writing—original draft preparation, H.L.; writing—review and editing, T.K. and M.H.; visualization, H.L.; supervision, T.K.; project administration, T.K.; funding acquisition, T.K. All authors have read and agreed to the published version of the manuscript.

Funding: This research was funded by the Telecommunications Advancement Foundation.

Institutional Review Board Statement: Not applicable.

Informed Consent Statement: Not applicable.

Data Availability Statement: Not applicable.

Acknowledgments: This study was supported by JAXA Research Announcement on Earth Observations (No. ER2A2N208). The airborne lidar data was provided by the Ibaraki Prefectural Government (permit No. RINSEI-264).

Conflicts of Interest: The authors declare no conflict of interest.

Appendix A

Figure A1 compares the data in the Japanese forestry register with predictions from the remote-sensing model for the targeted cities in Ibaraki Prefecture.

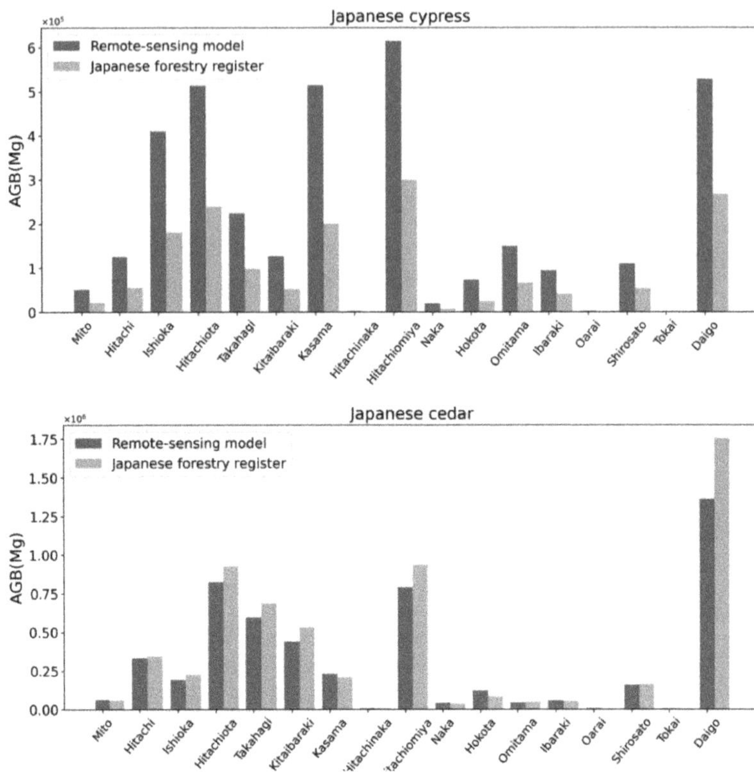

Figure A1. The aboveground biomass (AGB) estimated by the remote-sensing model and recorded in the Japanese forestry register for Japanese cedar and Japanese cypress in the 17 targeted cities in Ibaraki Prefecture.

References

1. Dong, J.; Kaufmann, R.K.; Myneni, R.B.; Tucker, C.J.; Kauppi, P.E.; Liski, J.; Buermann, W.; Alexeyev, V.; Hughes, M.K. Remote sensing estimates of boreal and temperate forest woody biomass: Carbon pools, sources, and sinks. *Remote Sens. Environ.* **2003**, *84*, 393–410. [CrossRef]
2. Gower, S.T. Patterns and Mechanisms of The Forest Carbon Cycle. *Annu. Rev. Environ. Resour.* **2003**, *28*, 169–204. [CrossRef]
3. Fahey, T.J.; Woodbury, P.B.; Battles, J.J.; Goodale, C.L.; Hamburg, S.P.; Ollinger, S.V.; Woodall, C.W. Forest carbon storage: Ecology, management, and policy. *Front. Ecol. Environ.* **2010**, *8*, 245–252. [CrossRef]
4. Goetz, S.; Dubayah, R. Advances in remote sensing technology and implications for measuring and monitoring forest carbon stocks and change. *Carbon Manag.* **2014**, *2*, 231–244. [CrossRef]
5. Chojnacky, D.C.; Heath, L.S.; Jenkins, J.C. Updated generalized biomass equations for North American tree species. *Forestry* **2013**, *87*, 129–151. [CrossRef]
6. Kuenzer, C.; Bluemel, A.; Gebhardt, S.; Quoc, T.V.; Dech, S. Remote Sensing of Mangrove Ecosystems: A Review. *Remote Sens.* **2011**, *3*, 878–928. [CrossRef]
7. Zolkos, S.G.; Goetz, S.J.; Dubayah, R. A meta-analysis of terrestrial aboveground biomass estimation using lidar remote sensing. *Remote Sens. Environ.* **2013**, *128*, 289–298. [CrossRef]
8. Lefsky, M.A. A global forest canopy height map from the Moderate Resolution Imaging Spectroradiometer and the Geoscience Laser Altimeter System. *Geophys. Res. Lett.* **2010**, *37*. [CrossRef]
9. Ioki, K.; Tsuyuki, S.; Hirata, Y.; Phua, M.-H.; Wong, W.V.C.; Ling, Z.-Y.; Saito, H.; Takao, G. Estimating above-ground biomass of tropical rainforest of different degradation levels in Northern Borneo using airborne LiDAR. *For. Ecol. Manag.* **2014**, *328*, 335–341. [CrossRef]
10. Jubanski, J.; Ballhorn, U.; Kronseder, K.; Siegert, F. Detection of large above-ground biomass variability in lowland forest ecosystems by airborne LiDAR. *Biogeosciences* **2013**, *10*, 3917–3930. [CrossRef]

11. Réjou-Méchain, M.; Tymen, B.; Blanc, L.; Fauset, S.; Feldpausch, T.R.; Monteagudo, A.; Phillips, O.L.; Richard, H.; Chave, J. Using repeated small-footprint LiDAR acquisitions to infer spatial and temporal variations of a high-biomass Neotropical forest. *Remote Sens. Environ.* **2015**, *169*, 93–101. [CrossRef]
12. Popescu, S.C. Estimating biomass of individual pine trees using airborne lidar. *Biomass Bioenergy* **2007**, *31*, 646–655. [CrossRef]
13. Wang, H.; Seaborn, T.; Wang, Z.; Caudill, C.C.; Link, T.E. Modeling tree canopy height using machine learning over mixed vegetation landscapes. *Int. J. Appl. Earth Obs. Geoinf.* **2021**, *101*, 102353. [CrossRef]
14. Nguyen, T.H.; Jones, S.; Soto-Berelov, M.; Haywood, A.; Hislop, S. Landsat Time-Series for Estimating Forest Aboveground Biomass and Its Dynamics across Space and Time: A Review. *Remote Sens.* **2019**, *12*, 98. [CrossRef]
15. Chauhan, S.; Darvishzadeh, R.; Boschetti, M.; Nelson, A. Estimation of crop angle of inclination for lodged wheat using multi-sensor SAR data. *Remote Sens. Environ.* **2020**, *236*, 111488. [CrossRef]
16. Le Hegarat-Mascle, S.; Zribi, M.; Alem, F.; Weisse, A.; Loumagne, C. Soil moisture estimation from ERS/SAR data: Toward an operational methodology. *IEEE Trans. Geosci. Remote Sens.* **2002**, *40*, 2647–2658. [CrossRef]
17. Neumann, M.; Ferro-Famil, L.; Reigber, A. Estimation of Forest Structure, Ground, and Canopy Layer Characteristics from Multibaseline Polarimetric Interferometric SAR Data. *IEEE Trans. Geosci. Remote Sens.* **2010**, *48*, 1086–1104. [CrossRef]
18. Fagan, M.; DeFries, R. *Measurement and Monitoring of the World's Forests: A Review and Summary of Remote Sensing Technical Capability, 2009–2015*; Resources of the Future: Washington, DC, USA, 2009; 131p.
19. Imhoff, M.L. Radar backscatter and biomass saturation: Ramifications for global biomass inventory. *IEEE Trans. Geosci. Remote Sens.* **1995**, *33*, 511–518. [CrossRef]
20. Amini, J.; Sumantyo, J.T.S. Employing a Method on SAR and Optical Images for Forest Biomass Estimation. *IEEE Trans. Geosci. Remote Sens.* **2009**, *47*, 4020–4026. [CrossRef]
21. Lillesand, T.; Kiefer, R.W.; Chipman, J. *Remote Sensing and Image Interpretation*; John Wiley & Sons: Hoboken, NJ, USA, 2015.
22. Pflugmacher, D.; Cohen, W.B.; Kennedy, R.E. Using Landsat-derived disturbance history (1972–2010) to predict current forest structure. *Remote. Sens. Environ.* **2012**, *122*, 146–165. [CrossRef]
23. Carreiras, J.M.B.; Vasconcelos, M.J.; Lucas, R.M. Understanding the relationship between aboveground biomass and ALOS PALSAR data in the forests of Guinea-Bissau (West Africa). *Remote Sens. Environ.* **2012**, *121*, 426–442. [CrossRef]
24. Peregon, A.; Yamagata, Y. The use of ALOS/PALSAR backscatter to estimate above-ground forest biomass: A case study in Western Siberia. *Remote Sens. Environ.* **2013**, *137*, 139–146. [CrossRef]
25. Lucas, R.; Armston, J.; Fairfax, R.; Fensham, R.; Accad, A.; Carreiras, J.; Kelley, J.; Bunting, P.; Clewley, D.; Bray, S.; et al. An Evaluation of the ALOS PALSAR L-Band Backscatter—Above Ground Biomass Relationship Queensland, Australia: Impacts of Surface Moisture Condition and Vegetation Structure. *IEEE J. Sel. Top. Appl. Earth Obs. Remote Sens.* **2010**, *3*, 576–593. [CrossRef]
26. Cartus, O.; Santoro, M.; Kellndorfer, J. Mapping Forest aboveground biomass in the Northeastern United States with ALOS PALSAR dual-polarization L-band. *Remote Sens. Environ.* **2012**, *124*, 466–478. [CrossRef]
27. Forkuor, G.; Benewinde Zoungrana, J.-B.; Dimobe, K.; Ouattara, B.; Vadrevu, K.P.; Tondoh, J.E. Above-ground biomass mapping in West African dryland forest using Sentinel-1 and 2 datasets—A case study. *Remote Sens. Environ.* **2020**, *236*, e111496. [CrossRef]
28. Dang, A.T.N.; Nandy, S.; Srinet, R.; Luong, N.V.; Ghosh, S.; Senthil Kumar, A. Forest aboveground biomass estimation using machine learning regression algorithm in Yok Don National Park, Vietnam. *Ecol. Inform.* **2019**, *50*, 24–32. [CrossRef]
29. Breiman, L. Random forests. *Mach. Learn.* **2001**, *45*, 5–32. [CrossRef]
30. Rodríguez-Veiga, P.; Quegan, S.; Carreiras, J.; Persson, H.J.; Fransson, J.E.S.; Hoscilo, A.; Ziółkowski, D.; Stereńczak, K.; Lohberger, S.; Stängel, M.; et al. Forest biomass retrieval approaches from earth observation in different biomes. *Int. J. Appl. Earth Obs. Geoinf.* **2019**, *77*, 53–68. [CrossRef]
31. Olden, J.D.; Lawler, J.J.; Poff, N.L. Machine learning methods without tears: A primer for ecologists. *Q. Rev. Biol.* **2008**, *83*, 171–193. [CrossRef] [PubMed]
32. Ahmed, O.S.; Franklin, S.E.; Wulder, M.A.; White, J.C. Characterizing stand-level forest canopy cover and height using Landsat time series, samples of airborne LiDAR, and the Random Forest algorithm. *ISPRS J. Photogramm. Remote Sens.* **2015**, *101*, 89–101. [CrossRef]
33. Wang, Y.; Zhang, X.; Guo, Z. Estimation of tree height and aboveground biomass of coniferous forests in North China using stereo ZY-3, multispectral Sentinel-2, and DEM data. *Ecol. Indic.* **2021**, *126*, 107645. [CrossRef]
34. Ye, Q.; Yu, S.; Liu, J.; Zhao, Q.; Zhao, Z. Aboveground biomass estimation of black locust planted forests with aspect variable using machine learning regression algorithms. *Ecol. Indic.* **2021**, *129*, 107948. [CrossRef]
35. Avitabile, V.; Herold, M.; Heuvelink, G.B.; Lewis, S.L.; Phillips, O.L.; Asner, G.P.; Armston, J.; Ashton, P.S.; Banin, L.; Bayol, N.; et al. An integrated pan-tropical biomass map using multiple reference datasets. *Glob. Change Biol.* **2016**, *22*, 1406–1420. [CrossRef]
36. Houghton, R.A. Aboveground Forest Biomass and the Global Carbon Balance. *Glob. Chang. Biol.* **2005**, *11*, 945–958. [CrossRef]
37. Mermoz, S.; Le Toan, T.; Villard, L.; Réjou-Méchain, M.; Seifert-Granzin, J. Biomass assessment in the Cameroon savanna using ALOS PALSAR data. *Remote Sens. Environ.* **2014**, *155*, 109–119. [CrossRef]
38. Rodríguez-Veiga, P.; Saatchi, S.; Tansey, K.; Balzter, H. Magnitude, spatial distribution and uncertainty of forest biomass stocks in Mexico. *Remote Sens. Environ.* **2016**, *183*, 265–281. [CrossRef]
39. Kellndorfer, J.; Walker, W.; Kirsch, K.; Fiske, G.; Bishop, J.; LaPoint, L.; Hoppus, M.; Westfall, J. NACP Aboveground Biomass and Carbon Baseline Data, V. 2 (NBCD 2000), USA. 2000. Available online: https://daac.ornl.gov/NACP/guides/NBCD_2000_V2.html (accessed on 5 April 2021).

40. Barbosa, J.M.; Broadbent, E.N.; Bitencourt, M.D. Remote Sensing of Aboveground Biomass in Tropical Secondary Forests: A Review. *Int. J. For. Res.* **2014**, *2014*, 715796. [CrossRef]
41. Japan Forestry Agency. *Data for Japanese Cedar and Cypress Plantations*; Japan Forestry Agency: Tokyo, Japan, 2014.
42. Japan Forestry Agency. *Table of Standing Tree Trunk Volume: Western Japan Version*; Forestry Survey Group: Tokyo, Japan, 1970.
43. Ministry of the Environment; Center for Global Environmental Research (CGER); National Institute for Environmental Studies (NIES). *National Greenhouse Gas Inventory Report of JAPAN*; National Institute for Environmental Studies: Onogawa, Japan, 2021.
44. Shimada, M.; Itoh, T.; Motooka, T.; Watanabe, M.; Shiraishi, T.; Thapa, R.; Lucas, R. New global forest/non-forest maps from ALOS PALSAR data (2007–2010). *Remote Sens. Environ.* **2014**, *155*, 13–31. [CrossRef]
45. Lee, J.S.; Jurkevich, L.; Dewaele, P.; Wambacq, P.; Oosterlinck, A. Speckle filtering of synthetic aperture radar images: A review. *Remote Sens. Rev.* **1994**, *8*, 313–340. [CrossRef]
46. Tian, X.; Su, Z.; Chen, E.; Li, Z.; van der Tol, C.; Guo, J.; He, Q. Estimation of forest above-ground biomass using multi-parameter remote sensing data over a cold and arid area. *Int. J. Appl. Earth Obs. Geoinf.* **2012**, *14*, 160–168. [CrossRef]
47. Haralick, R.M.; Shanmugam, K.; Dinstein, I.H. Textural features for image classification. *IEEE Trans. Syst. Man Cybern.* **1973**, *3*, 610–621. [CrossRef]
48. Thapa, R.B.; Watanabe, M.; Motohka, T.; Shimada, M. Potential of high-resolution ALOS–PALSAR mosaic texture for aboveground forest carbon tracking in tropical region. *Remote Sens. Environ.* **2015**, *160*, 122–133. [CrossRef]
49. Hayashi, M.; Motohka, T.; Sawada, Y. Aboveground Biomass Mapping Using ALOS-2/PALSAR-2 Time-Series Images for Borneo's Forest. *IEEE J. Sel. Top. Appl. Earth Obs. Remote Sens.* **2019**, *12*, 5167–5177. [CrossRef]
50. Avtar, R.; Sawada, H.; Takeuchi, W.; Singh, G. Characterization of forests and deforestation in Cambodia using ALOS/PALSAR observation. *Geocarto Int.* **2012**, *27*, 119–137. [CrossRef]
51. Lehmann, E.A.; Caccetta, P.A.; Zhou, Z.-S.; McNeill, S.J.; Wu, X.; Mitchell, A.L. Joint processing of Landsat and ALOS-PALSAR data for forest mapping and monitoring. *IEEE Trans. Geosci. Remote Sens.* **2011**, *50*, 55–67. [CrossRef]
52. Huggannavar, V.; Shetty, A. Biomass Estimation Using Synergy of ALOS-PALSAR and Landsat Data in Tropical Forests of Brazil. In *Applications of Geomatics in Civil Engineering*; Springer: Singapore, 2020.
53. Dong, J.; Xiao, X.; Sheldon, S.; Biradar, C.; Duong, N.D.; Hazarika, M. A comparison of forest cover maps in Mainland Southeast Asia from multiple sources: PALSAR, MERIS, MODIS and FRA. *Remote Sens. Environ.* **2012**, *127*, 60–73. [CrossRef]
54. Kim, Y.; Jackson, T.; Bindlish, R.; Lee, H.; Hong, S. Radar vegetation index for estimating the vegetation water content of rice and soybean. *IEEE Geosci. Remote Sens. Lett.* **2011**, *9*, 564–568.
55. Rouse, J.W.; Haas, R.H.; Schell, J.A.; Deering, D.W. Monitoring vegetation systems in the Great Plains with ERTS. *NASA Spec. Publ.* **1974**, *351*, 309.
56. Huete, A.; Didan, K.; Miura, T.; Rodriguez, E.P.; Gao, X.; Ferreira, L.G. Overview of the radiometric and biophysical performance of the MODIS vegetation indices. *Remote Sens. Environ.* **2002**, *83*, 195–213. [CrossRef]
57. Tucker, C.J. Red and photographic infrared linear combinations for monitoring vegetation. *Remote Sens. Environ.* **1979**, *8*, 127–150. [CrossRef]
58. Gitelson, A.A.; Kaufman, Y.J.; Merzlyak, M.N. Use of a green channel in remote sensing of global vegetation from EOS-MODIS. *Remote Sens. Environ.* **1996**, *58*, 289–298. [CrossRef]
59. Huete, A.R. A soil-adjusted vegetation index (SAVI). *Remote Sens. Environ.* **1988**, *25*, 295–309. [CrossRef]
60. Gitelson, A.A.; Merzlyak, M.N. Remote sensing of chlorophyll concentration in higher plant leaves. *Adv. Space Res.* **1998**, *22*, 689–692. [CrossRef]
61. Sripada, R.P. *Determining in-Season Nitrogen Requirements for Corn Using Aerial Color-Infrared Photography*; North Carolina State University: Raleigh, NC, USA, 2005.
62. Birth, G.S.; McVey, G.R. Measuring the color of growing turf with a reflectance spectrophotometer 1. *Agron. J.* **1968**, *60*, 640–643. [CrossRef]
63. Sripada, R.P.; Heiniger, R.W.; White, J.G.; Meijer, A.D. Aerial color infrared photography for determining early in-season nitrogen requirements in corn. *Agron. J.* **2006**, *98*, 968–977. [CrossRef]
64. Chai, T.; Draxler, R.R. Root mean square error (RMSE) or mean absolute error (MAE)?—Arguments against avoiding RMSE in the literature. *Geosci. Model Dev.* **2014**, *7*, 1247–1250. [CrossRef]
65. Bui, D.T.; Tuan, T.A.; Klempe, H.; Pradhan, B.; Revhaug, I. Spatial prediction models for shallow landslide hazards: A comparative assessment of the efficacy of support vector machines, artificial neural networks, kernel logistic regression, and logistic model tree. *Landslides* **2016**, *13*, 361–378.
66. Lu, D. The potential and challenge of remote sensing-based biomass estimation. *Int. J. Remote Sens.* **2007**, *27*, 1297–1328. [CrossRef]
67. Sinha, S.; Jeganathan, C.; Sharma, L.K.; Nathawat, M.S. A review of radar remote sensing for biomass estimation. *Int. J. Environ. Sci. Technol.* **2015**, *12*, 1779–1792. [CrossRef]
68. Watanabe, M.; Shimada, M.; Rosenqvist, A.; Tadono, T.; Matsuoka, M.; Romshoo, S.A.; Ohta, K.; Furuta, R.; Nakamura, K.; Moriyama, T. Forest Structure Dependency of the Relation between L-Band*sigma*^{0} and Biophysical Parameters. *IEEE Trans. Geosci. Remote Sens.* **2006**, *44*, 3154–3165. [CrossRef]
69. Suzuki, R.; Kim, Y.; Ishii, R. Sensitivity of the backscatter intensity of ALOS/PALSAR to the above-ground biomass and other biophysical parameters of boreal forest in Alaska. *Polar Sci.* **2013**, *7*, 100–112. [CrossRef]

70. Nesha, M.K.; Hussin, Y.A.; van Leeuwen, L.M.; Sulistioadi, Y.B. Modeling and mapping aboveground biomass of the restored mangroves using ALOS-2 PALSAR-2 in East Kalimantan, Indonesia. *Int. J. Appl. Earth Obs. Geoinf.* **2020**, *91*, 102158. [CrossRef]
71. Yu, Y.; Saatchi, S. Sensitivity of L-Band SAR Backscatter to Aboveground Biomass of Global Forests. *Remote Sens.* **2016**, *8*, 522. [CrossRef]
72. Government Ibaraki Prefecture. Forest Distribution Map. 2020. Available online: https://www.rinya.maff.go.jp/kanto/ibaraki/index.html (accessed on 5 April 2021).
73. Strobl, C.; Boulesteix, A.-L.; Zeileis, A.; Hothorn, T. Bias in random forest variable importance measures: Illustrations, sources and a solution. *BMC Bioinform.* **2007**, *8*, 25. [CrossRef] [PubMed]
74. Belgiu, M.; Drăguţ, L. Random forest in remote sensing: A review of applications and future directions. *ISPRS J. Photogramm. Remote Sens.* **2016**, *114*, 24–31. [CrossRef]
75. Zhao, P.; Lu, D.; Wang, G.; Liu, L.; Li, D.; Zhu, J.; Yu, S. Forest aboveground biomass estimation in Zhejiang Province using the integration of Landsat TM and ALOS PALSAR data. *Int. J. Appl. Earth Obs. Geoinf.* **2016**, *53*, 1–15. [CrossRef]
76. Jiang, X.; Li, G.; Lu, D.; Chen, E.; Wei, X. Stratification-Based Forest Aboveground Biomass Estimation in a Subtropical Region Using Airborne Lidar Data. *Remote Sens.* **2020**, *12*, 1101. [CrossRef]
77. Vafaei, S.; Soosani, J.; Adeli, K.; Fadaei, H.; Naghavi, H.; Pham, T.; Tien Bui, D. Improving Accuracy Estimation of Forest Aboveground Biomass Based on Incorporation of ALOS-2 PALSAR-2 and Sentinel-2A Imagery and Machine Learning: A Case Study of the Hyrcanian Forest Area (Iran). *Remote Sens.* **2018**, *10*, 172. [CrossRef]
78. TSITSI, B. Remote sensing of aboveground forest biomass: A review. *Trop. Ecol.* **2016**, *57*, 125–132.
79. Lu, D.; Chen, Q.; Wang, G.; Liu, L.; Li, G.; Moran, E. A survey of remote sensing-based aboveground biomass estimation methods in forest ecosystems. *Int. J. Digit. Earth* **2014**, *9*, 63–105. [CrossRef]
80. Campbell, M.J.; Dennison, P.E.; Kerr, K.L.; Brewer, S.C.; Anderegg, W.R.L. Scaled biomass estimation in woodland ecosystems: Testing the individual and combined capacities of satellite multispectral and lidar data. *Remote Sens. Environ.* **2021**, *262*, 112511. [CrossRef]
81. Egusa, T.; Kumagai, T.O.; Shiraishi, N. Carbon stock in Japanese forests has been greatly underestimated. *Sci. Rep.* **2020**, *10*, 7895. [CrossRef] [PubMed]
82. *Report On Program for Emergent Development of Forest Carbon Stocks Dataset: Fiscal Year 2003*; Forestry and Forest Products Research Institute: Matsunosato, Japan, 2004.
83. Tsuzuki, H.; Nelson, R.; Sweda, T. Estimating Timber Stock of Ehime Prefecture, Japan using Airborne Laser Profiling (<Special Issue> Silvilaser). *J. For. Plan.* **2008**, *13*, 259–265. [CrossRef]
84. Matsushita, K.; Yoshida, S. Analysis of the Resent Situation and Problems in Forestry Statistics in Japan. *J. For. Econ.* **1998**, *44*, 7–13.
85. Li, A.; Dhakal, S.; Glenn, N.F.; Spaete, L.P.; Shinneman, D.J.; Pilliod, D.S.; Arkle, R.S.; McIlroy, S.K. Lidar Aboveground Vegetation Biomass Estimates in Shrublands: Prediction, Uncertainties and Application to Coarser Scales. *Remote Sens.* **2017**, *9*, 903. [CrossRef]
86. Salas, W.A.; Ducey, M.J.; Rignot, E.; Skole, D. Assessment of JERS-1 SAR for monitoring secondary vegetation in Amazonia: I. Spatial and temporal variability in backscatter across a chrono-sequence of secondary vegetation stands in Rondonia. *Int. J. Remote Sens.* **2010**, *23*, 1357–1379. [CrossRef]

Article

Comparison of Field and SAR-Derived Descriptors in the Retrieval of Soil Moisture from Oil Palm Crops Using PALSAR-2

Veena Shashikant [1], Abdul Rashid Mohamed Shariff [1,2,3,*], Aimrun Wayayok [1,2,3], Md Rowshon Kamal [1], Yang Ping Lee [4] and Wataru Takeuchi [5]

1 Department of Biological and Agricultural Engineering, Faculty of Engineering, Universiti Putra Malaysia, Serdang 43400, Malaysia; gs52191@student.upm.edu.my (V.S.); aimrun@upm.edu.my (A.W.); rowshon@upm.edu.my (M.R.K.)
2 SMART Farming Technology Research Center, Faculty of Engineering, Universiti Putra Malaysia, Serdang 43400, Malaysia
3 Laboratory of Plantation System Technology and Mechanization, Institute of Plantation Studies, Universiti Putra Malaysia, Serdang 43400, Malaysia
4 FGV R&D Sdn Bhd, Level 9, Wisma FGV, Jalan Raja Laut, Kuala Lumpur 50350, Malaysia; yangp.lee@fgvholdings.com
5 Department of Human and Social Systems, Institute of Industrial Science, The University of Tokyo, Tokyo 153-8505, Japan; wataru@iis.u-tokyo.ac.jp
* Correspondence: rashidpls@upm.edu.my; Tel.: +60-39769-6414

Citation: Shashikant, V.; Mohamed Shariff, A.R.; Wayayok, A.; Kamal, M.R.; Lee, Y.P.; Takeuchi, W. Comparison of Field and SAR-Derived Descriptors in the Retrieval of Soil Moisture from Oil Palm Crops Using PALSAR-2. *Remote Sens.* **2021**, *13*, 4729. https://doi.org/10.3390/rs13234729

Academic Editors: Takeo Tadono and Masato Ohki

Received: 14 September 2021
Accepted: 5 November 2021
Published: 23 November 2021

Publisher's Note: MDPI stays neutral with regard to jurisdictional claims in published maps and institutional affiliations.

Abstract: Synthetic-aperture radar's (SAR's) capacity to resolve the cloud cover concerns encountered while gathering optical data has tremendous potential for soil moisture data retrieval using SAR data. It is possible to use SAR data to recover soil moisture because the backscatter coefficient is sensitive to both soil and vegetation by penetrating through the vegetation layer. This study investigated the feasibility of employing a SAR-derived radar vegetation index (RVI), the ratios of the backscatter coefficients using polarizations of HH/HV ($R_{HH/HV}$) and HV/HH ($R_{HH/HV}$) to an oil palm crops as vegetation indicators in the water cloud model (WCM) using phased-array L-band SAR-2 (PALSAR-2). These data were compared to the manual leaf area index (LAI) and a physical soil sampling method for computing soil moisture. The field data included the LAI input parameters and, more importantly, physical soil samples from which to calculate the soil moisture. The fieldwork was carried out in Chuping District, Perlis State, Malaysia. Corresponding PALSAR-2 data were collected on three observation dates in 2019: 17 January, 16 April, and 9 July. The results showed that the WCM modeled using the LAI under HV polarization demonstrated promising accuracy, with the root mean square error recorded as 0.033 m^3/m^3. This was comparable to the RVI and $R_{HH/HV}$ under HV polarization, which had accuracies of 0.031 and 0.049 m^3/m^3, respectively. The findings of this study suggest that SAR-based indicators, $R_{HH/HV}$ and RVI using PALSAR-2, can be used to reduce field-related input in the retrieval of soil moisture data using the WCM for oil palm crop.

Keywords: leaf area index; leave one out cross-validation; oil palm; radar vegetation index; synthetic aperture radar; soil moisture; vegetation descriptors; water cloud model

1. Introduction

Oil palm has long been recognized as a vital crop in tropical agricultural regions with a consistently increasing output rate, especially in Indonesia and Malaysia, which export significant amounts of crude palm oil to other countries [1]. In Malaysia, oil palm crop production occupies 71% of the agricultural land [2]. Oil palm crop is the second most important source of edible oil, behind soybean, in terms of production [3]. Beyond its core role as an edible oil, palm oil has spawned other palm-based sectors, such as specialized fats, cocoa-butter alternatives, oleochemicals, soaps, domestic detergents, nutritional supplements and, most recently, bioenergy [4]. Tropical regions like Malaysia that have sufficient rainfall and sunshine and appropriate soil conditions are ideal for oil palm cultivation [5]. Because of the increasing demand for palm oil, a major concern is maintaining crop yields at optimum levels and minimizing labor and fertilizer usage [6]. Due to the fact that the crops in oil palm plantations are linked directly with the ground, soil quality is an important factor when it comes to crop uptake and health [7]. Soil characteristics and climatic conditions are known to vary on a minute scale and are particularly site specific [8].

It has been well established that soil moisture and precipitation have the highest correlations in arid and dry regions and weaker correlations in wet regions, indicating that soil moisture and precipitation are more complex than what is viewed on the surface [9]. The intricate interaction between soil moisture and precipitation has been noted as important in the land-surface context. Correlations between precipitation and soil moisture are the strongest in areas with sparse vegetation, whereas forests and heavily vegetated areas have weaker correlations [10]. Understanding this enables study to focus on numerous specific scientific challenges such as subsurface recharge assessment and the identification of drought–flood cycles. Such studies are important for tropical countries—particularly agricultural nations where widespread applications are possible for scheduled irrigations and soil moisture modeling [11]. However, a lack of information on such topics makes it difficult for the farmers in those countries to take appropriate precautions to ensure the productivity of their crops. Furthermore, hydrological models are often developed for use under static conditions [12]. Additionally, in areas where oil palms are cultivated, soil moisture is equally important for supporting palm tree growth. Therefore, in order to estimate soil moisture by conventional means, highly reliable gravimetric measurements are taken, although this is regarded as time and resource intensive [13]. In response to this, time–domain reflectometry sensors are widely preferred [14], which can provide continuous measurements [15].

Soil moisture mapping is accomplished mostly through extensive point measurements, which can be expensive [16]. Numerous interpolation techniques have been used to produce gridded soil moisture data from field observations, including deterministic approaches such as inverse distance weighting (IDW), local polynomial interpolation (LPI) and radial basis function (RBF) as well as geostatistical methods such as ordinary kriging (OK) [17]. Deterministic methods can be examined using measured points evaluated based on their extent of similarity. It has been noted that model IDW, using soil moisture, is capable of investigating the distribution of drought conditions [18]. When precipitation is encountered on a catchment scale, it has been found that IDW, with the inverse distance to a power number, has a greater impact on simulated outcomes than the scale of grid sampling [19]. In a separate case, the evaluation of soil moisture using deterministic methods, such as IDW and RBF, using global polynomial interpolation, LPI and OK, have been examined, with OK being found to be more effective due to the fact of its use of geostatistical interpolation techniques that utilize the statistical properties of the measured points [20]. As it reduces the variance of estimate error, OK is the most used geostatistical interpolation approach and the best linear unbiased estimator [21]. In complicated terrains, OK is highly dependent on the homogeneity and density of the soil samples [22]. Recently, the topic of soil moisture has concentrated on the spatial and temporal variability of the moisture content in hillslopes and catchments. Fluctuations in soil moisture on slopes are more complicated because of the synergy and superposition effects of land use types, slope gradient, slope aspect, slope position, and elevation [23]. Hilly areas often face the problem of sparse rain gauge networks, which limits the accessibility of the data and affects the interpolation accuracy [24]. Therefore, using remote sensing as a tool, satellite imagery can provide useful information about the Earth's surface, with images being one of the most popular data sources for remote sensing.

Remote sensing is used in the oil palm industry in tree detection [25], monitoring for pests and disease mapping [26], and in nutrient detection [27]. Optical imaging gathers energy emitted from the surface of the Earth in the visible and near-infrared range [28], resulting in indices that represent the vegetation cover. The normalized difference vegetation index (NDVI), which is a normalized ratio of near-infrared to visible red, is the most commonly used metric [29]. It is a flexible and an effective indicator for distinguishing vegetation from non-vegetation and includes the ability to interpret the health of oil palm trees [30].

On the other hand, microwave remote sensing, or active remote sensing, can produce images regardless of weather or lighting conditions by using its own radiation for illumination, which can penetrate clouds and reach the Earth's surface. Microwave remote sensing has addressed the issue of cloud cover through optical sensors in remote sensing [31], clouds being a major impediment, particularly in tropical areas where oil palms are commonly cultivated [32]. Microwave remote sensing using PALSAR-2 generates data based on backscattered radiation from the ground, with a lengthier wavelength providing better penetrative capability [33]. As the radar has better penetrative capacity, it can be used to distinguish a smooth surface from a rough surface [34]. L-band SAR imagery provides the optimum diagnostic of oil palm canopies for growth monitoring [35]. As a result, the L band at a wavelength of 15–30 cm can penetrate tree canopies and offer information on sub-canopy structures [36]; hence, because of this capability, SAR can be employed in the categorization of oil palms.

Various types of information about the surface can be obtained from the vegetation cover by studying the polarization of the emitted and received radar signal. In HH, the signal is horizontally emitted and horizontally received; in HV, it is horizontally emitted and vertically received; in VH, it is vertically emitted and horizontally received; in VV, it is vertically emitted and vertically received [37]. Polarimetric SAR is a technique used to extract information from vegetation, with important information for oil palm crop categorization being carried by HH and HV signals [38]. In order to distinguish oil palm cover from natural forest and acacia plantations, both the C band and L band can be used to enhance the classification accuracy [39]. Moreover, using an optical sensor, object-based classification was used to improve classification accuracy in oil palm and acacia plantations [40]. Recently, SAR images have been shown to be capable of penetrating oil palm trunks, where basal root disease can be distinguished using a machine-learning model [41].

In the last decade, a better understanding of SAR has allowed the retrieval of soil moisture data from woody plants [42] and agricultural crops [43,44] using vegetation and soil parameters and the water cloud model (WCM). The WCM was proposed as a collection of similar spherical particles that are consistently distributed across the volumetric vegetation layer [45]. Originally, the WCM established an equation for the total backscatter coefficient as a function of soil volumetric moisture content, vegetation moisture content, and plant height [45]. Field-based vegetation parameters, such as the LAI [46–49] and vegetation water content [50,51], have been widely used in WCMs. The WCM has the advantage of being able to explain complicated scatter patterns in a vegetated area using simple bulk vegetation descriptors [52]. However, there is a lack of understanding or agreement on the best collection of vegetation descriptors. Recent studies have shown that using the NDVI [53], based on optical images and the radar vegetation index (RVI) [54,55], and the ratio of backscatter coefficient polarization (e.g., HH/VV [56] or VH/VV [56,57]) as descriptors provides successful soil moisture data retrieval in both the C and L bands. However, HV/HH has been used to understand the dynamics of soil moisture based on radar data, which lessens the effect of soil surface roughness [58]. The HV backscatter coefficient has been found to be sensitive, in the P and L bands, to plant biomass and plant water content [59].

In this study, the main goal was to extract soil moisture data from the oil palm cultivated site using the WCM and SAR-based vegetation descriptors, such as RVI and the ratio of the backscatter coefficients HH/HV ($R_{HH/HV}$) and HV/HH ($R_{HV/HH}$), and compare this with data from the LAI field-based vegetation descriptor.

2. Study Area and Materials

2.1. Study Area

Chuping, in Perlis State, is a flat-terrain oil palm growing location. For this study, the area of the oil palm crop was approximately 28 ha. As of the data collection date, it covered 4-year-old palm stands that had just begun to bear fresh oil palm fruits. The study area's central coordinates were 6°31′07.2″ N, 100°19′07.7″ E, located in the subdistrict called Kilang Gula Chuping. The area has a relatively flat terrain with a slope angle of 4–12% and an elevation of 21.6 m. The soil type was identified as Chuping and Dampar—sandy clay loam and clay loam. The study was conducted over three periods during weather conditions similar to those in which the SAR images were acquired (see Section 2.2.2). Early in the year, precipitation rates were quite low. This was particularly true in January–March, which are considered to be the driest months of the year, according to meteorological data from a previous study conducted in the same area [60]. The latter months of the year experienced sufficient precipitation, with the average precipitation being 1362.38 mm per year [61].

2.2. Data Collection

2.2.1. Field Data

In order to determine the soil moisture content at a depth of 0–5 cm, a soil gravimetric technique was used in a grid point shown in Figure 1. For this, fresh weights of soil were taken in the field, with their dry weights being calculated in the laboratory, following oven drying. The soil samples were obtained from 32 locations in the study area, resulting in 96 soil samples taken on three different dates. In addition, oil palm fronds were collected for estimation of the LAI. For oil palm crop, the standard approach of destructive sampling was used to determine the LAI. In addition to being an excellent predictor of a palm's nutritional condition, fronds are easy to identify and sample. In oil palm crop, using the 17th frond is widely accepted to estimate LAI [62]. According to the conventional method for evaluating LAI, which was developed specifically for oil palm crop, it was determined using the variables A_f as leaf area per frond in m^2 of the 17th frond from the palm crown, F_n as the total number of fronds per sampled tree, and P_{DEN} as the number of palm trees per hectare, using the following equation [63]:

$$\text{LAI}\left(\text{m}^2/\text{m}^2\right) = A_f \times F_n \times \frac{P_{DEN}}{10000} \quad (1)$$

The leaflet area was measured using an LI-3100C area meter (LI-COR Inc., Lincoln, NE, USA). The total leaflet area of each frond was calculated by multiplying one side of the leaflet area by two.

2.2.2. Remote Sensing Data

The backscatter coefficient of the oil palms was extracted using PALSAR-2 data. High-resolution PALSAR-2 images were collected through our participation with the Japanese Aerospace Exploration Agency (JAXA) using the Earth Observation Research Announcement 2 platform. Three 2019 PALSAR-2 images, from the HH and HV polarization on 17 January, 19 April, and 9 July, were used. The specifics of these SAR data are shown in Table 1, with all three images having been acquired in Strip Map 3 mode, in ascending order at 6.25 × 6.25 m resolution. All the PALSAR-2 images used in this study were constructed using a 16 bit data type with each pixel containing a digital number (DN). These DNs did not correspond to the radar signal of the ground features or objects. As a result, the DNs had to be converted into backscatter coefficients and expressed in decibels, as described in Equation (2). For the PALSAR-2 data provided by JAXA, the calibration factor (CF) was −83.0 dB [64]:

$$\sigma^0 = 10 \times \log_{10}\left(DN^2\right) + CF \quad (2)$$

Once the $\sigma^0{}_{HH}$ and $\sigma^0{}_{HV}$ for each field point were available, the images were radiometrically calibrated using the Shuttle Radar Topography Mission's digital elevation model (3 arc-second). Following that, the images were orthorectified with respect to geographic locations in order to eliminate speckles and noise from the PALSAR-2 images; a Lee filter was used with a 5 × 5 window size. It has been previously noted that the Lee filter works very well in terms of maintaining an image's spectral characteristics while decreasing speckling [65]. The open-source Sentinel Application Platform version 6.0.0 was used to commence all the SAR-related preprocessing presented in Table 1.

Table 1. PALSAR-2 satellite image acquisition and incident angle.

Date of Acquisition	Polarization	Incident Angle
17 January 2019	HH + HV	30.4–42.4°
19 April 2019	HH + HV	41.2–53.3°
9 July 2019	HH + HV	30.4–42.4°

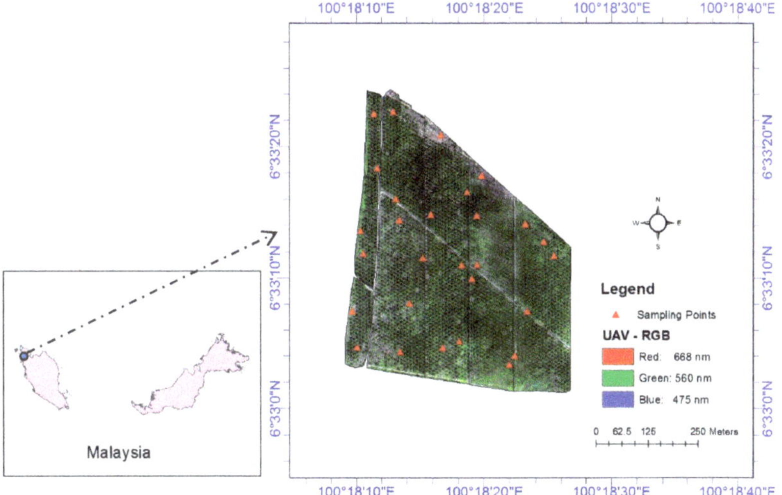

Figure 1. Oil palm cultivation area in Chuping, Perlis State, Malaysia.

In addition to the PALSAR-2 images, a DJI Phantom 4 Unmanned Aerial Vehicle (UAV), equipped with a Micasense® RedEdge camera (Micasense Inc., Seattle, WA, USA) multispectral sensor was employed to survey the study area on 17 January 2019. The Micasense® camera gathers information in five spectral bands, spanning the visible through red-edge and infrared spectrums. Specifically, red, green, blue, near-infrared, and red-edge images were captured at central wavelengths of 668, 560, 475, 840, and 717 nm, respectively. The sensor was calibrated on-site, prior to flight, using the reference panel for accurate ground reflectance calibration. The imagery from the UAV platform enabled us to compute the NDVI [66], as shown in Equation (3), in order to identify bare soil with NDVI values of less than 0.2. To confirm the classification was indeed bare soil, ground-truthing was performed.

$$\text{NDVI} = \frac{\rho_{840} - \rho_{668}}{\rho_{840} + \rho_{668}} \tag{3}$$

3. Methodology

The WCM was first developed by Attema and Ulaby [45] for alfalfa, corn, and wheat crops. It is a broadly applied model for vegetation-covered areas, because it is composed of two components: the direct contribution of vegetation and the attenuation component. Many studies have successfully applied the WCM to various crops, such as winter wheat [55], wheat and corn [50], multi-crop agriculture [54], and forests [67]. The WCM was established on the assumption that the canopy's "cloud" was composed of similar water droplets, scattered randomly throughout the canopy [68]. In this study, the WCM was used to retrieve soil moisture data from oil palm crop using PALSAR-2 data. Based on the assumption that the influence of soil surface roughness on observed backscatter is consistent over a short timespan at a given site, the temporal variation in SAR backscattering will be solely a reflection of changes in vegetation and soil moisture [54]. Consequently, in this study, a multi-temporal SAR data set was used in the WCM. With the input of SAR-derived indices and field-gathered vegetation descriptors (from the LAI), it was possible to compare both the vegetation descriptors to evaluate the WCM and retrieve the soil moisture parameter.

The WCM considered both soil moisture and vegetation characteristics, with Equation (4) showing the four empirical coefficients: A and B are vegetative characteristics and C and D are soil parameters [69]. In Equations (5) and (6), parameter A corresponds to the albedo of the vegetation, with B being an attenuation factor. Parameter D indicates the sensitivity of the radar signal to soil moisture, while C can be a calibration constant in Equation (7). Equation (5) shows the backscatter coefficient from the direct contribution of vegetation, whereas Equation (6) gives the attenuation component for the vegetation-covered surface. Hence, the equation is modified to:

$$\sigma^0{}_{tot} = \sigma^0{}_{veg} + \tau^2\sigma^0{}_{soil} \quad (4)$$

where

$$\sigma^0{}_{veg} = \mathrm{A}\ \times V_1 \times \cos\theta\left(1-\tau^2\right) \quad (5)$$

$$\tau^2 = \mathrm{Exp}\ (-2 \times\ \mathrm{B}\ \times V_2 \times \sec\theta) \quad (6)$$

$$\sigma^0{}_{soil} = \mathrm{C}M_v + \mathrm{D} \quad (7)$$

The moisture content held in the canopy and its geometry have an impact on the backscatter coefficient in terms of both V_1 and V_2. The soil moisture (M_v) is described in m^3/m^3 and θ represents the incidence angle of the SAR images. After solving for parameters C and D using a linear model fitting procedure, the values of C and D are replaced in Equations (5) and (6), allowing for the solution of parameters A and B using the nonlinear least squares method (NLSM) [51,53]. It has been reported that A and B can be estimated using Levenberg–Marquardt optimization in the NLSM [47]. However, descriptors relating to vegetation have varied implications for the WCM. Several experiments have been conducted, employing plant height, the LAI, the leaf–water area index (LWAI), and the normalized plant-water content (NPWC) as variables, to measure V_1 and V_2 [29,70,71]. In this study, the vegetation descriptors V_1 = 1 and V_2 = LAI were chosen because they have contributed to the best model performance using other field-based descriptors such as LWAI and NPWC [61]. This is referred to as Model 1 (see Table 2). The SAR-derived indices were used for modeling the oil palm WCM and are referred to as Models 2, 3, and 4. The RVI, being derived from dual polarization [72], was used as shown in Equation (8). The RVI equation was initially introduced by proposing the use of the four polarizations (i.e., HH, HV, VH, and VV) [73]. However, it has been found that the RVI provides a good approximation of surface scattering when only two polarizations are used [74].

$$\mathrm{RVI} = \frac{4\sigma^0{}_{HV}}{\sigma^0{}_{HH} + \sigma^0{}_{HV}} \quad (8)$$

It has been noted that RVI values range from 0 to 1, with 0 being associated with bare soil and 1 with higher vegetation [75]. In this study, along with the RVI, other vegetation descriptors, such as the calculated ratios $R_{HH/HV} = \frac{\sigma^0{}_{HH}}{\sigma^0{}_{HV}}$ and HV/HH as $R_{HV/HH} = \frac{\sigma^0{}_{HV}}{\sigma^0{}_{HH}}$, were used to evaluate the soil moisture (Table 2).

Table 2. Simplified WCM using modeled vegetation descriptors.

Model	Vegetation Descriptors, V_1 and V_2
1	V_1 = 1, V_2 = LAI
2	V_1 = V_2 = RVI
3	$V_1 = V_2 = R_{HH/HV}$
4	$V_1 = V_2 = R_{HV/HH}$

To evaluate the WCM for soil moisture data retrieval using the models listed in Table 2, the leave-one-out cross-validation (LOOCV) method was used—a deterministic validation procedure that enables accurate replication using the same data set [76]. Each time the model was evaluated, one of the data samples was omitted, with the remaining n − 1 data sample being used to train the model. The LOOCV method has been demonstrated as being superior to split-sample validation, especially when sample sizes are limited [77]. Model evaluation can be expressed in performance metrics, such as the coefficient of determination (R^2) and the root mean square error (RMSE) [78,79], calculated as shown in Equations (9) and (10), respectively. For each parameter combination, a pair of predicted and observed values were obtained.

$$R^2 = \left(\frac{\sum_{i=1}^{n} \left(X_{obs} - \overline{X_{obs}}\right)\left(X_{sim} - \overline{X_{sim}}\right)}{\sqrt{\sum_{i=1}^{n} \left(X_{obs} - \overline{X_{obs}}\right)^2 \sum_{i=1}^{n} \left(X_{sim} - \overline{X_{sim}}\right)^2}} \right)^2 \quad (9)$$

$$\text{RMSE} = \sqrt{\frac{\sum_{i=1}^{n}(X_{sim} - X_{obs})^2}{n}} \quad (10)$$

The RMSE was estimated using Equation (9), where X_{sim} is the simulated $\sigma^0{}_{tot}$ and X_{obs} is the observed $\sigma^0{}_{tot}$. The RMSE is widely accepted for assessing the gap between model predictions and actual observations from the environment in soil moisture-related studies [80,81]. Most scholars accept the RMSE for soil moisture data retrieval by referring to the Global Monitoring for Environment and Security (GMES) requirement from the European Space Agency for accuracy, with soil moisture values below 0.05 m^3/m^3 being considered as favoring the guidelines [82,83].

4. Results

4.1. WCM Parameterization

In the WCM approach, vegetation parameters describe the scattering from the vegetation cover on the ground. Estimation of the WCM parameters first requires calibration of the values for bare soil in order to obtain the soil-related parameters C and D from Equation (7), then correcting for the effects of vegetation on the backscattering coefficients. For parameters C and D, the input of field or SAR-based indicators, along with the incident angle and soil moisture, are required in order for the total backscatter to be calibrated. The WCM was calibrated differently for each model, for both $\sigma^0{}_{HH}$ and $\sigma^0{}_{HV}$, in order to localize the vegetation parameters as shown in Table 3 using the LOOCV approach for cross-validation. The WCM parameterization is important for obtaining a good fit with the field measurements, as described in Equations (4)–(7), enabling the retrieval of the soil moisture values. Using the LOOCV method to estimate the actual error in the developed model, all the steps in the algorithm, including parameter tuning, have to be repeated in each cross-validation loop [83]. For the SAR-based vegetation descriptors, the RVI was derived from the PALSAR-2 images, where it has been shown to describe the structural vegetation characteristics, and the RVI correlates with the vegetation water content and

LAI indicators [84]. The $R_{HH/HV}$ and $R_{HV/HH}$ were employed to evaluate the potential use of these simple ratios as vegetation descriptors because the latter has been reported as being able to distinguish fluctuations in soil moisture using SAR data, and also to identify areas where the influence of soil surface roughness can be mitigated [58]. To evaluate the model further, a comparison of the WCM-modeled backscatter coefficients was checked against the observed backscatter coefficients using the respective polarization, as indicated in Section 4.2.

Table 3. Fitting of the WCM using HH and HV polarization.

Vegetation Descriptor by Model		Model Coefficients									
		HH					HV				
V_1	V_2	A	B	C	D	n	A	B	C	D	n
1	LAI	0.012	0.001	−26.015	−2.864	96	0.317	0.013	22.207	−23.866	96
RVI	RVI	0.319	0.017	−13.648	−5.784	96	0.613	0.008	24.556	−23.894	96
$R_{HH/HV}$	$R_{HH/HV}$	0.181	0.016	−11.663	−6.462	96	0.450	0.133	21.874	−22.487	96
$R_{HV/HH}$	$R_{HV/HH}$	0.758	0.007	−15.200	−5.900	96	0.826	0.010	20.320	−23.500	96

4.2. Sensitivity Backscatter Coefficient vs. Vegetation Descriptors

To understand the suitability of vegetation descriptors in the retrieval of soil moisture data over oil palm crops, four WCMs were used to evaluate the potential use of SAR-based parameters. SAR backscatter coefficients are connected to vegetation features on the ground, such as crop form, height, size, geometric arrangement, and density, all of which vary per crop [85,86]. In this study, a simplified WCM was evaluated in terms of both $\sigma^0{}_{HH}$ and $\sigma^0{}_{HV}$ to understand its polarization sensitivity to the oil palm crop. The results were determined using the model metrics of R^2 and the RMSE between the observed and WCM-simulated backscatter coefficients as shown in Table 4. Overall, using the LOOCV method, R^2 ranged from 0.930 to 0.983 for the HH polarization and from 0.948 to 0.991 for the HV, with the RMSE being 0.425–2.257 dB and 0.635–1.282 dB, respectively. Using the LAI field vegetation descriptor for the palms produced, a low RMSE value of 0.635 dB under HV polarization with R^2 = 0.983 (Table 4, Figure 2). For the RVI, the SAR-derived descriptor $R_{HH/HV}$ and $R_{HV/HH}$ were evaluated for the same day as the LAI indicator using Equation (1). Under the same polarization, when the RVI was used in the WCM, the model showed a higher RMSE of 0.702 dB with an R^2 of 0.975 recorded. The modeled backscatter coefficient for the vegetation descriptor $R_{HH/HV}$ (Model 3, Table 4) had an R^2 of 0.982 and an RMSE of 0.828 dB. For $R_{HV/HH}$, the RMSE was higher than for $R_{HH/HV}$, at 1.282 dB for $\sigma^0{}_{HV}$, with an R^2 = 0.930.

Table 4. RSME values for the WCM-simulated and observed backscatter using PALSAR-2 with different vegetation descriptors.

Polarization	RMSE (dB)			
	Model 1	Model 2	Model 3	Model 4
HH	2.257	0.425	0.472	1.883
HV	0.635	0.702	0.828	1.282
Polarization	R^2			
	Model 1	Model 2	Model 3	Model 4
HH	0.948	0.990	0.991	0.964
HV	0.983	0.975	0.982	0.930

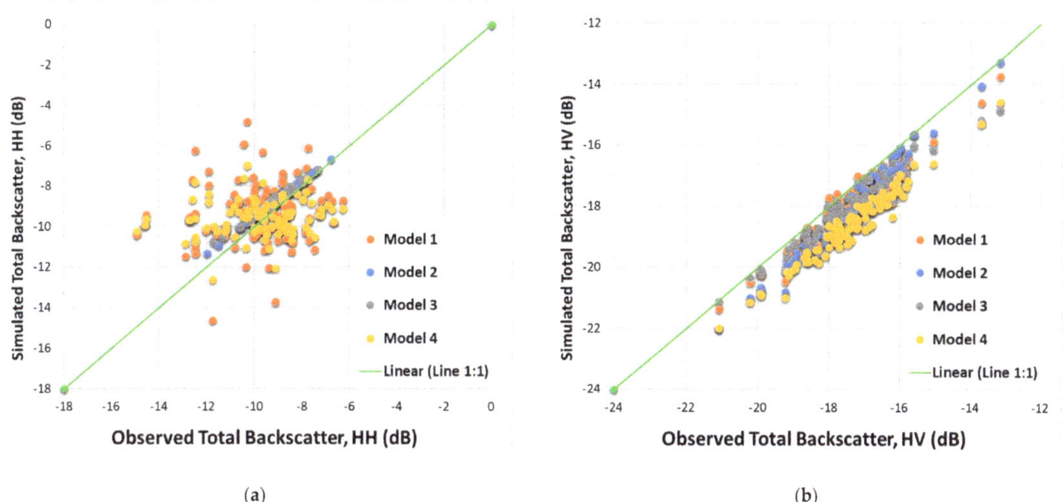

Figure 2. Scatterplots showing the simulated and observed backscatter coefficients under (**a**) HH and (**b**) HV polarization.

On the other hand, the HH polarization with the LAI vegetation descriptor had an R^2 of 0.948, with a higher RMSE, at 2.257 dB, than the HV polarization in the Model 1 (Table 4, Figure 2a). The Model 1 in HH polarization produced the highest RMSE values compared to the other models. In Model 3, the $R_{HH/HV}$, vegetation indicator was comparable with the RVI model with both being comparable to the LAI model under HH polarization. Both SAR-based model indicators showed a similar accuracy with RMSEs of 0.425 dB and 0.472 dB, respectively, as indicated in Table 4, and with an R^2 of 0.990 and 0.991. In the model using $R_{HV/HH}$, an R^2 value of 0.964 was observed with a higher RMSE of 1.883 dB.

4.3. Soil Moisture Data Retrieval

The purpose of this study was to retrieve soil moisture data from oil palm crops where soil moisture is an important indicator of the water requirements of the crop, being an important factor in crop development and yield [64,87]. Furthermore, the retrieval of soil moisture data is useful in seasonal or agricultural drought monitoring in terms of understanding the significant areas affected [28]. In this study, statistical metrics were employed in order to understand soil moisture data retrieval from the WCM used. Table 5 and Figure 3 show the data retrieval using Models 1–4 under HH and HV polarization. It was noted that, under both polarizations, the vegetation descriptors attempted to represent the vegetation layer as carefully as possible. Numerous studies have demonstrated that the type of vegetation, the geometric structure of its cover (including height, branch and leaf forms, and density distribution) and its water content have an effect on radar backscattering and radar wave transmittance in the plant canopy [88–90]. In order to minimize errors in the soil moisture content data, multiple angles, and multitemporal SAR data inversion were used to help to eradicate the consequences of the plant layer on the radar backscatter [90]. When the field-based LAI was used to retrieve the soil moisture data, the HV polarization showed a high R^2 of 0.949, with a low RMSE of 0.033 m^3/m^3. Under HH polarization, however, the LAI indicator showed a higher RMSE of 0.087 m^3/m^3.

Table 5. R^2 and the RMSEs for the soil moisture data retrieved and observed from an oil palm crops (in m^3/m^3) using PALSAR-2, given according to the proposed models.

Vegetation Descriptor by Model	Statistics Metrics			
	HH		HV	
	R^2	RMSE (m^3/m^3)	R^2	RMSE (m^3/m^3)
Model 1	0.901	0.087	0.949	0.033
Model 2	0.973	0.036	0.960	0.031
Model 3	0.946	0.049	0.974	0.049
Model 4	0.898	0.128	0.898	0.066

The main reason for evaluating the SAR-derived indicators was to avoid the cloud-cover concerns that arise from optical data, which mainly affects tropical regions [91]. From the SAR-derived Models 2–4 (Table 5), it was found that the HV polarization showed RMSEs ranging from 0.031 to 0.066 m^3/m^3. This suggested that the HV polarization was consistent in retrieving the soil moisture data. This is similar to the mentioned descriptors, which showed a lower RMSE from the backscatter model fit (Table 4). This finding correlated with the field evaluation of the WCM, with the HV polarization providing a more accurate estimation of soil moisture [92]. The RVI produced the lowest RMSE among the other SAR-derived models at 0.031 m^3/m^3. For the HH polarization, SAR-derived Models 2 and 3 had lower RMSE values of 0.036 and 0.049 m^3/m^3, and with comparable R^2 values (Table 5). The SAR-derived indicators performed better than the field-based vegetation descriptor, according to Model 1, under HH polarization. However, the $R_{HV/HH}$ showed low accuracy in the RMSE comparison for both the polarizations, being 0.128 and 0.066 m^3/m^3, respectively.

Our findings are in agreement with those of previous studies, in which it has been reported that RVI indicators in the WCM have been successfully evaluated to replace field indicators in order to overcome optical data concerns [41,62]. It was noted that the RVI model has been posited as a new descriptor that can be used to distinguish the backscattering from the crop canopy and the underlying soil surface in cases where the crop parameter cannot be obtained from the field, with the RVI being directly calculated from the SAR [55]. Overall, the soil moisture data retrieval in this study was successful, based on the parameterization of the WCM for the oil palm crop, with the use of the RVI and $R_{HH/HV}$ as vegetation descriptors proving as dependable as the LAI descriptors. However, the SAR-derived indicators were noted as producing lower RMSEs under HV polarization, similarly to the LAI descriptor under HV polarization. The scatterplots of the observed and retrieved soil moisture data, based on the polarization of each model, are shown in Figure 3.

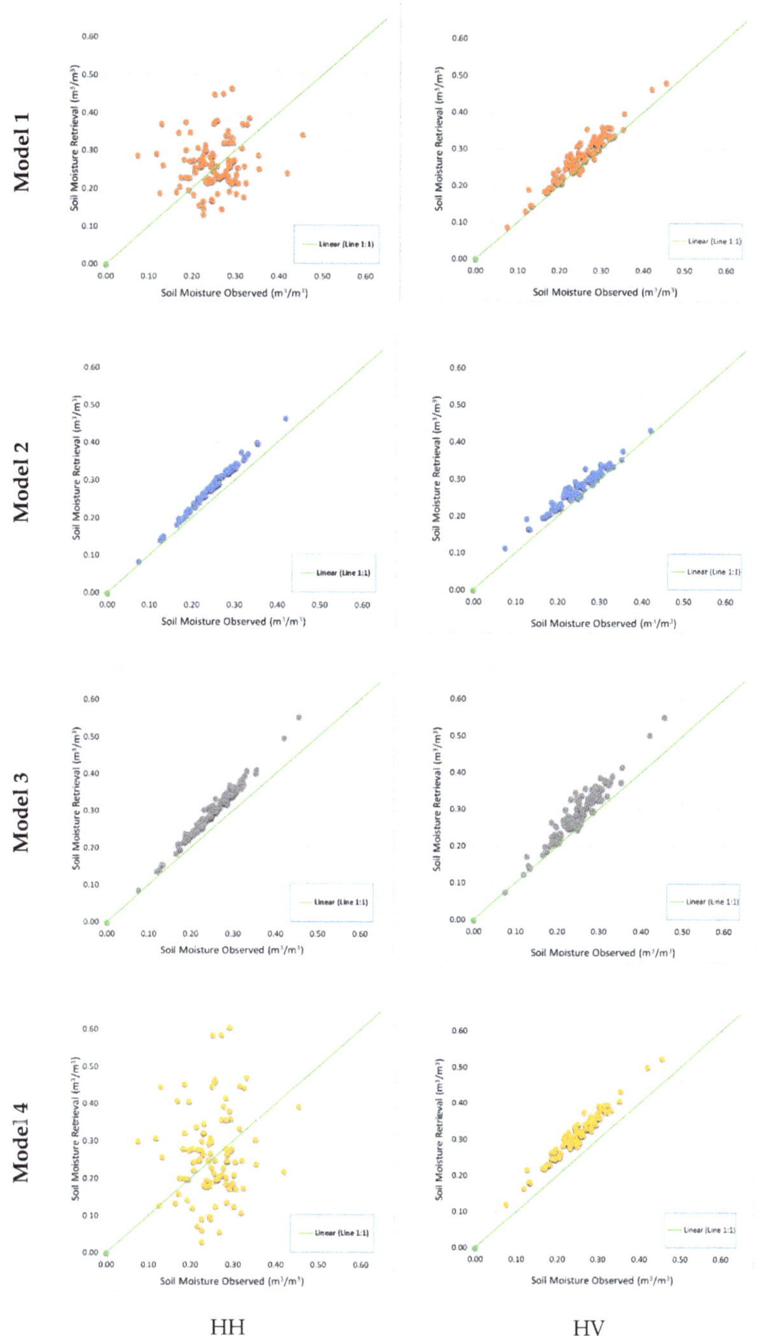

Figure 3. Scatterplots showing the observed and retrieved soil moisture data from an oil palm crop under both HH (**left**) and HV (**right**) polarization in different models in the L band.

5. Discussion

The WCM is a semi-empirical model, founded on theoretical ideas and relationships, but which employs a simplified method based on field- and SA-based parameters. In order to build the WCM, a calibration process was performed using variables, including LAI from the oil palm crop, soil moisture data from the field data collection as well as the backscattering coefficients, the RVI, $R_{HH/HV}$ and $R_{HV/HH}$, and the incidence angle from PALSAR-2. Using these variables, the parameters A, B, C, and D were considered in fine-tuning the WCM—important steps specific to each crop and location [49]. Adding on, to improve the fine-tuning estimation of the parameters mentioned, LOOCV was implemented using the concept of iteration. This fine-tuning of the parameters (Table 3) is dependent on the sensor configuration, vegetation cover and soil characteristics. In this instance, the terrain was relatively flat and, therefore, the oil palm backscattering contributed to the radar signal as shown in Figure 2. It is important to note that the vegetation parameters $V_1 = 1$ and $V_2 =$ LAI were used for comparison to the ground vegetation cover in this study, as these have previously been found to be the best soil moisture indicators, among other vegetation parameters, such as the LWAI and NPWC, for oil palm crop [61]. The results were in agreement with those of previous studies on other crops, with the LAI variable being superior in sugarcane, cherry, rice [46], and wheat [93]. The accuracy obtained in the retrieval of soil moisture data using the LAI (Table 5) showed that the HV polarization RMSE of 0.033 m^3/m^3 using the L band fulfilled the GMES requirement of RMSE < 0.05 m^3/m^3. By contrast, the HH polarization produced a higher RMSE in this study than in another study that used PALSAR-2, where the soil moisture was variable, giving a retrieval accuracy of approximately 6.0% [94]. In relation to this, the HV polarization in the L band is more sensitive to the vegetation structure and biomass of oil palm when compared to HH polarization in peninsular Malaysia [95]. However, comparable results were found under VV polarization using the LAI in wheat, with an RMSE of 4.19% using the advanced SAR (ASAR) C-band sensor. For oil palm crop, using LAI in the field is a destructive, manual method [96], but it is widely regarded as the most accurate method for estimating the true LAI [97]. However, estimating LAI using this direct method is time-consuming, tedious, and labor-intensive [63].

L-band backscatter interacts at the top of the canopy as well at the soil. Using this capability, the L-band SAR-derived descriptors were considered worthy of evaluation in order to obtain an understanding of the possibility of reducing this field-based variable into the WCM to allow for simplified model fine-tuning and soil moisture data retrieval. Positive correlations were found between all SAR-based descriptors and soil moisture in oil palm under HH polarization using the RVI and $R_{HH/HV}$ ranging from 0.036 to 0.049 m^3/m^3 RMSE, followed by $R_{HV/HH}$ with a RMSE accuracy of 0.128 m^3/m^3. For the RVI, $R_{HH/HV}$ and $R_{HV/HH}$ were employed where greater accuracy was found under the HV polarization than the HH polarization. Under HV polarization, the RVI vegetation descriptors used in multiple crops have demonstrated an accuracy of 0.085 m^3/m^3 [54], which was improved in this study at 0.031 m^3/m^3.

Similarly, using $R_{HH/VV}$, $R_{VH/VV}$, RVI, and the generalized volume scattering model based radar vegetation index, employed in a recent study, showed similar accuracies to the findings of this study [56]. In addition, WCM studies using optical-based descriptors (commonly the NDVI) have also been found to be accurate to within the GMES standards. In comparing the NDVI with $R_{VH/VV}$, crop phenology and crop growth changes have been demonstrated and found to have an of accuracy of 0.12 cm^3/cm^3 in corn at the growth level [57]. It was noted that the accuracy of soil moisture data retrieval can be affected by the preprocessing and filtering process; hence, some consideration must be given to evaluating the filtering window size, incident angle, and the SAR imaging resolution. Soil moisture data retrieved from multi-polarized and multi-angled RADARSAT-2 images have produced WCMs with accuracies of RMSE = 5.9% and 6.6%, respectively [98]. Using the WCM, Zribi et al. [99] obtained comparable results for a semi-arid environment, at RMSE = 0.06 m^3/m^3, using ASAR data. However, because crop structures vary in time and space, and radar interactions between the soil and vegetation are complex, the proposed approach's spatial and temporal transferability requires more measurements of soil and vegetation properties, and corresponding radar observations, to provide more robust results. On the other hand, cross-validation in this study was achieved using the LOOCV method to enable fine-tuning of the parameters and error reduction in the evaluated data set.

Based on the outcome of this study, we envisage that the WCM approach can be embedded into crop automated irrigation systems, particularly in oil palm, where appropriate soil moisture must be accessible, since insufficient or excessive moisture will have a detrimental effect on nutrient uptake and yields. On the other hand, soil moisture retrieval from PALSAR-2 can reduce laborious soil sampling work and result in time and cost savings. Using the findings from this study, we were able to successfully reduce field-based parameters, allowing the WCM approach to be evaluated further to develop an efficient soil moisture model for the oil palm industry, particularly in rural plantation areas with limited physical access for conventional soil sampling.

6. Conclusions

In this study, the WCM model was calibrated using L band SAR data, with the field-based LAI indicator and SAR-derived RVI, $R_{HH/HV}$, and $R_{HV/HH}$ as input vegetation descriptors for an oil palm crop with in-field soil moisture. The aim was to evaluate the SAR-derived indicators from PALSAR-2 for their suitability in reducing the need for field-based parameter data collection. Our findings allow a simplification of the WCM that enables SAR benefits to be adapted for soil moisture data retrieval in oil palm. The model fit showed that with HV polarization, the RVI and $R_{HH/HV}$ produced a good replication backscatter coefficient compared to using the LAI as the vegetation parameter. The WCM modeled using the RVI and $R_{HH/HV}$ had accuracies of 0.425 and 0.472 dB RMSE. With HV polarization, the field-based LAI indicator showed the model fit with an R^2 of 0.983 and RMSE of 0.635 dB, using PALSAR-2 data. Our results showed that the soil moisture data retrieval was successful with an RMSE ranging as low as 0.033 m^3/m^3 using the field-based LAI indicator under HV polarization. The SAR-based RVI indicator, however, gave better accuracy with HV polarization at 0.031 m^3/m^3. The $R_{HH/HV}$ polarization demonstrated an equally good capability of soil moisture data retrieval, at an RMSE of 0.049 m^3/m^3 with the same polarization.

Based on these results, it was demonstrated that the WCM is applicable to oil palm crop, with the performance of the model being evaluated using different vegetation descriptors, providing an understanding of the potential use of SAR-derived vegetation descriptors using PALSAR-2. It is suggested that full polarization of the L band to be used for exploiting the SAR-based indicators in oil palm WCMs, and also to examine the impact of the VH and VV polarization effects. For future work, C band backscattering from the oil palm trees crown canopies can be more thoroughly evaluated to be implemented in the WCM for biophysical estimation of vegetation cover. Investigation of the C band, using field-based vegetation water content measurements in the oil palm canopy can be explored using Equation (6) to study the accuracy of retrieving vegetation variable, e.g., LAI.

Author Contributions: Conceptualization, V.S. and A.R.M.S.; data curation, V.S.; formal analysis, V.S. and A.R.M.S.; investigation, V.S., A.R.M.S. and A.W.; methodology, V.S., A.R.M.S., A.W., M.R.K., Y.P.L. and W.T.; resources, A.R.M.S., A.W., M.R.K., Y.P.L. and W.T.; software, V.S. and M.R.K.; supervision, A.R.M.S., A.W., M.R.K., Y.P.L. and W.T.; validation, V.S., A.R.M.S., A.W., M.R.K., Y.P.L. and W.T.; visualization, A.R.M.S. and A.W.; writing—original draft, V.S., A.R.M.S., and Y.P.L.; writing—review and editing, V.S., A.R.M.S., A.W., M.R.K., Y.P.L. and W.T. All authors have read and agreed to the published version of the manuscript.

Funding: The PALSAR-2 data were gathered with funding from the Japanese Aerospace Agency as part of the Earth Observation Research Announcement 2 partnership (Principal Investigator No. ER2A2N180).

Institutional Review Board Statement: Not applicable.

Informed Consent Statement: Not applicable.

Data Availability Statement: The PALSAR-2 data obtained are subject to restrictions, as a special request for high-resolution data has to be made.

Acknowledgments: The authors would like to express their gratitude to Rajah Selvarajah for providing financial support for the acquisition of the UAV data and for the fieldwork logistics necessary to finish the research. The authors thank the Soil and Water Conservation Laboratory, Department of Biological and Agricultural Engineering, Faculty of Engineering, Universiti Putra Malaysia (UPM), and especially Mohamad Hafis bin Ramli, for providing the equipment to extract the soil samples and the oven for drying them. All authors are grateful to Ghazali bin Kassim for arranging for us to borrow the GPS equipment used for locating the soil samples in the field from the Spatial Information System Laboratory, Biological and Agricultural Engineering, Faculty of Engineering, UPM. FGV R&D Sdn Bhd are thanked for facilitating the collection of data in the field, particularly Mohd Shahkhirat bin Norizan, Syahidah Abu Hassan, Muhamad Mazuan Bin Yahaya, Haryati Abidin, Noorsusilawati bt. Mandangan, Mohd Mahfuz Roslan, Muhammad Farid Abdul Rahim, and Mohd Na'aim bin Samad.

Conflicts of Interest: The authors declare no conflict of interest.

References

1. Quezada, J.C.; Etter, A.; Ghazoul, J.; Buttler, A.; Guillaume, T. Carbon neutral expansion of oil palm plantations in the Neotropics. *Sci. Adv.* **2019**, *5*, eaaw4418. [CrossRef] [PubMed]
2. Alam, A.S.A.F.; Er, A.C.; Begum, H. Malaysian oil palm industry: Prospect and problem. *J. Food Agric. Environ.* **2015**, *13*, 143–148.
3. Arunachalam, V. Oil Palm. In *Genomics of Cultivated Palms*, 1st ed.; Elsevier Inc Publications: Amsterdam, The Netherlands, 2012; pp. 29–48.
4. Voora, V.; Larrea, C.; Bermúdez, S.; Baliño, S. *Global Market Report: Palm Oil*; International Institute for Sustainable Development: Winnipeg, MB, Canada, 2020.
5. Lim, C.H.; Cheah, Z.H.; Lee, X.H.; How, B.S.; Ng, W.P.Q.; Ngan, S.L.; Lim, S.; Lam, H.L. Harvesting and evacuation route optimisation model for fresh fruit bunch in the oil palm plantation site. *J. Clean. Prod.* **2021**, *307*, 127238. [CrossRef]
6. Suharjito; Elwirehardja, G.N.; Prayoga, J.S. Oil palm fresh fruit bunch ripeness classification on mobile devices using deep learning approaches. *Comput. Electron. Agric.* **2021**, *188*, 106359. [CrossRef]
7. Jaroenkietkajorn, U.; Gheewala, S.H. Land suitability assessment for oil palm plantations in Thailand. *Sustain. Prod. Consum.* **2021**, *28*, 1104–1113. [CrossRef]
8. Bos, U.; Maier, S.D.; Horn, R.; Leistner, P.; Finkbeiner, M. A GIS based method to calculate regionalized land use characterization factors for life cycle impact assessment using LANCA®. *Int. J. Life Cycle Assess.* **2020**, *25*, 1259–1277. [CrossRef]
9. Sehler, R.; Li, J.; Reager, J.; Ye, H. Investigating Relationship Between Soil Moisture and Precipitation Globally Using Remote Sensing Observations. *J. Contemp. Water Res. Educ.* **2019**, *168*, 106–118. [CrossRef]
10. Li, J.; Sehler, R.; Reager, J.T., II; Ye, H. Using Satellite-based Observations to Investigate the Soil Moisture—Precipitation Relationship. *AGU Fall Meet. Abstr.* **2019**, *2019*, 1910.
11. Ali, G. Climate change and associated spatial heterogeneity of Pakistan: Empirical evidence using multidisciplinary approach. *Sci. Total Environ.* **2018**, *634*, 95–108. [CrossRef] [PubMed]
12. Fowler, H.J.; Blenkinsop, S.; Tebaldi, C. Linking climate change modelling to impacts studies: Recent advances in downscaling techniques for hydrological modelling. *Int. J. Clim.* **2007**, *27*, 1547–1578. [CrossRef]
13. Shukla, A.; Panchal, H.; Mishra, D.M.; Patel, P.R.; Srivastava, H.S.; Patel, P. Soil Moisture Estimation using Gravimetric Technique and FDR Probe Technique: A Comparative Analysis. *Am. Int. J. Res. Form. Appl. Nat. Sci.* **2014**, *8*, 89–92.
14. Degirmenci, H.; Gonen, E.; Boyaci, S.; Resources, W.; Location, T. A comparison of the gravimetric and TDR methods in terms of determining the soil water content of the corn plant. *Sci. Pap.-Ser. A-Agron.* **2016**, *59*, 153–158.
15. Teixeira, W.G.; Huwe, B.; Schroth, G.; Marques, J.D. Sampling and TDR probe insertion in the determination of the volumetric soil water content. *Rev. Bras. Ciênc. Solo* **2003**, *27*, 575–582. [CrossRef]
16. Kaleita, A.L.; Tian, L.F.; Hirschi, M.C. Relationship between Soil Moisture Content and Soil Surface Reflectance. *Trans. ASAE* **2005**, *48*, 1979–1986. [CrossRef]
17. Al-Mamoori, S.K.; Al-Maliki, L.A.; Al-Sulttani, A.H.; El-Tawil, K.; Al-Ansari, N. Statistical analysis of the best GIS interpolation method for bearing capacity estimation in An-Najaf City, Iraq. *Environ. Earth Sci.* **2021**, *80*, 683. [CrossRef]
18. Chen, H.; Fan, L.; Wu, W.; Liu, H.-B. Comparison of spatial interpolation methods for soil moisture and its application for monitoring drought. *Environ. Monit. Assess.* **2017**, *189*, 525. [CrossRef]
19. Gilewski, P. Impact of the Grid Resolution and Deterministic Interpolation of Precipitation on Rainfall-Runoff Modeling in a Sparsely Gauged Mountainous Catchment. *Water* **2021**, *13*, 230. [CrossRef]

20. İmamoğlu, M.Z.; Sertel, E. Analysis of Different Interpolation Methods for Soil Moisture Mapping Using Field Measurements and Remotely Sensed Data. *Int. J. Environ. Geoinformatics* **2016**, *3*, 11–25. [CrossRef]
21. Oliver, M.A.; Webster, R. Geostatistical Prediction: Kriging. In *Basic Steps in Geostatistics: The Variogram and Kriging*; Springer Nature: Basingstoke, UK, 2015; pp. 43–69. [CrossRef]
22. Shen, Q.; Wang, Y.; Wang, X.; Liu, X.; Zhang, X.; Zhang, S. Comparing interpolation methods to predict soil total phosphorus in the Mollisol area of Northeast China. *CATENA* **2019**, *174*, 59–72. [CrossRef]
23. Guo, X.; Fu, Q.; Hang, Y.; Lu, H.; Gao, F.; Si, J. Spatial variability of soil moisture in relation to land use types and topographic features on hillslopes in the black soil (mollisols) area of northeast China. *Sustainability* **2020**, *12*, 3552. [CrossRef]
24. Degré, A.; Ly, S.; Charles, C. Different methods for spatial interpolation of rainfall data for operational hydrology and hydrological modeling at watershed scale: A review. *Biotechnol. Agron. Société Environ.* **2013**, *17*, 392–406. Available online: https://popups.uliege.be/1780-4507 (accessed on 11 October 2021).
25. Yarak, K.; Witayangkurn, A.; Kritiyutanont, K.; Arunplod, C.; Shibasaki, R. Oil palm tree detection and health classification on high-resolution imagery using deep learning. *Agriculture* **2021**, *11*, 183. [CrossRef]
26. Malinee, R.; Stratoulias, D.; Nuthammachot, N. Detection of oil palm disease in plantations in krabi province, thailand with high spatial resolution satellite imagery. *Agriculture* **2021**, *11*, 251. [CrossRef]
27. Ibrahim, A.L.; Hashim, M.; AIi, M.I.; Rasib, W.; Kadir, W.H.W.; Sumari, M.R.; Haron, K. Detecting and Mapping Nutrients Concentration in Oil Palm Plantation Using Remote Sensing and Geographic Information System. In Proceedings of the PIPOC 2003 International Palm Oil Congress (Agriculture), Kuala Lumpur, Malaysia, 24–28 August 2003; pp. 261–271.
28. Zhu, L.; Suomalainen, J.; Liu, J.; Hyyppä, J.; Kaartinen, H.; Haggren, H. A Review: Remote Sensing Sensors. Available online: https://books.google.co.jp/books?id=8m6QDwAAQBAJ&printsec=frontcover&hl=zh-CN#v=onepage&q&f=false (accessed on 3 November 2021).
29. Xue, J.; Su, B. Significant Remote Sensing Vegetation Indices: A Review of Developments and Applications. *J. Sens.* **2017**, *2017*, 1353691. [CrossRef]
30. Shashikant, V.; Shariff, A.R.M.; Nordin, L.; Pradhan, B. NDVI of Oil Palm Trees by Landsat-5 Imagery. In Proceedings of the 33rd Conference on Remote Sensing, Pattaya, Thailand, 26–30 November 2012; p. 5.
31. Lehmann, E.A.; Caccetta, P.; Lowell, K.; Mitchell, A.; Zhou, Z.-S.; Held, A.; Milne, T.; Tapley, I. SAR and optical remote sensing: Assessment of complementarity and interoperability in the context of a large-scale operational forest monitoring system. *Remote Sens. Environ.* **2015**, *156*, 335–348. [CrossRef]
32. Saatchi, S.S. Application of SAR Remote Sensing in Land Surface Processes Over Tropical Region. In Proceedings of the Anais VIII Simposio Brasileiro de Sensoriamento Remoto, Salvador, Brasil, 14–19 April 1996; pp. 369–374.
33. Ottinger, M.; Kuenzer, C. Spaceborne L-Band Synthetic Aperture Radar Data for Geoscientific Analyses in Coastal Land Applications: A Review. *Remote Sens.* **2020**, *12*, 2228. [CrossRef]
34. Daliman, S.; Rahman, S.A.; Abu Bakar, S.; Busu, I. Segmentation of oil palm area based on GLCM-SVM and NDVI. In Proceedings of the 2014 IEEE REGION 10 SYMPOSIUM, Kuala Lumpur, Malaysia, 14–16 April 2014; pp. 645–650. [CrossRef]
35. Teng, K.C.; Koay, J.Y.; Tey, S.H.; Lim, K.S.; Ewe, H.T.; Chuah, H.T. A Dense Medium Microwave Backscattering Model for the Remote Sensing of Oil Palm. *IEEE Trans. Geosci. Remote Sens.* **2014**, *53*, 3250–3259. [CrossRef]
36. Ibharim, N.; Mustapha, M.; Lihan, T.; Mazlan, A. Mapping mangrove changes in the Matang Mangrove Forest using multi temporal satellite imageries. *Ocean Coast. Manag.* **2015**, *114*, 64–76. [CrossRef]
37. Woodhouse, I.H. *Introduction to Microwave Remote Sensing*; CRC Press: Boca Raton, FL, USA, 2017.
38. Li, L.; Dong, J.; Tenku, S.N.; Xiao, X. Mapping Oil Palm Plantations in Cameroon Using PALSAR 50-m Orthorectified Mosaic Images. *Remote Sens.* **2015**, *7*, 1206–1224. [CrossRef]
39. Dong, X.; Quegan, S.; Yumiko, U.; Hu, C.; Zeng, T. Feasibility Study of C- and L-band SAR Time Series Data in Tracking Indonesian Plantation and Natural Forest Cover Changes. *IEEE J. Sel. Top. Appl. Earth Obs. Remote Sens.* **2015**, *8*, 3692–3699. [CrossRef]
40. Fadaei, H.; Ishii, R.; Suzuki, R.; Kendawang, J.J. Detection of oil palm and acacia plantation areas using object based classification in sarawak, Malaysia. In Proceedings of the 34th Asian Conference on Remote Sensing (ACRS), Bali, Indonesia, 20–24 October 2013; pp. 1–7.
41. Hashim, I.; Shariff, A.; Bejo, S.; Muharam, F.; Ahmad, K. Machine-Learning Approach Using SAR Data for the Classification of Oil Palm Trees That Are Non-Infected and Infected with the Basal Stem Rot Disease. *Agronomy* **2021**, *11*, 532. [CrossRef]
42. Reiche, J.; Lucas, R.; Mitchell, A.L.; Verbesselt, J.; Hoekman, D.H.; Haarpaintner, J.; Kellndorfer, J.M.; Rosenqvist, A.; Lehmann, E.A.; Woodcook, C.E.; et al. Combining satellite data for better tropical forest monitoring. *Nat. Clim. Chang.* **2016**, *6*, 120–122. [CrossRef]
43. Kumar, P.; Prasad, R.; Choudhary, A.; Gupta, D.K.; Mishra, V.N.; Vishwakarma, A.K.; Singh, A.K.; Srivastava, P.K. Comprehensive evaluation of soil moisture retrieval models under different crop cover types using C-band synthetic aperture radar data. *Geocarto Int.* **2018**, *34*, 1022–1041. [CrossRef]
44. Petropoulos, G.P.; Ireland, G.; Barrett, B. Surface soil moisture retrievals from remote sensing: Current status, products & future trends. *Phys. Chem. Earth Parts A/B/C* **2015**, *83–84*, 36–56.
45. Attema, E.P.W.; Ulaby, F.T. Vegetation modeled as a water cloud. *Radio Sci.* **1978**, *13*, 357–364. [CrossRef]

46. Said, S.; Kothyari, U.; Arora, M. Vegetation effects on soil moisture estimation from ERS-2 SAR images. *Hydrol. Sci. J.* **2012**, *57*, 517–534. [CrossRef]
47. Chauhan, S.; Srivastava, H.S.; Patel, P. Improved parameterization of water cloud model for hybrid-polarized backscatter simulation using interaction factor. *ISPRS—Int. Arch. Photogramm. Remote Sens. Spat. Inf. Sci.* **2017**, *XLII-4/W2*, 61–66. [CrossRef]
48. Léonard, A.; Bériaux, E.; Defourny, P. Complementarity of linear polarizations in C-band SAR imagery to estimate leaf area index for maize and winter wheat. *ESA Living Planet Symp.* **2013**, *12*, 957–964.
49. Beriaux, E.; Lucau-Danila, C.; Auquiere, E.; Defourny, P. Multiyear independent validation of the water cloud model for retrieving maize leaf area index from SAR time series. *Int. J. Remote Sens.* **2013**, *34*, 4156–4181. [CrossRef]
50. Zhang, L.; Meng, Q.; Yao, S.; Wang, Q.; Zeng, J.; Zhao, S.; Ma, J. Soil Moisture Retrieval from the Chinese GF-3 Satellite and Optical Data over Agricultural Fields. *Sensors* **2018**, *18*, 2675. [CrossRef]
51. Tao, L.; Wang, G.; Chen, X.; Li, J.; Cai, Q. Estimation of soil moisture using a vegetation scattering model in wheat fields. *J. Appl. Remote Sens.* **2019**, *13*, 044503. [CrossRef]
52. Park, S.-E.; Jung, Y.T.; Cho, J.-H.; Moon, H.; Han, S.-H. Theoretical Evaluation of Water Cloud Model Vegetation Parameters. *Remote Sens.* **2019**, *11*, 894. [CrossRef]
53. Khabazan, S.; Hosseini, M.; Saradjian, M.R.; Motagh, M.; Magagi, R. Accuracy Assessment of IWCM Soil Moisture Estimation Model in Different Frequency and Polarization Bands. *J. Indian Soc. Remote Sens.* **2015**, *43*, 859–865. [CrossRef]
54. Li, J.; Wang, S. Using SAR-Derived Vegetation Descriptors in a Water Cloud Model to Improve Soil Moisture Retrieval. *Remote Sens.* **2018**, *10*, 1370. [CrossRef]
55. Yue, J.; Yang, G.; Qi, X.; Wang, Y. Soil moisture retrieval in well covered farmland by Radarsat-2 SAR data. *Int. Geosci. Remote Sens. Symp.* **2016**, 1699–1702. [CrossRef]
56. Bhogapurapu, N.; Mandal, D.; Rao, Y.S.; Bhattacharya, A. Soil Moisture Retrieval Using SAR Derived Vegetation Descriptors in Water Cloud Model. In Proceedings of the IGARSS 2020—2020 IEEE International Geoscience and Remote Sensing Symposium, Waikoloa, HI, USA, 26 September–2 October 2020; pp. 4696–4699. [CrossRef]
57. Wang, Z.; Zhao, T.; Qiu, J.; Zhao, X.; Li, R.; Wang, S. Microwave-based vegetation descriptors in the parameterization of water cloud model at L-band for soil moisture retrieval over croplands. *GIScience Remote Sens.* **2020**, *58*, 48–67. [CrossRef]
58. Jacome, A.; Bernier, M.; Chokmani, K.; Gauthier, Y.; Poulin, J.; De Sève, D. Monitoring Volumetric Surface Soil Moisture Content at the La Grande Basin Boreal Wetland by Radar Multi Polarization Data. *Remote Sens.* **2013**, *5*, 4919–4941. [CrossRef]
59. Baronti, S.; del Frate, F.; Ferrazzoli, P.; Paloscia, S.; Pampaloni, P.; Schiavon, G. Sar Polarimetric features of agricultural areas. *Int. J. Remote Sens.* **2010**, *16*, 2639–2656. [CrossRef]
60. Shashikant, V.; Shariff, A.M.; Wayayok, A.; Kamal, R.; Lee, Y.; Takeuchi, W. Utilizing TVDI and NDWI to Classify Severity of Agricultural Drought in Chuping, Malaysia. *Agronomy* **2021**, *11*, 1243. [CrossRef]
61. Shashikant, V.; Shariff, A.R.M.; Wayayok, A.; Kamal, R.; Lee, Y.P.; Takeuchi, W. Vegetation Effects on Soil Moisture Retrieval from Water Cloud Model Using PALSAR-2 for Oil Palm Trees. *Remote Sens.* **2021**, *13*, 4023. [CrossRef]
62. Kok, Z.H.; Shariff, A.R.B.M.; Khairunniza-Bejo, S.; Kim, H.-T.; Ahamed, T.; Cheah, S.S.; Wahid, S.A.A. Plot-Based Classification of Macronutrient Levels in Oil Palm Trees with Landsat-8 Images and Machine Learning. *Remote Sens.* **2021**, *13*, 2029. [CrossRef]
63. Awal, M.A.; Wan Ishak, W.I. Measurement of Oil Palm LAI by Manual and LAI-2000 Method. *Asian J. Sci. Res.* **2007**, *1*, 49–56. [CrossRef]
64. Motohka, T.; Isoguchi, O.; Sakashita, M.; Shimada, M. ALOS-2 PALSAR-2 Cal/Val Updates. In Proceedings of the JAXA/EORC Joint PI Meeting of Global Environment Observation Mission FY2017, Tokyo, Japan, 22–26 January 2018.
65. Ozdarici, A.; Conference, Z.A. A comparison of SAR filtering techniques on agricultural area identification. In Proceedings of the Imaging and Geospatial Information Society (ASPRS), San Diego, CA, USA, 26–30 April 2010.
66. Tucker, C.J. Red and photographic infrared linear combinations for monitoring vegetation. *Remote Sens. Environ.* **1979**, *8*, 127–150. [CrossRef]
67. Pulliainen, J.T.; Heiska, K.; Hyyppa, J.; Hallikainen, M.T. Backscattering Properties of Boreal Forests at the C and X-Bands. *IEEE Trans. Geosci. Remote Sens.* **1994**, *32*, 1041–1050. [CrossRef]
68. Ulaby, F.; Allen, C.; Eger, G.; Kanemasu, E. Relating the microwave backscattering coefficient to leaf area index. *Remote Sens. Environ.* **1984**, *14*, 113–133. [CrossRef]
69. Prévot, L.; Champion, I.; Guyot, G. Estimating surface soil moisture and leaf area index of a wheat canopy using a dual-frequency (C and X bands) scatterometer. *Remote Sens. Environ.* **1993**, *46*, 331–339. [CrossRef]
70. Xu, H.; Steven, M.D.; Jaggard, K.W. Monitoring leaf area of sugar beet using ERS-1 SAR data. *Int. J. Remote Sens.* **1996**, *17*, 3401–3410. [CrossRef]
71. Dabrowska-Zielinska, K.; Inoue, Y.; Kowalik, W.; Gruszczynska, M. Inferring the effect of plant and soil variables on C- and L-band SAR backscatter over agricultural fields, based on model analysis. *Adv. Space Res.* **2007**, *39*, 139–148. [CrossRef]
72. Charbonneau, F.; Trudel, M.; Fernandes, R. Use of Dual Polarization and Multi-Incidence SAR for soil permeability mapping. In Proceedings of the 2005 Advanced Synthetic Aperture Radar (ASAR) Workshop, St-Hubert, QC, Canada, 15–17 November 2005.
73. Kim, Y.; Van Zyl, J.J. A Time-Series Approach to Estimate Soil Moisture Using Polarimetric Radar Data. *IEEE Trans. Geosci. Remote Sens.* **2009**, *47*, 2519–2527. [CrossRef]
74. Trudel, M.; Charbonneau, F.; Leconte, R. Using RADARSAT-2 polarimetric and ENVISAT-ASAR dual-polarization data for estimating soil moisture over agricultural fields. *Can. J. Remote Sens.* **2012**, *38*, 514–527.

75. Kim, Y.-H.; Oh, J. Comparative Analysis of the Multispectral Vegetation Indices and the Radar Vegetation Index. *J. Korean Soc. Surv. Geodesy, Photogramm. Cartogr.* **2014**, *32*, 607–615. [CrossRef]
76. Akhtar, A.M.; Qazi, W.A.; Ahmad, S.R.; Gilani, H.; Mahmood, S.A.; Rasool, A. Integration of high-resolution optical and SAR satellite remote sensing datasets for aboveground biomass estimation in subtropical pine forest, Pakistan. *Environ. Monit. Assess.* **2020**, *192*, 584. [CrossRef] [PubMed]
77. Schönbrodt-Stitt, S.; Ahmadian, N.; Kurtenbach, M.; Conrad, C.; Romano, N.; Bogena, H.R.; Vereecken, H.; Nasta, P. Statistical Exploration of SENTINEL-1 Data, Terrain Parameters, and in-situ Data for Estimating the Near-Surface Soil Moisture in a Mediterranean Agroecosystem. *Front. Water* **2021**, *3*, 75. [CrossRef]
78. Wang, H.; Allain-Bailhache, S.; Méric, S.; Pottier, É.; Allain, S.; Pottier, E. Soil Parameter Retrievals Over Bare Agricultural Fields Using Multi-angular RADARSAT-2 Dataset. *IEEE J. Sel. Top. Appl. Earth Obs. Remote Sens.* **2016**, *9*, 5666–5676. [CrossRef]
79. Han, D.; Wang, P.; Tansey, K.; Zhou, X.; Zhang, S.; Tian, H.; Zhang, J.; Li, H. Linking an agro-meteorological model and a water cloud model for estimating soil water content over wheat fields. *Comput. Electron. Agric.* **2020**, *179*, 105833. [CrossRef]
80. Li, Y.; Zhang, C.; Heng, W. Retrieving Surface Soil Moisture over Wheat-Covered Areas Using Data from Sentinel-1 and Sentinel-2. *Water* **2021**, *13*, 1981. [CrossRef]
81. Paloscia, S.; Pettinato, S.; Santi, E.; Notarnicola, C.; Pasolli, L.; Reppucci, A. Soil moisture mapping using Sentinel-1 images: Algorithm and preliminary validation. *Remote Sens. Environ.* **2013**, *134*, 234–248. [CrossRef]
82. Santi, E.; Paloscia, S.; Pettinato, S.; Notarnicola, C.; Pasolli, L.; Pistocchi, A. Remote Sensing Comparison between SAR Soil Moisture Estimates and Hydrological Model Simulations over the Scrivia Test Site. *Remote Sens.* **2013**, *5*, 4961–4976. [CrossRef]
83. Varma, S.; Simon, R. Bias in error estimation when using cross-validation for model selection. *BMC Bioinform.* **2006**, *7*, 91. [CrossRef]
84. Kim, Y.; Jackson, T.; Bindlish, R.; Lee, H.; Hong, S. Radar Vegetation Index for Estimating the Vegetation Water Content of Rice and Soybean. *IEEE Geosci. Remote Sens. Lett.* **2011**, *9*, 564–568. [CrossRef]
85. Joshi, N.P.; Mitchard, E.T.A.; Schumacher, J.; Johannsen, V.K.; Saatchi, S.; Fensholt, R. L-Band SAR Backscatter Related to Forest Cover, Height and Aboveground Biomass at Multiple Spatial Scales across Denmark. *Remote Sens.* **2015**, *7*, 4442–4472. [CrossRef]
86. Goh, K.J.; Teo, C.B. Agronomic Principles and Practices of Fertilizer Management Of Oil Palm. *Agron. Princ. Pract. Oil Palm Cultiv.* **2011**, 241–389.
87. Disney, M.; Lewis, P.; Saich, P. 3D modelling of forest canopy structure for remote sensing simulations in the optical and microwave domains. *Remote Sens. Environ.* **2006**, *100*, 114–132. [CrossRef]
88. Pizaña, J.M.G.; Hernández, J.M.N.; Romero, N.C. Remote Sensing-Based Biomass Estimation. Available online: https://books.google.co.jp/books?hl=zh-CN&lr=&id=Wm-QDwAAQBAJ&oi=fnd&pg=PA3&dq=Remote+Sensing-Based+Biomass+Estimation.+Environ.+Appl.+Remote+Sens.+2016&ots=CkzgeLf5eT&sig=zPDDULDOZN6dR55GnHBQjxYY76M&redir_esc=y#v=onepage&q&f=false (accessed on 3 November 2021).
89. Ghasemi, N.; Sahebi, M.R.; Mohammadzadeh, A. A review on biomass estimation methods using synthetic aperture radar data. *Int. J. Ofgeomatics Geosci.* **2011**, *1*, 776–788.
90. Srivastava, H.S.; Patel, P.; Sharma, Y.; Navalgund, R.R. Large-Area Soil Moisture Estimation Using Multi-Incidence-Angle RADARSAT-1 SAR Data. *IEEE Trans. Geosci. Remote Sens.* **2009**, *47*, 2528–2535. [CrossRef]
91. Hoekman, D.; Quinones, M.; Vissers, M. K&C Science Report—Phase 1 Tropical Forest and Wetlands Mapping, Case Study Borneo. Available online: https://library.wur.nl/WebQuery/wurpubs/fulltext/50976 (accessed on 3 November 2021).
92. Saradjian, M.; Hosseini, M. Soil moisture estimation by using multipolarization SAR image. *Adv. Space Res.* **2011**, *48*, 278–286. [CrossRef]
93. Yadav, V.P.; Prasad, R.; Bala, R. Leaf area index estimation of wheat crop using modified water cloud model from the time-series SAR and optical satellite data. *Geocarto Int.* **2019**, *36*, 791–802. [CrossRef]
94. Satalino, G.; Mattia, F.; Balenzano, A.; Panciera, R.; Walker, J.; Walker, J. Soil moisture maps from time series of PALSAR-1 scansar data over Australia. In Proceedings of the 2013 IEEE International Geoscience and Remote Sensing Symposium—IGARSS, Melbourne, VIC, Australia, 21–26 July 2013; pp. 719–722. [CrossRef]
95. Najib, N.E.M.; Kanniah, K.D.; Cracknell, A.P.; Yu, L. Synergy of Active and Passive Remote Sensing Data for Effective Mapping of Oil Palm Plantation in Malaysia. *Forests* **2020**, *11*, 858. [CrossRef]
96. Mathur, R.K.; Suresh, K.; Bhanusri, A.; Naveen, K.P.; Prasad, M.V.; Rao, B.N.; Kalidas, P. *Research Highlights 2011–2015*; ICAR-Indian Institute of Oil Palm Research: Andhra Pradesh, India, 2017; pp. 1–64.
97. Beets, P.N.; Reutebuch, S.; Kimberley, M.O.; Oliver, G.R.; Pearce, S.H.; McGaughey, R.J. Leaf Area Index, Biomass Carbon and Growth Rate of Radiata Pine Genetic Types and Relationships with LiDAR. *Forests* **2011**, *2*, 637–659. [CrossRef]
98. Gherboudj, I.; Magagi, R.; Berg, A.A.; Toth, B. Soil moisture retrieval over agricultural fields from multi-polarized and multi-angular RADARSAT-2 SAR data. *Remote Sens. Environ.* **2011**, *115*, 33–43. [CrossRef]
99. Zribi, M.; Baghdadi, N.; Holah, N.; Fafin, O.; Guérin, C. Evaluation of a rough soil surface description with ASAR-ENVISAT radar data. *Remote Sens. Environ.* **2005**, *95*, 67–76. [CrossRef]

remote sensing MDPI

Technical Note

L-Band SAR Co-Polarized Phase Difference Modeling for Corn Fields

Matías Ernesto Barber [1,2,*], David Sebastián Rava [1] and Carlos López-Martínez [3]

1 Quantitative Remote Sensing Group, Institute of Astronomy and Space Physics (IAFE), Buenos Aires 1428, Argentina; drava@iafe.uba.ar
2 Department of Physics, Engineering School, University of Buenos Aires (UBA), Buenos Aires 1428, Argentina
3 Signal Theory and Communications Department (TSC), Universitat Politècnica de Catalunya (UPC), 08034 Barcelona, Spain; carlos.lopezmartinez@upc.edu
* Correspondence: mbarber@iafe.uba.ar

Abstract: This research aims at modeling the microwave backscatter of corn fields by coupling an incoherent, interaction-based scattering model with a semi-empirical bulk vegetation dielectric model. The scattering model is fitted to co-polarized phase difference measurements over several corn fields imaged with fully polarimetric synthetic aperture radar (SAR) images with incidence angles ranging from 20° to 60°. The dataset comprised two field campaigns, one over Canada with the Uninhabited Aerial Vehicle Synthetic Aperture Radar (UAVSAR, 1.258 GHz) and the other one over Argentina with Advanced Land Observing Satellite 2 (ALOS-2) Phased Array type L-band Synthetic Aperture Radar (PALSAR-2) (ALOS-2/PALSAR-2, 1.236 GHz), totaling 60 data measurements over 28 grown corn fields at peak biomass with stalk gravimetric moisture larger than 0.8 g/g. Co-polarized phase differences were computed using a maximum likelihood estimation technique from each field's measured speckled sample histograms. After minimizing the difference between the model and data measurements for varying incidence angles by a nonlinear least-squares fitting, well agreement was found with a root mean squared error of 24.3° for co-polarized phase difference measurements in the range of −170.3° to −19.13°. Model parameterization by stalk gravimetric moisture instead of its complex dielectric constant is also addressed. Further validation was undertaken for the UAVSAR dataset on earlier corn stages, where overall sensitivity to stalk height, stalk gravimetric moisture, and stalk area density agreed with ground data, with the sensitivity to stalk diameter being the weakest. This study provides a new perspective on the use of co-polarized phase differences in retrieving corn stalk features through inverse modeling techniques from space.

Keywords: synthetic aperture radar; polarimetric radar; co-polarized phase difference; radar scattering; vegetation; radar applications; agriculture

Citation: Barber, M.E.; Rava, D.S.; López-Martínez, C. L-Band SAR Co-Polarized Phase Difference Modeling for Corn Fields. *Remote Sens.* **2021**, *13*, 4593. https://doi.org/10.3390/rs13224593

Academic Editors: Takeo Tadono, Masato Ohki and Klaus Scipal

Received: 29 August 2021
Accepted: 11 November 2021
Published: 15 November 2021

1. Introduction

The potential of active microwaves to monitor agricultural areas is recognized as a key feature for supporting application-oriented approaches such as crop classification schemes (e.g., [1–3]), crop height estimation (e.g., [4–6]), soil moisture estimation (e.g., [7,8]), among others, and to aid decision-makers in managing and assessing agricultural resources. Towards this goal, the NASA/JPL's UAVSAR airborne L-band mission was deployed to support several soil moisture and vegetation features inversion strategies [9–11]. In this respect, the systematic use of polarimetric SAR data from orbiting sensors at L-band over croplands was almost limited to JAXA's Advanced Land Observing Satellite 2 (ALOS-2) Phased Array type L-band Synthetic Aperture Radar (PALSAR-2) mission (global.jaxa.jp/projects/sat/alos2) over the years. However, this situation has recently improved with the successful launch of the Argentinean L-band SAR constellation mission SAOCOM-1A and 1B (saocom.invap.com.ar) on 7 October 2018, and 30 August 2020, respectively. Both sensors have a lifespan of 5.5 years and were designed with interferometric and polarimetric capabilities. Within its goals, the SAOCOM constellation will provide fully polarimetric acquisitions dedicated to monitoring large cropland areas in Argentina, representing an important contribution to agriculture and hydrology worldwide.

The NASA-ISRO Synthetic Aperture Radar (NISAR) mission, which is planned to be launched in 2023, will provide L- and S-band full-polarized data over vegetated terrain, adding up its polarimetric capabilities to existing imagery [12]. In addition, the European Space Agency has recently signed the contract to develop the new high-priority Copernicus Radar Observation System for Europe in L-band (ROSE-L) as part of Europe's Copernicus program. With a launch planned in 2028, this system will present polarimetric capabilities and its main product types and formats will be aligned as much as possible with the ones of Sentinel-1, for enhanced continuity [13].

Among major crops, corn is the most cultivated cereal worldwide according to the latest Food and Agriculture Organization (FAO) data [14], with a total production of 1149 Mt in 2019, followed by wheat (765.8 Mt), paddy rice (755.5 Mt), soybeans (333.7 Mt), and barley (159.0 Mt) in the same year. Following the significant SAR missions mentioned, amplitude and phase measurements will be systematically delivered to cover most of these major crops, among which corn fields have unique features: corn plants have the largest dimensions with stalk heights up to 3 m, stalk diameters up to 2.5 cm and large moisture contents up to 0.90 g/g [11,15,16]. Furthermore, corn seeds are usually planted in a regular pattern of 7 to 9 plants per square meter onto rows separated 75 cm apart [11,16,17]. This pattern and the unique plant features, often in the resonant regime for wavelengths at the L-band, make the interaction of electromagnetic waves with corn fields very complex to model.

Efforts in this direction were made on computing the scattering of a collection of randomly distributed vertical cylinders, thus modeling the plant stalks over a dielectric half-space. Smaller plant elements such as leaves and cobs were usually disregarded. High order solutions involving multiple interactions among the cylinders and the underlying dielectric half-space were obtained by Monte Carlo simulation or by radiative transfer theory ([18,19]). However, for an application-oriented approach, a Monte Carlo simulation is of limited practical use because of the ensemble-based statistical nature of its solution. In the radiative transfer approach, solutions for modeling large dielectric structures such as corn stalks should deal with an overestimation of phase and extinction matrices [18].

A more straightforward approach that incorporates much of the interaction complexity with few input parameters is the model developed by Ulaby et al. [17]. This model relied on previous experimental measurements to treat a corn canopy as a low-loss medium, thus allowing for a description in terms of an equivalent dielectric medium characterized by a complex index of refraction. With the noticeably uneven distribution of volumetric moisture content between leaves and stalks during much of the growth stages, the contribution of the plant leaves to total scattering can be disregarded for longer wavelengths, such as in L-band.

Ulaby's model was experimentally validated in [17] using an image-based relative phase calibration, where near-range azimuth rows were assumed to have a co-polarized phase difference near zero, and thus converting relative values to absolute values in the remaining image. An ad hoc 180° phase shift added to the model ([17], Equation (5)) should be disregarded on properly absolute calibrated images such as that of the aforementioned SAR missions. The dataset used in Ulaby's research for validation involved relatively mature, dried vegetation with low stalk volumetric moisture [17]. No validation is reported for other conditions, nor was further research in this respect found elsewhere. Moreover, research on L-band co-polarized phase differences on crops is scarce (e.g., [20]). Most of the research using polarimetric SAR data relied on higher frequencies (C- and X-band) [21–23] or multi-polarization intensity-only studies [24]. These shortcomings will be addressed in this manuscript, which turns out to be a novel contribution of this work.

When corn plants reach their peak biomass, vegetation water content is near maximum, and canopy attenuation and stalk's coherent effects are significant. The empirical fitting used in [17] to compute the dielectric constants of stalks from their gravimetric water content was limited by its upper bound ([25], Chapter 4–9.2). For larger water contents, the model developed by Mätzler in [26] will be considered.

In this research, a validation of Ulaby's incoherent multi-parameter model with experimental data on grown corn fields is shown. Mätzler's model for a bulk dielectric constant is coupled with Ulaby's model to account for the large water content found in the stalks of grown corn plants and to avoid time-consuming, laboratory-based dielectric constant measurements. Two datasets were used, (1) fields in Canada imaged by the airborne sensor UAVSAR and (2) fields in Argentina imaged by the satellite-borne ALOS-2/PALSAR-2 sensor. Good agreement is made, which enables us to consider this model for retrieving purposes through inverse modeling techniques.

The outline of this paper is given as follows: Section 2 states a brief review of Ulaby's fitting model, a sensitivity assessment of its model parameters with stalk features, and a description of Mätzler's model to estimate stalk dielectric constant from gravimetric measurements. Then, details of data used in this paper are introduced, including SAR data quality and the method for estimating co-polarized phase differences. The SAR data statistics and the fitting of the Ulaby–Mätzler's model to remotely-sensed co-polarized phase differences are analyzed in Section 3. This coupled model and its implications for corn parameter retrieval are discussed in Section 4. Concluding remarks are stated in Section 5.

2. Materials and Methods

2.1. Incoherent Multi-Parameter Fitting Model

The co-polarized phase difference ϕ is defined as the difference in the absolute phase between the linearly polarized HH and VV complex scattering amplitudes

$$\phi = \phi_{HH} - \phi_{VV}. \tag{1}$$

In lossy media, ϕ accounts for many scattering mechanisms and contributions. On a grown corn canopy, ϕ is modeled as a result of the sum of three single contributions

$$\phi = \phi_p + \phi_{st} + \phi_s, \tag{2}$$

where ϕ_p accounts for the phase term due to wave propagation through the canopy, ϕ_{st} for the forward scattering by the soil surface followed by bistatic scattering by the stalks, or the reverse process, and ϕ_s for the specular reflection on the soil. Each one of these scattering mechanisms was evaluated following Ulaby's model [17] and compared to SAR data.

Ulaby's model [17] to be fitted accounts for the scattering interaction between the plant stalk and an underlying rough, moistened surface to compute (2). Corn plants were modeled as vertical dielectric cylinders, long enough relative to the wavelength to rely on the infinity cylinder scattering solution, which is given in the form of a series [27]

$$T_{H,V}(\theta_i, \theta_s; k, a_0, \varepsilon_{st}) = \sum_{n=-\infty}^{+\infty} (-1)^n e^{in\theta_s} C_n^{H,V}(\theta_i; k, a_0, \varepsilon_{st}), \tag{3}$$

where $T_{H,V}$ is the normalized far-field scattering amplitude, the subscript states the polarization of the impinging wave onto a linear basis (H or V), θ_i is the incidence angle relative to the plane containing the cylinder's axis, and θ_s is the azimuth scattered angle. The dependence of the functions $C_n^{H,V}$ on the wavenumber k of the impinging wave, the radius a_0 and the complex dielectric constant ε_{st} of the cylinder is cumbersome and the reader is referred to [27] for their analytical expressions.

The solution given by (3) is applied two-fold. Firstly, Ulaby et al. [17] have shown that propagation in a layer comprising identical vertical cylinders randomly positioned on the ground may be modeled in terms of an equivalent dielectric medium characterized by a polarization-dependent complex index of refraction. The model assumed stalks are arranged with N cylinder per unit area and are far away enough such that multiple scattering is negligible. Hence, the phase constant of the index of refraction is used to compute the co-polarized phase difference for two-way propagation ($\theta_s = \pi$ in (3)). Secondly, the scattering solution in (3) is used to compute the phase difference between waves bistatically reflected by the stalks by considering specular scattering only ($\theta_s = 0$ in (3)).

The first term on the right side in (2) computes the phase term due to the two-way, slanted propagation through the canopy,

$$\phi_p = \frac{4Nh}{k} \tan\theta[\mathrm{Im}\{T_H(\theta_i, \pi)\} - \mathrm{Im}\{T_V(\theta_i, \pi)\}], \tag{4}$$

where h is stalk height. In (4), the scattering features of the stalks are accounted for in the $T_{H,V}$ amplitudes, where canopy bulk features are accounted for in the stalk density N and in h. The scattered angle is evaluated at the forward direction ($\theta_s = \pi$) [27].

The second term in (2) accounts for the phase term resulting from forward scattering by the soil surface followed by bistatic scattering by the stalks, or the reverse process,

$$\phi_{st} = \tan^{-1}\left(\frac{\mathrm{Im}\{T_H(\theta_i, 0)/T_V(\theta_i, 0)\}}{\mathrm{Re}\{T_H(\theta_i, 0)/T_V(\theta_i, 0)\}}\right), \tag{5}$$

where the solution should be sought in the domain $(-\pi, \pi]$. Here, $\theta_s = 0$ accounted for the specular direction.

The third term in (2) is the contribution from specular reflection on the soil through Fresnel reflection coefficients R_H and R_V [25]

$$\phi_s = \tan^{-1}\left(\frac{\mathrm{Im}\{R_H(\theta_i, \varepsilon_s)/R_V(\theta_i, \varepsilon_s)\}}{\mathrm{Re}\{R_H(\theta_i, \varepsilon_s)/R_V(\theta_i, \varepsilon_s)\}}\right), \tag{6}$$

where ε_s is the complex dielectric constant of the soil surface underlying the canopy. The contribution of this term is about $-180°$ due to the small imaginary part of ε_s in typical soils and the difference in sign between R_H and R_V. Because of this term, total co-polarized phase difference ϕ, over grown corn canopies yields negative values on absolute calibrated polarimetric images.

2.2. Sensitivity Analysis of the Model Parameters

The three phase terms defined from (4) to (6) account respectively for the phase difference by propagation through the stalks, by the bistatic reflection, and by the soil. Each of these terms has different contributions to the total co-polarized phase difference in (2). In what follows, a sensitivity analysis will be carried out, where frequency will be fixed at an intermediate 1.25 GHz, that is, between those of UAVSAR and ALOS-2/PALSAR-2.

Among the three terms, the soil term ϕ_s has a simple dependency on the soil's complex dielectric constant $\varepsilon_s = \varepsilon_s' + i\varepsilon_s''$. A typical imaginary-to-real ratio of ε_s is 0.10, and commonly used empirical models predict this ratio to be as large as 0.25 [28–30]. Then, it follows from Figure 1 that the ϕ_s dynamic range is less than 16° for an imaginary-to-real ratio of 0.10 (black lines) and 0.25 (blue lines). Three incidence angles 20°, 40°, and 60° are evaluated. Then, for a typical observation geometry at 40° incidence angle, the sensitivity to the dielectric constant shown by ϕ_s is of little relevance to the total phase difference.

The propagation term in (4) has a linear dependence with the stalk density N and with the stalk height h. Moreover, since they are of the same order of magnitude, the effect of varying N or h on ϕ_p will be equivalent. Conversely, the a_0 and ε_{st} are nonlinear model parameters through $T_{H,V}$ and the following sensitivity analysis will be focused on them. First, contour levels depicting the dependence of ϕ_p on a_0 and h at $\theta_i = 40°$ are shown in Figure 2a. The contours are variations of ϕ_p computed as

$$\Delta\phi_p = \frac{\partial\phi_p}{\partial a_0}\Delta a_0 + \frac{\partial\phi_p}{\partial h}\Delta h, \tag{7}$$

where it is understood that the other three terms involving derivatives (on ε'_{st}, ε''_{st}, and N) were computed and evaluated from the mean values collected on the ground, and these are indicated in the inset in Figure 2a.

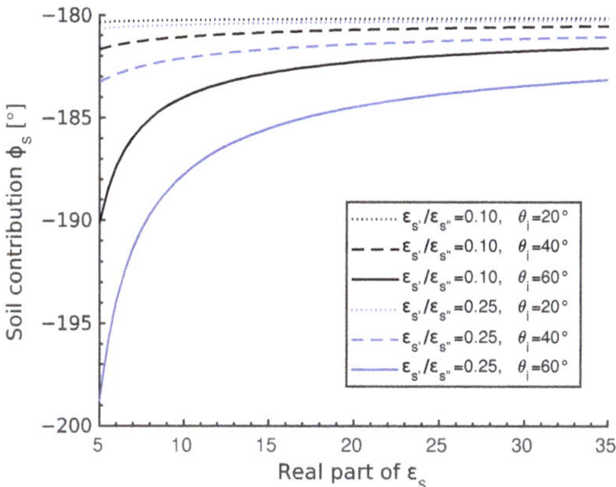

Figure 1. Sensitivity of soil term on the real part of the dielectric constant. The imaginary part is assumed to be 0.10 (black lines) and 0.25 (blue lines) of the real part.

A significant gradient indicated a high sensitivity to stalk height. This related to the linear term h in (4). Conversely, a small sensitivity on a_0 is related to a cancellation effect due to the difference operator in (4), since both T_H and T_V depend on a_0. The model exhibits $\Delta\phi_p \sim 20^\circ$ when evaluated at the ground measurements (white '+'-mark in Figure 2a). For a better comparison to ground measurements, stalk diameter $d = 2a_0$ instead of stalk radius a_0 is shown. Since N and h are of the same order, contours for ϕ_p varying N instead of h will result in sensitivities similar, slightly smaller though, to the ones depicted in Figure 2a.

The sensitivity analysis on ϕ_p for the real ε'_{st} and imaginary ε''_{st} parts of ε_{st} is shown in Figure 2b, where the inset indicates the parameters the model is evaluated at. Here, the contours range from around 0° to 20°, accounting for a larger sensitivity on ε_{st} in relation to that on h. However, $\Delta\phi_p \sim 18^\circ$ when evaluated at the ground measurements, similar to the sensitivity found in Figure 2a.

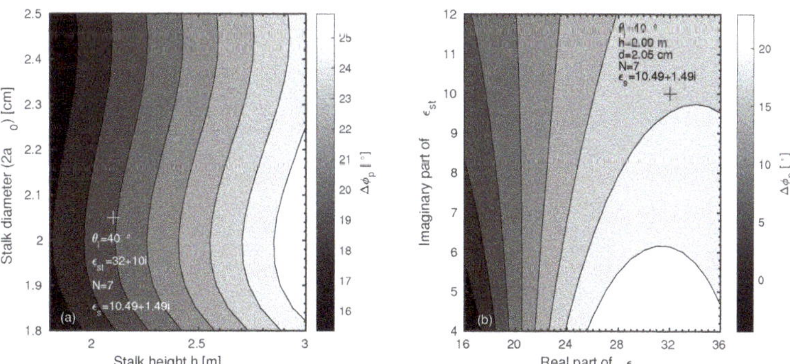

Figure 2. Sensitivity analysis of model parameters. (**a**) Sensitivity of the propagation term on stalk height and diameter $2a_0$. The '+'-mark indicates the average values for the dataset. (**b**) Sensitivity of propagation term on real ε'_{st} and imaginary ε''_{st} parts of ε_{st}.

The bistatic term ϕ_{st} does not depend on h nor N. Moreover, overall the variation of ϕ_{st} on a_0 ranges from -7° to -2° with about -5° of variation with the model evaluated at the ground measurements. Hence, the contribution of the bistatic term to the overall model sensitivity is negligible. Similar results are reached for the sensitivity of ϕ_{st} on ε'_{st} and ε''_{st}.

The aforementioned analysis for $\Delta\phi_p$ and $\Delta\phi_{st}$ considered a fixed incidence angle $\theta_i = 40°$. Computing the contour levels at different angles showed that:

1. At $\theta_i = 20°$, the ϕ_p variations evaluated at the ground measurements are $\Delta\phi_p = 0°$ on the $2a_0, h$-space and $\Delta\phi_p = 2°$ on $\varepsilon'_{st}, \varepsilon''_{st}$-space;
2. Similarly, at $\theta_i = 60°$ resulted in $\Delta\phi_p = 26°$ and $\Delta\phi_p = 28°$;
3. At both $\theta_i = 20°$ and $\theta_i = 60°$, $\Delta\phi_{st}$ is bounded between $-6°$ and $-3°$.

From these remarks, it turns out that sensitivity improved with increasing incidence angles for ϕ_p, whereas the contribution to the overall sensitivity of the ϕ_{st} term is negligible.

2.3. Microwave Dielectric Constant of Stalk from Gravimetric Measurements

The propagation and bistatic terms in (4) and (5) depend on stalk diameter and dielectric constant through the far-field solution in (3). Whereas the collection of a_0 is straightforward from the ground, ε_{st} requires tuned laboratory measurement techniques [31]. From a large set of dedicated measurements, semi-empirical models relating bulk vegetation dielectric properties with vegetation moisture were developed by Ulaby and El Rayes [32], and Mätzer [26]. These are shown in Figure 3 for the vegetation moisture within the stalks. Concerning the range of validity, a slight drawback in Ulaby and El Rayes' model is the upper bound of vegetation moisture used to fit the data. In effect, the model was fitted for gravimetric moisture m_g in the range 0.0–0.7 g/g ([25], Chapter 4-9.2). On the other hand, Mätzer's model [26] comprised measurements with larger moisture contents, which allowed setting an empirical fitting with m_g in the range 0.5–0.9 g/g. Since typical gravimetric moisture for growth corn stalks is larger than 0.6 g/g, Mätzler's model is better suited and will be used here to estimate the dielectric constant of vegetation material within corn stalks. Its input parameters are gravimetric moisture of plant stalks and frequency of the impinging wave. In Figure 3, note the trend in Ulaby and El Rayes' model of larger dielectric values with respect to those in Mätzer's model.

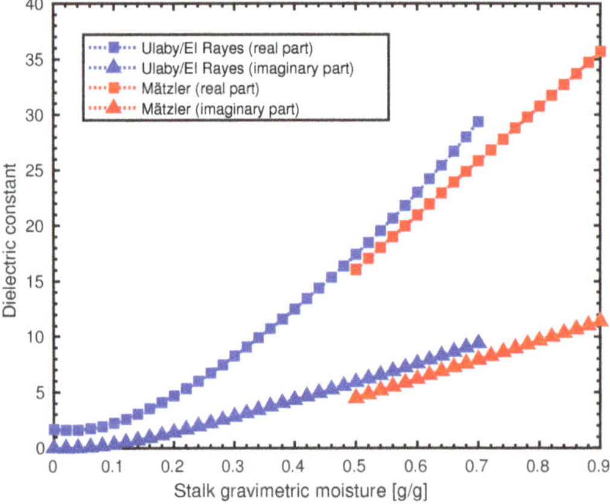

Figure 3. Microwave dielectric model for bulk vegetation from Ulaby and El Rayes [25,32] and from Mätzler [26].

2.4. Study Area and Ground Data Collection

The dataset to fit the Ulaby's incoherent multi-parameter model was taken from the Soil Moisture Active Passive Validation Experiment 2012 (SMAPVEx12) over southwest of Winnipeg, Manitoba, Canada, centered on the town of Elm Creek (98°0′23″ W, 49°40′48″ N) [11,33] and from intensive campaigns led by the SAOCOM mission's science team near the town Monte Buey (32°55′11″ W, 62°27′22″ S), located in central Argentina over Pampas Plain. With these datasets, co-polarized phase measurements on grown corn fields covered incidence angles roughly from 20° to 60°.

In Canada, eight corn fields were imaged by UAVSAR at peak biomass on 17 July 2012. Each acquisition of the UAVSAR comprised four main flight lines with different incidence angles totaling 32 data points. Vegetation was characterized within a one-day window from the flight day: stalk height and diameter were measured. The former was measured with a meter tape and the latter with a caliper over 10 plants within the field selected at random positions. Then, the 10-plant average is computed. Stalk gravimetric moisture was also measured on a less frequent basis, though, due to time-consuming laboratory procedures. Five plants along two rows (ten in total) were collected, bagged appropriately, and then weighted before and after the samples were placed in drying facilities to quantify their water content [11]. A summary of the stalk features for the eight fields in Canada is shown in Table 1.

In Argentina, CONAE has the largest instrumented site over croplands dedicated to calibrating the soil moisture retrieval algorithm for the SAOCOM 1A and 1B mission. In March, April, and June 2017, intensive campaigns over a 140 × 100 km region were conducted and basic ground information over 20 corn fields among other crop types was gathered. Ground information included 0–5-cm soil moisture, stalk height, and till status. Stalk height measurements were taken at convenient positions while the plant was standing in the field. Measurements involving the removal of the plant were disregarded due to time-constraints. The stalk height is summarized in Table 1. These fields were imaged by ALOS-2/PALSAR-2 on several dates totaling 30 data points.

Table 1. Corn stalk features from the ground data collection for two field campaigns in Canada and Argentina.

Feature	Canada (SMAPVex12)	Argentina (CONAE)
# Fields [-]	8	20
# Data points [-]	32	30
Stalk height h range [m]	1.93–2.53	1.80–3.00
Stalk diameter d range [cm]	1.85–2.35	-
Stalk moisture m_g range [g/g]	0.811–0.834	-
Stalk density N [1/m^2]	7.0–8.2	-

2.5. SAR Data and Its Quality and Processing Chain

Airborne UAVSAR provided full polarimetric imagery over Canada with local incidence angles ranging from 20° to 60°. It measured complex scattering coefficients at a frequency of 1.258 GHz. Co-polarized phase measurements are given with a root mean squared phase error ~5.3° and always smaller than 10° [34]. The pixel size on the ground projected image is 5.0 × 7.2 m onto a swath of 20 km.

As read from its metadata, UAVSAR imagery has the coherence matrix as a native image format where $S_{HH}S_{VV}^*$ is readily extracted from. Multi-looked (12 pixels in azimuth by 3 pixels in range) and ground range projected data were used. The ground projection method was nearest neighbor. With the $S_{HH}S_{VV}^*$-images, local incidence angle bands were also provided.

Concurrent with the ground measurements over Argentina, fully polarimetric images were acquired by satellite-borne ALOS-2/PALSAR-2 sensor at 1.236 GHz in High-sensitive Full Polarimetry mode with a 50-km swath width at two incidence angle ranges: 25–30° and 30–35°. This sensor delivered co-polarized phase difference measurements with an imbalance better than 0.618° ([35], Table 3).

For ALOS-2/PALSAR-2, the processing chain started with radiometric calibration from Single Look Complex (SLC) scenes. Subsequently, multilooking was applied (4 pixels in azimuth by 2 pixels in range) to obtain an approximate square pixel and improve the images' radiometric quality. Coherence matrices were computed and then geocoded to a 12 × 12 m ground pixel size using bilinear resampling. As the final product, output bands for complex scattering product $S_{HH}S_{VV}^*$ and for local incidence angle were generated.

2.6. Polarimetric Observable ϕ

With the above-mentioned phase-calibrated images, the derivation of the absolute co-polarized phase difference defined in (1) is given by

$$\phi = arg(S_{HH}S_{VV}^*), \tag{8}$$

where S_{HH} and S_{VV} are the co-polarized complex scattering amplitudes, and * denotes a complex conjugate. In (8), ϕ is defined in the range $-\pi < \phi \leq \pi$. The statistical distribution of ϕ for a speckled image is known, and its closed-form expression is [36]

$$\begin{aligned} P_{\phi}^{(n)}(\phi;\rho_0,\phi_0) = & \frac{\Gamma(n+1/2)(1-\rho_0^2)^n\beta}{2\sqrt{\pi}\Gamma(n)(1-\beta^2)}+ \\ & + \frac{(1-\rho_0^2)^n}{2\pi}{}_2F_1\left(n,1;1/2;\beta^2\right) \end{aligned} \tag{9}$$

with $\beta = \rho_0\cos(\phi - \phi_0)$ where ${}_2F_1\left(n,1;1/2;\beta^2\right)$ is a Gauss hypergeometric function. In (9), ρ_0 is the correlation between S_{HH} and S_{VV}^*, also known as coherence, ϕ_0 is the phase difference of the sample, $\Gamma(\cdot)$ is the gamma function, and n is the equivalent number of looks, which is estimated by means of a matrix-variate estimator based on the trace of the product of the covariance matrix C with itself (tr(CC)), thus using all polarimetric information [37].

3. Results

3.1. Co-Polarized Phase Difference ϕ_0 Estimation

The parameters ϕ_0 and ρ_0 in (9) were estimated using a Maximum Likelihood Estimation (MLE) [38], where (9) is the likelihood function to be maximized constrained to the observed SAR data. The MLE technique applied to multilooked histograms led to the fittings shown in Figure 4. Here, Figure 4a,b display the histogram for a 2.27 m-height corn field imaged with UAVSAR, and a 2.00 m-height corn field imaged with ALOS-2/PALSAR-2, respectively. The number of looks n estimated from the above matrix-variate estimator is also shown. Thus, the co-polarized phase difference estimator ϕ_0 is computed for each sampling site on each acquisition day. Furthermore, uncertainties in the estimates are computed with a 95% confidence level.

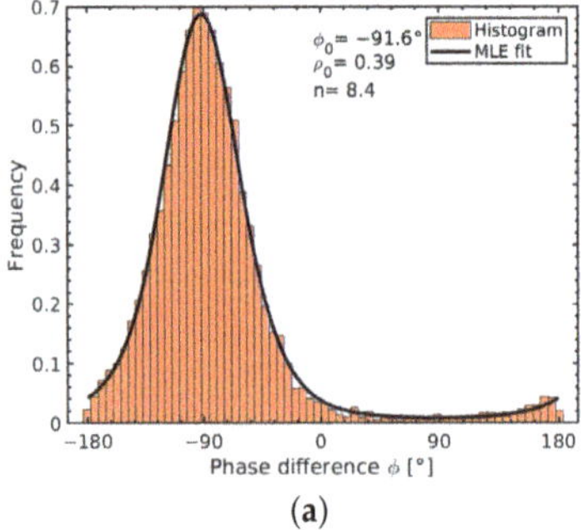

(a)

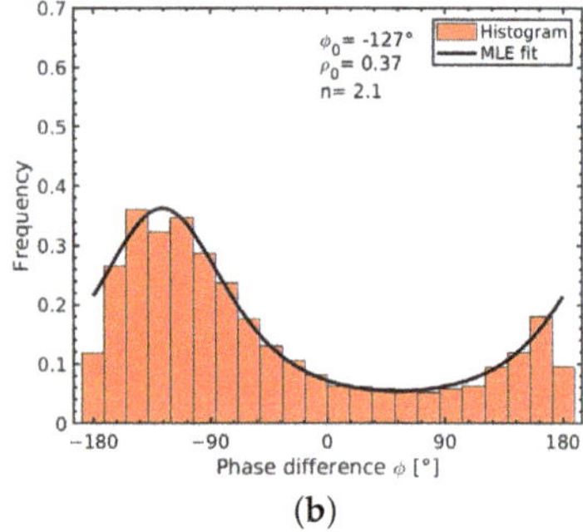

(b)

Figure 4. MLE fitting for speckled co-polarized phase difference histograms. (**a**) A 2.27-m-height corn field imaged by UAVSAR at incidence angle 49.98°. (**b**) A 2.00-m-eight corn field imaged by ALOS-2/PALSAR-2 at incidence angle 26.67°.

3.2. Ulaby's Model Fitting to SAR Data

With the model described in Section 2.1 and the HH-VV phase estimation methodology explained in Section 3.1, a nonlinear least-squares fitting of the measurements against the model is performed, as shown in Figure 5. The upper panel shows the estimated coherence ρ_0 and its uncertainties as bars resulting from the MLE technique. The middle panel shows the fitting itself with the thick black as the best-fitted total co-polarized phase difference ϕ_0. The dotted bands represent the interval defined by the root mean squared error (rmse). Fitted model parameters are also shown. Each term ϕ_p, ϕ_{st}, and ϕ_s is depicted separately in Figure 5c.

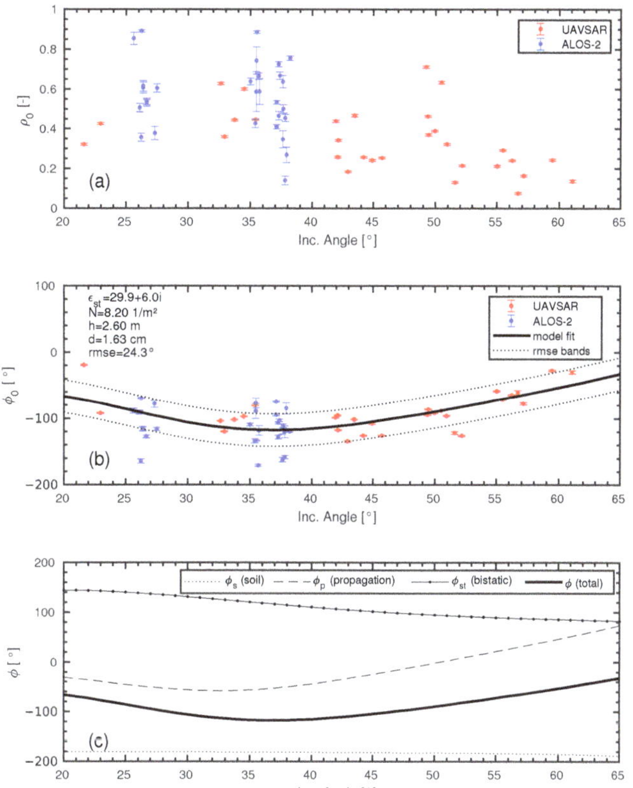

Figure 5. Model fitting by nonlinear least-squares and estimated parameters. (**a**) Coherence ρ_0. (**b**) Co-polarized phase difference ϕ_0 and model fitting. The fitted parameters are indicated. (**c**) Each contribution to the total phase difference is shown separately.

Overall, a good agreement is shown in the view of the dispersion found in the ground measurements, most remarkably in stalk height (see Table 1). A slight overestimation is expected since the corn plant developed above the stalk, resulting in an overall plant structure taller than the stalk itself. Furthermore, the vegetation material within the stalks occupied a smaller volume within the stalk rind, thus leading to an underestimation in the fitted diameter since the outer layer comprising the rind is almost dry. By means of Mätzler's vegetation model, shown in Figure 3, the fitted real part $\varepsilon'_{st} = 29.9$ corresponds to a $m_g = 0.78$ g/g, close to the laboratory-measured $m_g = (0.82 \pm 0.01)$ g/g. Moreover, the ϕ_0 estimates with lower ρ_0 (corresponding to 50° and 65° in Figure 5b) fitted well, indicating that using an MLE technique over the whole histogram resulted in sound, reliable estimates.

With regards to the dependency on the incidence angle shown in Figure 5c, the propagation term becomes relevant for large θ_i due to increasing interaction with dielectric stalks for slanted propagation paths through the corn plants. The soil term sets a reference level almost insensitive to incidence angle variations.

The Canada campaign collected corn parameters regularly, covering most of the development of the corn plants. For further assessment, the same procedure described above was applied to the corresponding UAVSAR dataset on dates before peak biomass on 17 July 2012. Table 2 summarizes the fitted model parameters and the root mean squared error on 5, 8, 14, and 17 July 2012. The corresponding ground measurements are grouped by dates to compensate for missing data, thus covering all eight sites.

While the trend in stalk height seemed to correspond to the plant's growth, the fitted ϕ_0 is somewhat insensitive to stalk diameter, likely due to the sensitivity of the underlying model to these parameters and the dispersion in model parameters. Again, an underestimation of the diameter is expected due to the smaller volume of the vegetation material within the stalks.

On the other hand, the stalk dielectric constant showed a significant sensitivity. With the measured stalk gravimetric moisture shown in Table 2, a straightforward comparison of fitted ε_{st} with m_g is made with the aid of Mätzler's model shown in Figure 3. In effect, as followed from Figure 3, the range 0.834–0.847 g/g corresponded to the ε'_{st} range 32–34, which is in reasonable agreement with the corresponding fitted parameter. Similarly, the range 0.811–0.834 g/g corresponded to ε'_{st} in the range 31–32. The complex dielectric part is governed by the salinity of the vegetation bulk material. Hence, it can be regarded as a second-order effect in relation to the real part dielectric constant and, therefore, its low sensitivity.

Table 2. Stalk features as compared to the fitting and to the ground data from dates prior to peak biomass. Field campaign in Canada with UAVSAR. The dielectric constant on ground data is estimated from stalk moisture by means of Mätzler's model shown in Figure 3.

	Date			
	5 July 2012	**8 July 2012**	**14 July 2012**	**17 July 2012**
Fitted pars.				
Height h [m]	1.42	1.83	2.56	2.60
Diameter d [cm]	1.80	1.80	1.80	1.80
Dielectric constant ε_{st} [-]	30.6 + 6.0i	31.4 + 6.0i	32.0 + 6.0i	24.9 + 6.0i
Density N [1/m^2]	7.15	7.39	8.16	8.20
Root mean sq. error [°]	16.3	20.8	21.8	22.3
Ground data				
Height h range [m]	1.19–1.77		1.93–2.53	
Diameter d range [cm]	2.00–2.29		1.85–2.35	
Moisture m_g range [g/g]	0.834–0.847		0.811–0.834	
Dielectric constant (real part) ε'_{st} [-]	32–34		31–32	

Finally, as mentioned in the introduction, some techniques make the estimation of crop height available. Hence, stalk height might be regarded as a known parameter in specific applications. On the other hand, stalk diameter and gravimetric moisture are plant features that involve time-consuming gathering procedures. If copolarized phase measurements are available over a known corn field at some late stage, Figure 6 can aid in parameter retrieval provided some guess in stalk gravimetric moisture or diameter is at hand. Usually, relationships between diameter and height are available for corn and maize elsewhere (e.g., [39]). The contour levels in Figure 6 were evaluated for several stalk heights by parameterizing the stalk dielectric constant ε_{st} with gravimetric moisture m_g using Mätzler's model.

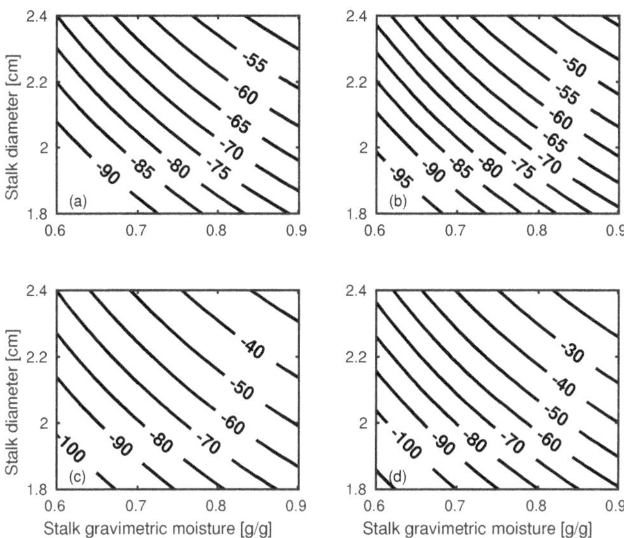

Figure 6. ϕ-contours resulting from the evaluation of (2) when coupling Mätzler's model with Ulaby's. Frequency is fixed at 1.25 GHz and incidence angle is 40°. Model parameters now include m_g instead of ε_{st}. Stalk height h is indicated. (**a**) h = 1.80 m. (**b**) h = 2.00 m. (**c**) h = 2.40 m. (**d**) h = 2.80 m. Contours are in degrees.

4. Discussion

Availability of a fully polarimetric dataset involving airborne and satellite-borne images and stalk dielectric, structural, and spatial parameters enabled a multi-parameter modeling over corn fields. The model considered here for the co-polarized phase difference comprised three incoherent contributions with different sensitivities. Whereas the soil term set an almost constant reference level of around -180°, propagation and bistatic terms had a marked dependency with height, diameter, and moisture of the stalks. By adding them up, the incoherent, interaction-based model fitting showed good agreement with UAVSAR and ALOS-2/PALSAR-2 acquisitions, provided the dispersion in the ground measurements be accounted for. By separating each of the contributions, a more accurate understanding of crop interaction is made, advancing previous research where a full explanation of observation data could not be given since considerable modeling efforts were required [20,24]. Moreover, a number of dedicated radar experiments [15,16,40] with detailed field measurements collected on corn fields can benefit from incorporating a co-polarized phase model to extend their findings to phase-related observables, since modeling efforts associated with these experiments were limited to intensity-related observables only.

More accurate crop scattering models will likely include detailed canopy physical attributes, other than only stalk height and width, such as leaf area index, leaf orientation distribution, and leaf size [41], among others. As a result, a direct relationship of the scattering with plant biophysical parameters might not be easy to develop. On the other hand, scattering models with interaction at higher orders for randomly distributed vertical cylinders rely on Monte Carlo simulations or iterative methods [18]. Thus, the few parameters implied in the Ulaby's model and its straightforward analytical expression highlight its usefulness.

From the sensitivity analysis on Ulaby's model described in Section 2.2, the stalk height resulted in the highest sensitivity on the propagation term ϕ_p for all the incidence angles. This goes in line with the application mentioned at the end of Section 3, where the contours shown in Figure 6 leverage the stalk height retrieval from other remotely-sensed techniques (i.e., [21]) through the improved sensitivity of the term ϕ_p in the total ϕ_0. In this regard, corn height estimates with a root mean square error around 40–50 cm over a growing season were demonstrated with machine learning techniques over a dataset of polarimetric SAR observables at the C-band [21]. This study also highlighted the relevance of polarimetric features related to double-bounce scattering (i.e., ϕ_{st}) [21]. Moreover, model parameterization by stalk gravimetric moisture content instead of its complex dielectric constant using Mätzler's model demonstrated a potential resource for dimensionality reduction, thus helping future application-oriented developments.

Several techniques are usually validated with data from airborne campaigns and then expected to be readily applied with similar levels of accuracy to imagery acquired by orbiting sensors. In the case analyzed in this research, field-based estimates from satellite-borne acquisitions such as those of ALOS-2/PALSAR-2 were clearly constrained by histograms with fewer data points since the larger pixel sizes involved were compared to airborne acquisitions. With the sound histogram-based, matrix-variate Maximum Likelihood Estimation technique described in Section 3.1, the estimates from ALOS-2/PALSAR-2 resulted in slightly larger, otherwise reasonably bounded, uncertainties than UAVSAR ones.

With the increasing availability of L-band space-borne SAR missions adding to existing C-band SAR resources (e.g., European Space Agency's Sentinel-1), multi-frequency methodologies may become fully operational in the near future. The multi-frequency approach exploits different penetration capabilities into the vegetation canopy. For instance, these enhanced capabilities can potentially circumvent typical issues regarding the classification of crops with similar architectures such as corn and sorghum, the latter widely spread in America and Africa. To a greater extent, if multi-frequency polarimetric SAR data become available, polarimetric modeling such as the Ulaby–Mätzler model can enhance further corn plant parameter retrieval.

5. Conclusions

Research on crop scattering processes can primarily benefit from fully polarimetric data. In addition to usual power scattering coefficients, a promising polarimetric observable for crop monitoring is the phase difference between the co-polarized complex scattering amplitudes. By leveraging the penetration capabilities at the L-band, fully polarimetric SAR missions become worthwhile over croplands. This study presents a scattering model coupled with a semi-empirical dielectric model for co-polarized phase differences resulting from the interaction of microwaves with grown corn canopies. The dataset included airborne and space-borne fully polarimetric SAR data with incidence angles ranging from 20° to 60°. A set of 60 data points was analyzed and used to perform an experimental data fitting with a nonlinear least-squares technique. The results showed a satisfactory agreement for corn co-polarized phase differences at the field scale, with an RMSE of around 24.3° considering airborne and space-borne acquisitions. Compared with available studies on corn phase differences with SAR data, this research provides a new perspective on using phase-related observables on fully polarimetric SAR data over corn fields.

Author Contributions: Conceptualization, M.E.B. and C.L.-M.; formal analysis, M.E.B.; funding acquisition, C.L.-M.; investigation, M.E.B.; methodology, M.E.B.; resources, M.E.B.; software, D.S.R.; validation, M.E.B. and D.S.R.; visualization, M.E.B.; writing—original draft, M.E.B. and D.S.R.; writing—review and editing, C.L.-M. All authors have read and agreed to the published version of the manuscript.

Funding: This research was partially funded by the Argentinean National Scientific and Technical Research Council (CONICET, project PICT 2015 N°3689), by the Spanish Ministry of Science and Innovation (project CICYT RTI2018-099008-B-C21/AEI/10.13039/501100011033 "Sensing with pioneering opportunistic techniques") and by the grant to "CommSensLab-UPC" Excellence Research Unit Maria de Maeztu (MINECO grant).

Institutional Review Board Statement: Not applicable.

Informed Consent Statement: Not applicable.

Data Availability Statement: The data are not publicly available due to license restrictions.

Acknowledgments: Special thanks to Heather McNairn and CONAE for sharing part of the Canada and Argentina ground data, respectively. The authors acknowledged Avik Bhattacharya for revising the manuscript and for his valuable comments.

Conflicts of Interest: The authors declare no conflict of interest.

References

1. McNairn, H.; Brisco, B. The application of C-band polarimetric SAR for agriculture: A review. *Can. J. Remote Sens.* **2004**, *30*, 525–542. [CrossRef]
2. Li, H.; Zhang, C.; Zhang, S.; Atkinson, P.M. Crop classification from full-year fully-polarimetric L-band UAVSAR time-series using the Random Forest algorithm. *Int. J. Appl. Earth Obs. Geoinf.* **2020**, *87*, 102032. [CrossRef]
3. Larrañaga, A.; Álvarez Mozos, J. On the Added Value of Quad-Pol Data in a Multi-Temporal Crop Classification Framework Based on RADARSAT-2 Imagery. *Remote Sens.* **2016**, *8*, 335. [CrossRef]
4. Erten, E.; Lopez-Sanchez, J.M.; Yuzugullu, O.; Hajnsek, I. Retrieval of agricultural crop height from space: A comparison of SAR techniques. *Remote Sens. Environ.* **2016**, *187*, 130–144. [CrossRef]
5. Lopez-Sanchez, J.M.; Vicente-Guijalba, F.; Erten, E.; Campos-Taberner, M.; Garcia-Haro, F.J. Retrieval of vegetation height in rice fields using polarimetric SAR interferometry with TanDEM-X data. *Remote Sens. Environ.* **2017**, *192*, 30–44. [CrossRef]
6. Ndikumana, E.; Ho Tong Minh, D.; Dang Nguyen, H.T.; Baghdadi, N.; Courault, D.; Hossard, L.; El Moussawi, I. Estimation of Rice Height and Biomass Using Multitemporal SAR Sentinel-1 for Camargue, Southern France. *Remote Sens.* **2018**, *10*, 1394. [CrossRef]
7. Barber, M.; Bruscantini, C.; Grings, F.; Karszenbaum, H. Bayesian Combined Active/Passive (B-CAP) Soil Moisture Retrieval Algorithm. *IEEE J. Sel. Top. Appl. Earth Obs. Remote Sens.* **2016**, *9*, 5449–5460. [CrossRef]
8. Das, N.N.; Entekhabi, D.; Dunbar, R.S.; Colliander, A.; Chen, F.; Crow, W.; Jackson, T.J.; Berg, A.; Bosch, D.D.; Caldwell, T.; et al. The SMAP mission combined active-passive soil moisture product at 9 km and 3 km spatial resolutions. *Remote Sens. Environ.* **2018**, *211*, 204–217. [CrossRef]
9. Hensley, S.; Michel, T.; Van Zyl, J.; Muellerschoen, R.; Chapman, B.; Oveisgharan, S.; Haddad, Z.S.; Jackson, T.; Mladenova, I. Effect of Soil Moisture on polarimetric-interferometric repeat pass observations by UAVSAR during 2010 Canadian Soil Moisture campaign. In Proceedings of the 2011 IEEE International Geoscience and Remote Sensing Symposium, Brussels, Belgium, 11–16 July 2011; pp. 1063–1066. [CrossRef]
10. Morandeira, N.S.; Franco, M.; Barber, M.; Grings, F. Modeling Bare Soil L-Band Polarimetric H-α Values Using a Second-Order SPM Model. *IEEE Geosci. Remote Sens. Lett.* **2016**, *13*, 399–403. [CrossRef]
11. McNairn, H.; Jackson, T.J.; Wiseman, G.; Bélair, S.; Berg, A.; Bullock, P.; Colliander, A.; Cosh, M.H.; Kim, S.B.; Magagi, R.; et al. The Soil Moisture Active Passive Validation Experiment 2012 (SMAPVEX12): Prelaunch Calibration and Validation of the SMAP Soil Moisture Algorithms. *IEEE Trans. Geosci. Remote Sens.* **2015**, *53*, 2784–2801. [CrossRef]
12. NASA Jet Propulsion Laboratory. *NASA-ISRO SAR (NISAR) Mission Science Users' Handbook*; NASA Jet Propulsion Laboratory: Pasadena, CA, USA, 2018; p. 261.
13. ESA. Copernicus L-Band SAR Mission Requirements Document. Available online: https://esamultimedia.esa.int/docs/EarthObservation/Copernicus_L-band_SAR_mission_ROSE-L_MRD_v2.0_issued.pdf (accessed on 8 November 2021).
14. FAOSTAT: The Food and Agriculture Organization (FAO) Statistics. Available online: http://www.fao.org/faostat/en/#data/QC (accessed on 25 February 2021).
15. Joseph, A.T.; van der Velde, R.; O'Neill, P.E.; Lang, R.H.; Gish, T. Soil Moisture Retrieval During a Corn Growth Cycle Using L-Band (1.6 GHz) Radar Observations. *IEEE Trans. Geosci. Remote Sens.* **2008**, *46*, 2365–2374. [CrossRef]
16. Sharma, A.; Lang, R.H.; Kurum, M.; O'neill, P.E.; Cosh, M.H. L-Band Radar Experiment and Modeling of a Corn Canopy Over a Full Growing Season. *IEEE Trans. Geosci. Remote Sens.* **2020**, *58*, 5821–5835. [CrossRef]
17. Ulaby, F.T.; Held, D.; Dobson, M.C.; McDonald, K.C.; Senior, T.B.A. Relating Polarization Phase Difference of SAR Signals to Scene Properties. *IEEE Trans. Geosci. Remote Sens.* **1987**, *GE-25*, 83–92. [CrossRef]
18. Sarabandi, K.; Polatin, P.F.; Ulaby, F.T. Monte Carlo simulation of scattering from a layer of vertical cylinders. *IEEE Trans. Antennas Propag.* **1993**, *41*, 465–475. [CrossRef]
19. Karam, M.A.; Fung, A.K. Electromagnetic scattering from a layer of finite length, randomly oriented, dielectric, circular cylinders over a rough interface with application to vegetation. *Int. J. Remote Sens.* **1988**, *9*, 1109–1134. [CrossRef]
20. Skriver, H.; Svendsen, M.T.; Thomsen, A.G. Multitemporal C- and L-band polarimetric signatures of crops. *IEEE Trans. Geosci. Remote Sens.* **1999**, *37*, 2413–2429. [CrossRef]
21. Xie, Q.; Wang, J.; Lopez-Sanchez, J.M.; Peng, X.; Liao, C.; Shang, J.; Zhu, J.; Fu, H.; Ballester-Berman, J.D. Crop Height Estimation of Corn from Multi-Year RADARSAT-2 Polarimetric Observables Using Machine Learning. *Remote Sens.* **2021**, *13*, 392. [CrossRef]
22. Cable, J.W.; Kovacs, J.M.; Jiao, X.; Shang, J. Agricultural Monitoring in Northeastern Ontario, Canada, Using Multi-Temporal Polarimetric RADARSAT-2 Data. *Remote Sens.* **2014**, *6*, 2343–2371. [CrossRef]
23. Lopez-Sanchez, J.M.; Ballester-Berman, J.D.; Hajnsek, I. First Results of Rice Monitoring Practices in Spain by Means of Time Series of TerraSAR-X Dual-Pol Images. *IEEE J. Sel. Top. Appl. Earth Obs. Remote Sens.* **2011**, *4*, 412–422. [CrossRef]
24. Balenzano, A.; Mattia, F.; Satalino, G.; Davidson, M. Dense Temporal Series of C- and L-band SAR Data for Soil Moisture Retrieval Over Agricultural Crops. *Sel. Top. Appl. Earth Obs. Remote Sens. IEEE J.* **2011**, *4*, 439–450. [CrossRef]

25. Ulaby, F.T.; Long, D.; Blackwell, W.; Elachi, C.; Fung, A.; Ruf, C.; Sarabandi, K.; van Zyl, J.; Zebker, H. *Microwave Radar and Radiometric Remote Sensing*; The University of Michigan Press: Ann Arbor, MI, USA, 2014.
26. Mätzler, C. Microwave (1–100 GHz) dielectric model of leaves. *IEEE Trans. Geosci. Remote Sens.* **1994**, *32*, 947–949. [CrossRef]
27. Ruck, G. *Radar Cross Section Handbook*; Number v. 2 in Radar Cross Section Handbook; Plenum Press: New York, NY, USA, 1970.
28. Hallikainen, M.; Ulaby, F.; Dobson, M.; El-Rayes, M.; Wu, L.K. Microwave Dielectric Behavior of Wet Soil-Part I: Empirical Models and Experimental Observations. *IEEE Trans. Geosci. Remote Sens.* **1985**, *GE-23*, 25–34. [CrossRef]
29. Peplinski, N.R.; Ulaby, F.T.; Dobson, M.C. Dielectric properties of soils in the 0.3–1.3-GHz range. *IEEE Trans. Geosci. Remote Sens.* **1995**, *33*, 803–807. [CrossRef]
30. Mironov, V.L.; Dobson, M.C.; Kaupp, V.H.; Komarov, S.A.; Kleshchenko, V.N. Generalized refractive mixing dielectric model for moist soils. *IEEE Trans. Geosci. Remote Sens.* **2004**, *42*, 773–785. [CrossRef]
31. El-Rayes, M.A.; Ulaby, F.T. Microwave Dielectric Spectrum of Vegetation—Part I: Experimental Observations. *IEEE Trans. Geosci. Remote Sens.* **1987**, *GE-25*, 541–549. [CrossRef]
32. Ulaby, F.T.; El-Rayes, M.A. Microwave Dielectric Spectrum of Vegetation—Part II: Dual-Dispersion Model. *IEEE Trans. Geosci. Remote Sens.* **1987**, *GE-25*, 550–557. [CrossRef]
33. SMAP Validation Experiment 2012, Experimental Plan. Available online: https://smapvex12.espaceweb.usherbrooke.ca (accessed on 22 September 2021).
34. Fore, A.; Chapman, B.; Hawkins, B.; Hensley, S.; Jones, C.; Michel, T.; Muellerschoen, R. UAVSAR Polarimetric Calibration. *Geosci. Remote Sens. IEEE Trans.* **2015**, *53*, 3481–3491. [CrossRef]
35. Motohka, T.; Isoguchi, O.; Sakashita, M.; Shimada, M. Results of ALOS-2 PALSAR-2 Calibration and Validation After 3 Years of Operation. In Proceedings of the IGARSS 2018—2018 IEEE International Geoscience and Remote Sensing Symposium, Valencia, Spain, 22–27 July 2018; pp. 4169–4170.
36. Lee, J.S.; Pottier, E. *Polarimetric Radar Imaging: From Basics to Applications*; CRC Press: Boca Raton, FL, USA, 2009. [CrossRef]
37. Anfinsen, S.; Doulgeris, A.; Eltoft, T. Estimation of the Equivalent Number of Looks in Polarimetric Synthetic Aperture Radar Imagery. *Geosci. Remote Sens. IEEE Trans.* **2009**, *47*, 3795–3809. [CrossRef]
38. Wasserman, L. *All of Statistics: A Concise Course in Statistical Inference*; Springer Texts in Statistics; Springer: New York, NY, USA, 2004.
39. Umburanas, R.C.; Kawakami, J.; Eschemback, V.; Ribeiro Oliari, I.C. Effect of green manure and sources of phosphorus on the morphology and yield of maize plants. *Braz. J. Appl. Technol. Agric. Sci.* **2013**, *6*, 49–56. [CrossRef]
40. Monsivais-Huertero, A.; Liu, P.W.; Judge, J. Phenology-Based Backscattering Model for Corn at L-Band. *IEEE Trans. Geosci. Remote Sens.* **2018**, *56*, 4989–5005. [CrossRef]
41. Bracaglia, M.; Ferrazzoli, P.; Guerriero, L. A fully polarimetric multiple scattering model for crops. *Remote Sens. Environ.* **1995**, *54*, 170–179. [CrossRef]

MDPI

Article

SAR-Based Flood Monitoring for Flatland with Frequently Fluctuating Water Surfaces: Proposal for the Normalized Backscatter Amplitude Difference Index (NoBADI)

Hiroto Nagai [1,*], Takahiro Abe [2] and Masato Ohki [3]

[1] School of Education, Waseda University, 1-6-1 Nishiwaseda, Shinjuku-ku, Tokyo 169-0051, Japan
[2] Graduate School of Bioresources, Mie University, 1577 Kurimamachiya-cho, Tsu-shi 514-8507, Mie Prefecture, Japan; abe@bio.mie-u.ac.jp
[3] Earth Observation Research Center, Japan Aerospace Exploration Agency, 2-1-1 Sengen, Tsukuba-shi 305-8505, Ibaraki Prefecture, Japan; ohki.masato@jaxa.jp
* Correspondence: nagai.hiroto@aoni.waseda.jp; Tel.: +81-3-3208-8370

Abstract: Space-based synthetic aperture radar (SAR) is a powerful tool for monitoring flood conditions over large areas without the influence of clouds and daylight. Permanent water surfaces can be excluded by comparing SAR images with pre-flood images, but fluctuating water surfaces, such as those found in flat wetlands, introduce uncertainty into flood mapping results. In order to reduce this uncertainty, a simple method called Normalized Backscatter Amplitude Difference Index (NoBADI) is proposed in this study. The NoBADI is calculated from a post-flood SAR image of backscatter amplitude and multiple images on non-flooding conditions. Preliminary analysis conducted in the US state of Florida, which was affected by Hurricane Irma in September 2017, shows that surfaces frequently covered by water (more than 20% of available data) have been successfully excluded by means of C-/L-band SAR (HH, HV, VV, and VH polarizations). Although a simple comparison of pre-flood and post-flood images is greatly affected by the spatial distribution of the water surface in the pre-flood image, the NoBADI method reduces the uncertainty of the reference water surface. This advantage will contribute in making quicker decisions during crisis management.

Keywords: SAR; ALOS-2; PALSAR-2; Sentinel-1; flood; NoBADI; Florida; Hurricane Irma

Citation: Nagai, H.; Abe, T.; Ohki, M. SAR-Based Flood Monitoring for Flatland with Frequently Fluctuating Water Surfaces: Proposal for the Normalized Backscatter Amplitude Difference Index (NoBADI). *Remote Sens.* **2021**, *13*, 4136. https://doi.org/10.3390/rs13204136

Academic Editor: Renaud Hostache

Received: 31 August 2021
Accepted: 12 October 2021
Published: 15 October 2021

Publisher's Note: MDPI stays neutral with regard to jurisdictional claims in published maps and institutional affiliations.

1. Introduction

Serious flooding events are caused by extreme weather conditions and seasonal tropical cyclones, and it is important for affected societies to respond quickly to disasters based on spatial awareness of the observed flood extent. Space-borne Synthetic Aperture Radar (SAR), a remote sensing technology from space, is an advanced solution for monitoring large-scale flood disasters extensively and quickly that is not hampered by cloud cover. The easiest method for identifying a flood surface is to extract low backscatter pixels, which in this case includes the permanent water surface. To exclude the permanent water surfaces such as rivers and lakes, significant backscatter reduction from a pre-flood SAR image would be extracted as flood-derived water surfaces [1–4]. Advanced methods analyze phase information based on the interferometric SAR (InSAR) technology. By means of L-band, C-band, and X-band SAR, not only basic experiments but also operational uses for actual crisis responses have been carried out [5,6].

When the spatial distribution of the water surface is not stable, the difference between the pre-flood and post-flood images becomes uncertain. In particular, low-elevation and flat terrains in the tropics tend to form numerous ponds and lakes with fluctuating water bodies due to intensive monsoon precipitation and other seasonal features. In such a case, different results of flood extraction would be obtained according to the acquisition dates of pre-flood images. This uncertainty might cause difficulties in rapid image analysis during an actual operation of crisis response.

One idea for reducing the uncertainty is to quantify the frequency of water extent in each pixel by considering the average and variability during non-flooding periods. Calculations of these statistical values are common in SAR-related studies, and examples include detecting anomaly pixels of the forest [7]. In this study, therefore, we focus on a simple method calculated from the mean and standard deviation values of multiple non-flooding conditions in addition to a post-flood image, which we call the Normalized Backscatter Amplitude Difference Index (NoBADI) in this study. The objective of this study is to evaluate the benefits of NoBADI applied to hurricane-derived flood areas formed around fluctuating water distributions. The obtained results are assessed by optically derived reference maps of water bodies and other SAR-based results, then possibilities and challenges of NoBADI are discussed.

2. Study Site

In this study, we chose the Florida peninsula in the United States as our research site. The study site is rectangular in shape (8.219–8.227° W; 27.281–27.346° N) located 30 km east of Sarasota, upstream of the Myakka River State Park. The site contains the Lettuce Lake (Figure 1). Another rectangle is defined as a validation site in order to assess classification accuracy.

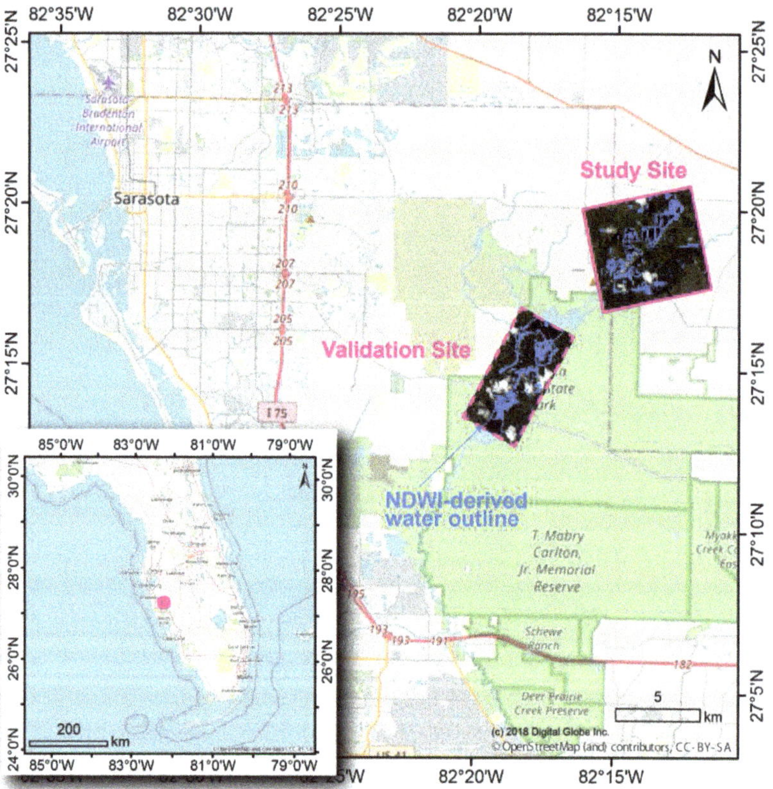

Figure 1. Location of the study site with a validation site shown as pink rectangles. Basemap is derived from OpenStreetMap.

This peninsula is located between the Gulf of Mexico and the Atlantic Ocean. The general climate of the region is characterized by a humid subtropical climate, with rainy seasons from May to October and a dry season from October to the end of April. Average annual precipitation reaches 1137 mm, of which 193 mm is the maximum monthly precipitation in August. Surface temperatures range from 11 to 32 degrees Celsius in Tampa [8].

The topography is dominated by flat terrain and is characterized by the longest coastline in the states of the U.S. (2170 km), where more than half of that is classified as sand beach. Most surfaces in the Florida peninsula have elevations below 5 m. The highest place is the Sugarloaf Mountain (95 m) in the Lake Wales Ridge, located in the center of the peninsula.

The state of Florida, with a population of 19.6 million, was severely damaged by Hurricane Irma. The hurricane formed on the West coast of Africa on 27 August 2017 [9]. It landed in Florida as a Category 4 hurricane on September 10 [10]. Power outages due to the hurricane affected 6.7 million people [11,12], which is equivalent to 36 % of the total population in Florida [12]. Six point five million people were ordered to evacuate and there was a serious problem of traffic congestion [13]. Social disruption caused by the hurricane caused the direct death of 11 people, whereas 115 deaths were indirectly caused by deterioration of the medical environment [14]. The hurricane disappeared and became a tropical storm on 11 September west of Gainesville, Florida [9].

3. Data

3.1. Reference Optical Imagery

Optical satellite imagery collected by WorldView-2 (WV2) derives spatial distribution of post-flood water bodies. The satellite is equipped with a high-resolution optical sensor called the WorldView-110 camera. Satellite operation started in 2009 by DigitalGlobe and is still continued by MAXAR for mostly commercial use and those related to activities for national security. The optical sensor contains eight multispectral bands (coastal, blue, green, yellow, red, red edge, and near-infrared (2)) and one visible panchromatic instrument, with the spatial resolution of 1.85 m and 0.46 m at the nadir angle, respectively [15].

A post-flood WV2 imagery after Hurricane Irma passed was purchased from a Japanese sales agent, Japan Space Imaging. Imagery collection was carried out at 15:59 (UTC), 12 September 2017 under 3% cloud cover condition. The given data consist of a part of multi-spectral bands (blue: 450–510 nm; green 510–580 nm; red 630–690 nm; near-infrared 770–895 nm). Those of product level 2A (standard) after radiometric correction were resampled to 2 m pixel spacing (nearest neighbor) for a WGS84/UTM projection as a 16 bit digital number.

The possible range of water distribution which was not influenced by flood hazards is determined by another optical satellite imagery collected by Sentinel-2 (S2). Identical satellites of Sentinel-2A and Sentinel-2B are operated by the European Space Agency in the Copernicus program to distribute collected data free of charge. The main sensor is a Multi Spectral Instrument (MSI), which contains 12 multispectral bands from coastal to short-wave infrared with the spatial resolutions of 10, 20, and 60 m [16]. Single satellite operations began on 23 June 2015, and dual operations began on 7 March 2017. The revisit period for a single satellite is 10 days. The approximate observation time is 10:30 a.m. local time.

We collected and processed all available S2 data since 27 October 2015 (~325 scenes) in a cloud platform for remote sensing using Google Earth Engine (GEE) (Table S1). These data were observed along the relative orbit number of 97, which was adequate for the study site. Those in product level 2A, surface reflectance at the bottom of atmosphere (after atmospheric correction), were resampled in 10 m (equivalent pixel sizes in a geographic coordination system).

3.2. SAR Data

First, L-band SAR data were collected by the Phased Array type L-band Synthetic Aperture Radar-2 (PALSAR-2) aboard the Advanced Land Observing Satellite-2 (ALOS-2). ALOS-2 was launched on 24 May 2014 and is operated by the Japan Aerospace Exploration Agency (JAXA) for distributing SAR imagery for commercial use or free of charge. Four scenes before the flood (20 January 2015, 1 September 2015, 19 January 2016, and 30 August 2016) and one scene after the flood (12 September 2017) were observed at 5:32 (UTC) from Path 47 (Frame 0530) with an off-nadir angle of 36.6 degrees. Stripmap (fine beam) mode at 5.3/9.1 m spatial resolutions in azimuth/range directions with HH-HV dual polarization (SM3) was selected for these observations.

For the same five scenes, data in two product levels are separately and independently selected for different processes. One is backscatter amplitude image which is ortho-rectified and geocoded by means of SRTM-90 m (v4.1) (i.e., JAXA's standard product, level 2.1) and resampled (bi-linear) to a pixel spacing of 6.25 m at an adequate WGS84/UTM (zone 17N) projection. Another includes Single Look Complex (SLC) data (i.e., JAXA's standard product, level 1.1) in which phase information is included for an interferogram. We obtained amplitude (L2.1) data as a 16-bit digital number using the Geotiff format and SLC (L1.1) data in CEOS format.

Second, C-band SAR data were collected by Sentinel-1A/B satellites (S1). These are identical two satellites launched on 3 April 2014 (A) and 25 April 2016 (B) for constellational operation by ESA in the Copernicus program. We collected and processed all available data (~143 scenes) in GEE free of charge (Table S2), which contains data observed on the closest day (15 September 2017) to the flood event (10 September 2017). These data were observed along the relative orbit of 121 (ascending direction) adequate for the study site. ScanSAR (IWS: Interferometric Wide Swath) mode at 20/22m spatial resolutions in azimuth/range directions with VV-VH dual polarization was selected for these S1 observations. The acquired amplitude data were ortho-rectified and resampled to pixel spacing of 10 m (equivalent pixel sizes in a geographic coordination system) as Level-1 Ground Range Detected (GRD) products before our process.

4. Methodology

4.1. Water Surface during the Flood

The post-flood water bodies as references are extracted from the WV2 image (Figure 2). Surface reflectance is calculated from the digital number using conversion parameters explained in [17]. The Normalized Difference Water Index (NDWI) for open water surface proposed by [18] is calculated as follows:

$$\mathrm{NDWI} = \frac{\mathrm{G} - \mathrm{N}}{\mathrm{G} + \mathrm{N}}, \tag{1}$$

where G and N are the surface reflectance in the green and near-infrared bands. NDWI is then binarized with a threshold of 0.0, where higher values correspond to the water surface.

4.2. Water Surface Not/Highly Related to the Flood

We generated a water extent map with non-flood status. In the study site, a part of numerous potential water bodies may change their extent due to low-and-flat topography and tropical precipitation. Therefore, only the fixed map of water extent was difficult to determine. Instead, the spatial distribution of the frequency of water cover is quantified in this study. A large number of NDWI images were stacked to quantify water-inundated opportunity at 0 to 100%.

Cloud-covered pixels are excluded with an attached quality band "QA60". NDWI is then calculated from the green (Band 3) and the near-infrared (Band 8) bands. Binarization of NDWI values with the threshold of 0.0 reclassifies water pixels from other land covers. In each location of the pixels, the number of water pixels divided by total available scene number denotes water frequency (WF). Pixels with permanent water will have a value of 1.0, and pixels with permanent land will have a value of 0.0. WF ranges from 0.0 to 1.0, with larger values for pixels that are covered by water more frequently.

The unique water extent caused by the hurricane Irma (HWE: Hurricane-induced Water Extent) is defined as the difference of WV2-derived water extent and S2-derived frequent water extent with a certain WF value. In the results, HWE is generated based on pre-flood water surface of WF > 0.2.

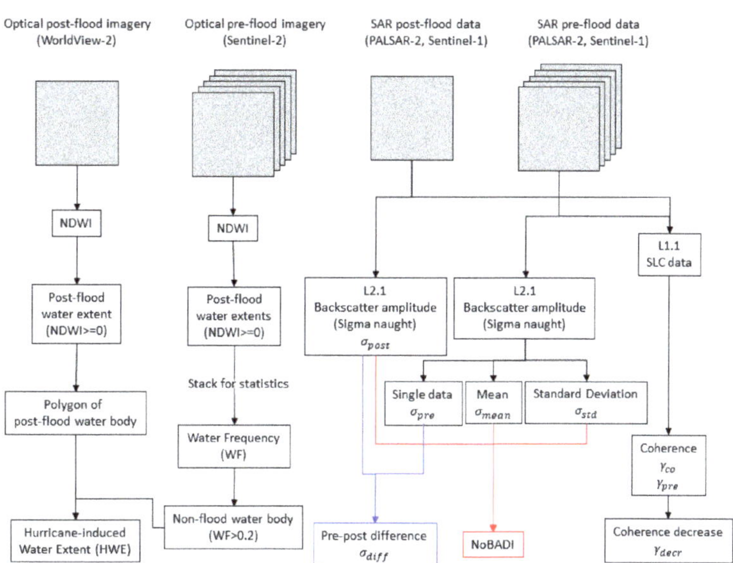

Figure 2. Overall data processing flow in the methodology.

4.3. L-Band SAR Processing

We calculate NoBADI and other outputs from PALSAR-2 data for flood mapping. As initial processing, backscatter amplitude (i.e., sigma naught) was calculated from the given digital number (DN) in L2.1 data as follows.

$$\sigma = 10\log_{10} \mathrm{DN}^2 - 83. \tag{2}$$

Low-pass filter (mean of 3 × 3 pixels) for speckle noise reduction is conducted, then the pixels are resampled (Nearest Neighbor) relative to the projection of WGS84/UTM (Zone 17N).

This study proposes NoBADI, which expresses how rare the water cover is in each pixel. This value should be intensive/weakened when water covers hardly/frequently inundated places. Non-flood mean value and standard deviation value are calculated from four pre-flood images, which are all of available Stripmap-mode scenes with an equal off-nadir angle in this location. They are combined with post-flood pixel value in order to calculate NoBADI with the following equation:

$$\mathrm{NoBADI} = \frac{\sigma_{\mathrm{post}} - \sigma_{\mathrm{mean}}}{\sigma_{\mathrm{std}}}, \tag{3}$$

where σ_{post} is the post-flood backscatter amplitude (i.e., sigma naught), σ_{mean} is the mean of the non-flood backscatter amplitudes, and σ_{std} is the standard deviation of the non-flood backscatter amplitudes (Figure 2).

The standard deviation shows smaller values for stable land surfaces and larger values for frequently fluctuating surfaces. This method moderates the usual water extent and enhances irregular water extent, even if these have similar values in terms of amplitude decrease over the hurricane-induced flood.

In order to evaluate NoBADI results, general methods of water body extraction are also performed. The simple difference of backscatter amplitude is calculated as follows:

$$\sigma_{diff} = \sigma_{post} - \sigma_{pre}, \tag{4}$$

where σ_{pre} is one of four scenes. Four results based on different pre-flood dates are calculated and compared.

In addition, coherence and coherence decreases are calculated by means of SLC (L1.1) data. Interferometric coherence, γ, can be calculated as follows:

$$\gamma = \frac{|\langle s_1 s_2^* \rangle|}{\sqrt{\langle s_1 s_1^* \rangle \langle s_2 s_2^* \rangle}}, \tag{5}$$

where s1 and s2 are a pair of single look complex (SLC) SAR data, s* represents the complex conjugate of s, and the bracket ⟨ ⟩ indicates averaging in a local window (4 × 4 pixels in this study). Coherence represents the similarity between two SAR images, and it decreases as the state of the ground surface changes.

To cancel non-hazardous changes from the coherence output, two coherence images including a pre-flood pair (19 January and 30 August 2016), coherence γ_{pre}, and a pre-flood and post-flood pair (12 September 2017 and 30 August 2016), coherence γ_{co}, were used. The coherence decrease, γ_{decr}, is calculated as follows.

$$\gamma_{decr} = \gamma_{co} - \gamma_{pre}. \tag{6}$$

4.4. C-Band SAR Processing

Furthermore, we calculated NoBADI from S1 data for flood mapping. Backscatter amplitude (i.e., sigma naught) was ready in GEE. NoBADI is calculated as proposed above, where all available scenes in GEE are stacked in order to calculate the mean and the standard deviation. After the processing, pixels are resampled (Nearest Neighbor) relative to the projection of WGS84/UTM (Zone 17N).

5. Results

5.1. Optical Water Mapping for Reference

Most parts of the post-flood water bodies in the study site are correctly extracted from the WV2-derived NDWI image and can be observed in visual comparison with the WV2 true color image (Figure 3a). More than six relatively large water bodies (ID numbers attached) and many smaller areas have been extracted. Some rather smaller portions (~10 m) of the water bodies are excluded, probably due to the mixed-pixel effect. Subsequent S2 and SAR data have larger pixel sizes than compared to these small water bodies; therefore, this exclusion does not affect the research objectives.

Potential water bodies, which are not related to hazardous conditions, are obtained by stacking the S2-NDWI images and quantifying the frequency of the presence of water bodies per pixel (0–100%) (Figure 3b). There are several locations with high WF values above 0.2 (indicated by the black outlines in Figure 3b), and slightly lower values are distributed around them. These locations are also covered by water during floods and correspond to the water areas of ID:1–6 (Figure 3a). Therefore, these locations are likely to be potential water bodies (lakes and/or ponds) with some area fluctuation. Scattered portions showing 0–0.01% values of WF (in darker gray) are distributed homogeneously in the study site. These do not correspond to post-flood water-body distribution and potential land cover.

5.2. Single Post-Flood SAR Imagery

Simple visual interpretation of a post-flood SAR image suggests spatial distribution of post-flood water bodies, including potential water extent. Color composite images of PALSAR-2(Figure 4a) and S1 (Figure 4b) dual polarizations denote multiple domains in blue, which is emphasized by their flat surface with smaller volumetric scatter as same as smaller backscatter. Some of these extents correspond to WV2-derived post-flood water distributions (blue outlines as same as Figure 3). The PALSAR-2 image has higher contrast and has easy to visually identify flat surfaces (in blue) than compared to the S1 image. On the other hand, the S1 image emphasizes water bodies that have been identified by WV2 data (blue outlines).

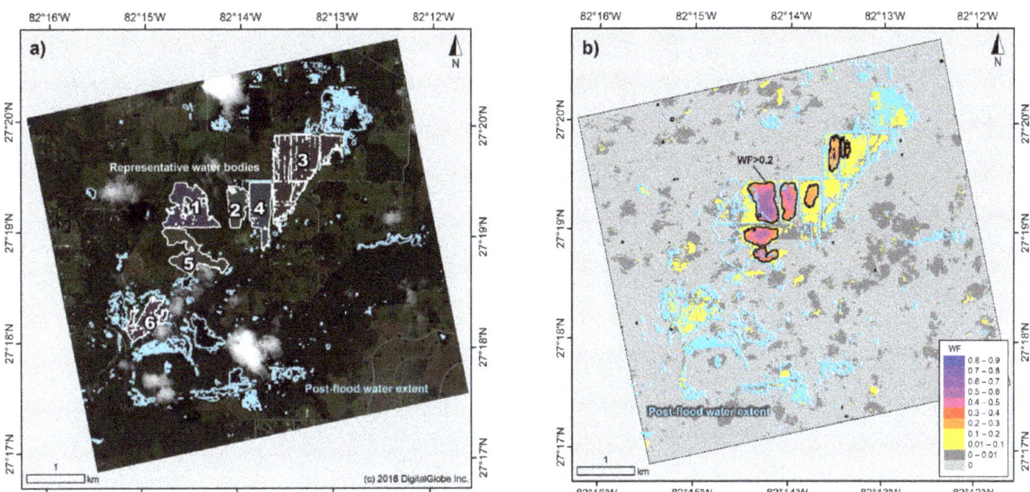

Figure 3. Optically derived water extent (**a**) after flood (12 September 2017) and (**b**) in general status shown in WF. ID numbers are assigned in representative six post-flood water bodies. Contour lines in black denote an example coastline of usual water extent (>0.2).

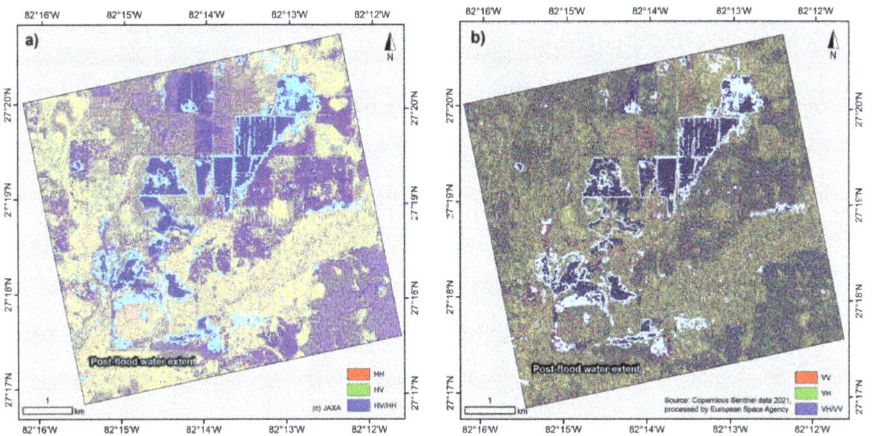

Figure 4. Post-flood amplitude imagery with RGB composite of dual polarization by (**a**) PALSAR-2 (12 September 2017) and (**b**) S1 (15 September 2015).

5.3. NoBADI

Spatial distributions of NoBADI in the study site were calculated from PALSAR-2 HH (Figure 5a) and HV (Figure 5b) polarizations as well as from S1 VV (Figure 5c) and VH (Figure 5d) polarizations, respectively. All images show higher values (zero or greater) that are mostly outside the WV2-derived flood portions (in blue/white outlines), whereas the majority of negative values are distributed inside the flood portions. These trends are not influenced by the difference of SAR wavelength (C-/L-bands), polarization (HH, HV, VV, and VH), and stacked number (4 or 143 scenes) of pre-flood images.

In the close-up panels (Figure 5e–h) centered on the representative major flood portions, it can be observed that high values close to zero are distributed in the frequent water areas (>20%) (in black outlines), while lower negative values (<-1) are distributed in the peripheral areas (green/blue) within the post-flood water bodies. These results indicate that NoBADI values generally have lower negative values only in HWE, avoiding potential water extents.

Several parts located at the central north edge of the study site denoted significantly lower values (in light blue) in terms of PALSAR-2 results (Figure 5a,b), which are not shown in S1 results (Figure 5c,d). These locations are not optically referenced in the WV2 true-color image (Figure 3a) because they correspond to cloud shadows. The presence of the water bodies was expected in the PALSAR-2 image (Figure 4a) but not in the S1 image (Figure 3b). Based on this limited information, it is possible that the water body here has decreased between the two observation dates, which may be the reason for the difference in NoBADI results.

5.4. Pre–Post Difference

In order to evaluate the NoBADI results, simple before/after differences in the amplitude of SAR backscatter were calculated for each polarization of PALSAR-2 HH (Figure 6a), HV (Figure 6b), S1 VV (Figure 6c), and VH (Figure 6d). Backscatter amplitude is small in flat surfaces such as water and paved roads. Darker pixels would indicate a severe decrease in backscatter (e.g., flood inundation), while brighter pixels would indicate a severe increase in backscatter (e.g., loss of water extent).

Dark regions should, therefore, correspond to flood-derived water extents. However, their distribution does not correspond to potential water extents (outline in yellow). Fundamentally, their spatial distribution is not homogeneous in the pre-flood water extent (blue/white outlines) among the four images that are visually identifiable in the closed-up panels (Figure 6e–h). Inside a representative water body of ID-3, for example, flood-derived water extent was suggested in the largest area based on the conditions of 19 January 2016 (Figure 6g) and 30 August 2016 (Figure 6h), whereas it was slightly suggested based on the condition of 20 January 2015 (Figure 6e). Based on the condition on 1 September 2015 (Figure 6f), relatively moderate and partial flooding is expected.

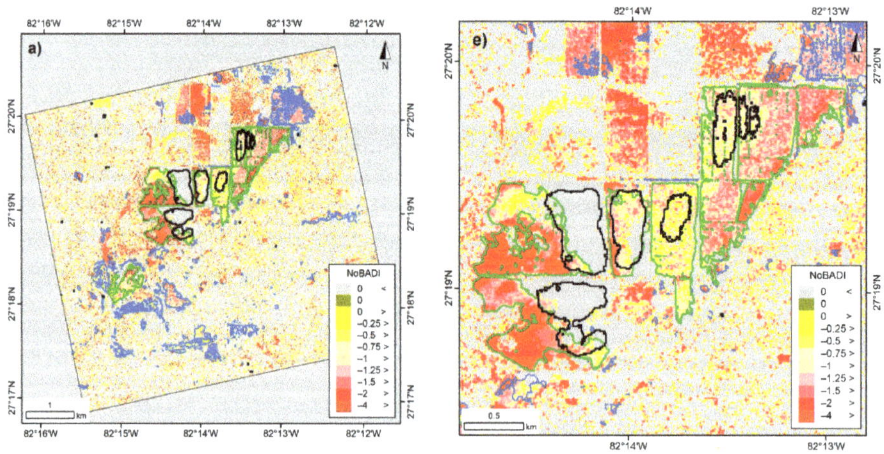

Figure 5. *Cont.*

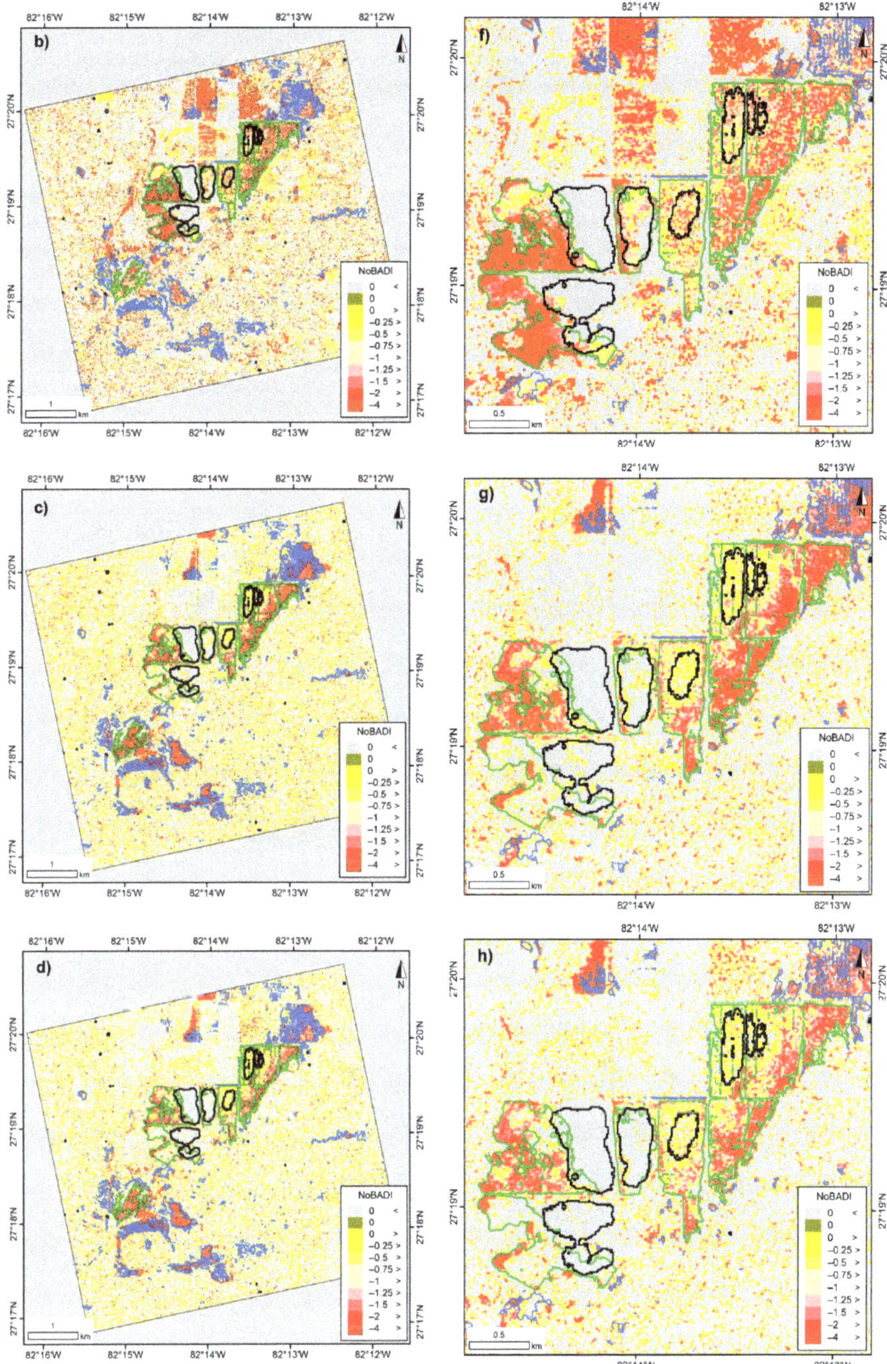

Figure 5. Spatial distribution of NoBADI calculated from PALSAR-2 (**a**) HH and (**b**) HV polarizations. Others are calculated by S1 (**c**) VV and (**d**) VH polarization. Optically identified water bodies are outlined in blue, in which ID-attached six portions are in green. Optically estimated frequent water extents (>20%) are outlined in black. Closed-up panels corresponding to the four results are in the same order (**e**–**h**).

The comparison of the most recent month does not ideally match the distribution of the reference water extent. Irregular changes in potential water bodies and not seasonal differences are dominant in the study area. This result suggests that selecting the most recent month's data as a reference does not always result in accurate water extraction from simple before/after differences.

5.5. Coherence

Multiple previous studies focused on coherence and other InSAR-derived results for flood mapping. The spatial distributions of PALSAR-2 coherence between post-flood (12 September 2017) and pre-flood (30 August 2016) scenes (Figure 7a) and coherence decrease versus two pre-flood scenes (30 August 2016 and 19 January 2016) (Figure 7b) were calculated. The two results do not show any visually identifiable features, and even pre-flood water bodies are difficult to extract. These results suggest that background coherence is generally low in this study area. This may be due to the relatively active growth of vegetation in the tropics and the disturbance of the physical properties of the ground by frequent rainfall over one year.

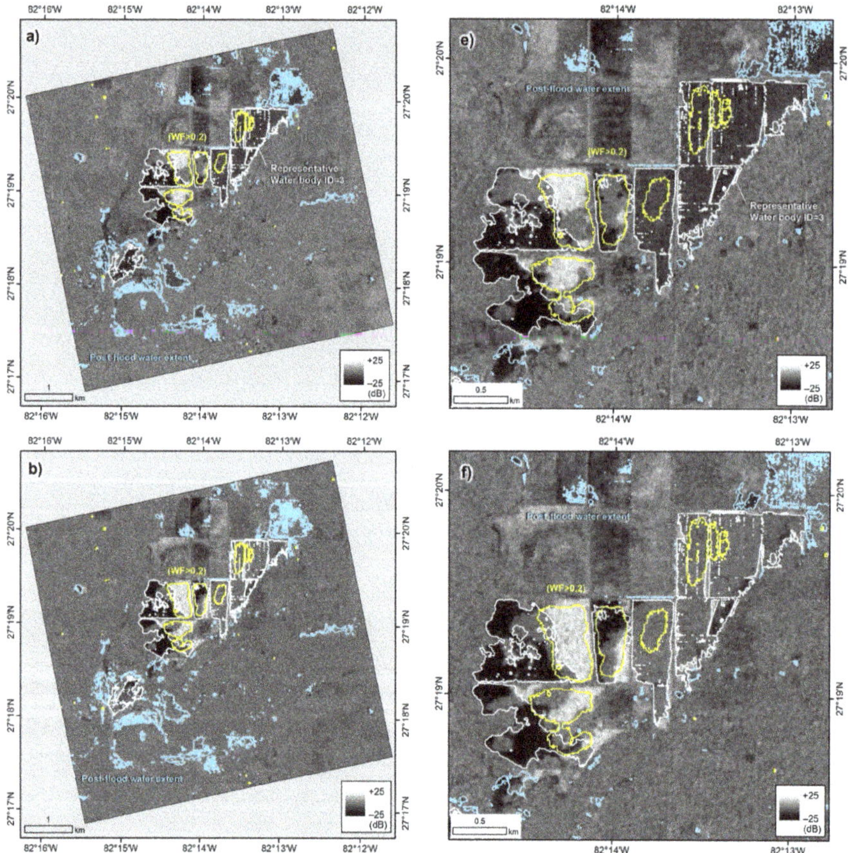

Figure 6. *Cont.*

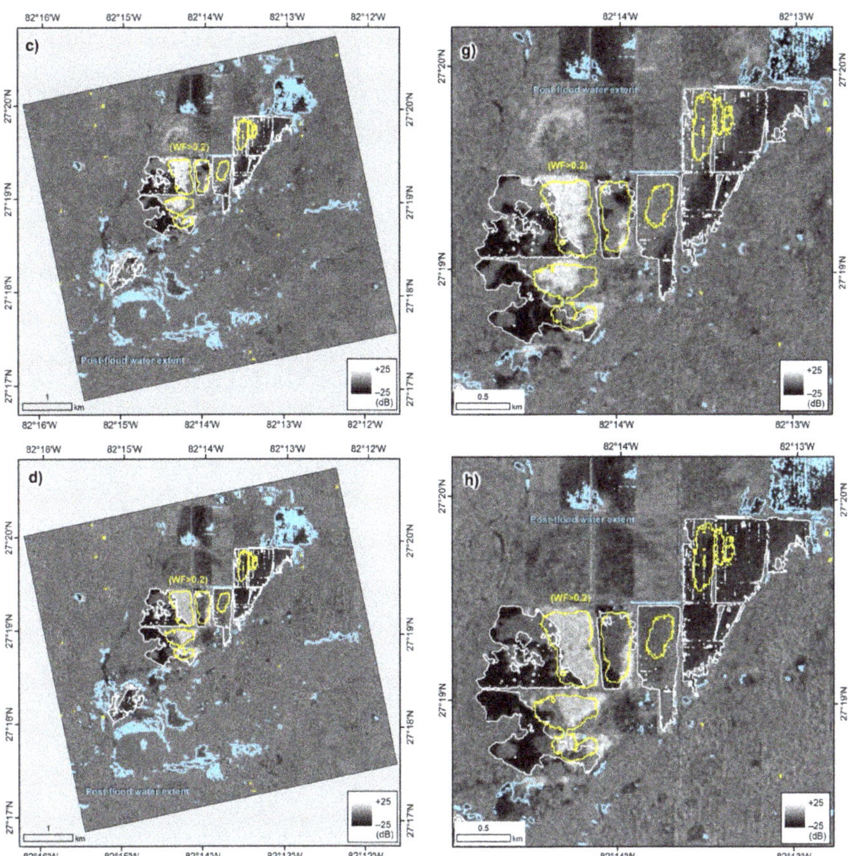

Figure 6. Pre–post differences of backscatter amplitude by PALSAR-2 over the flood event. Based on (**a**) 20 January 2015; (**b**) 1 September 2015; (**c**) 19 January 2016; and (**d**) 30 August 2016. Closed-up panels are shown (**e**–**h**) in the same temporal order. White outlines denote representative water bodies (in Figure 3a), and yellow outlines denotes the example of regular water extent (in Figure 3b).

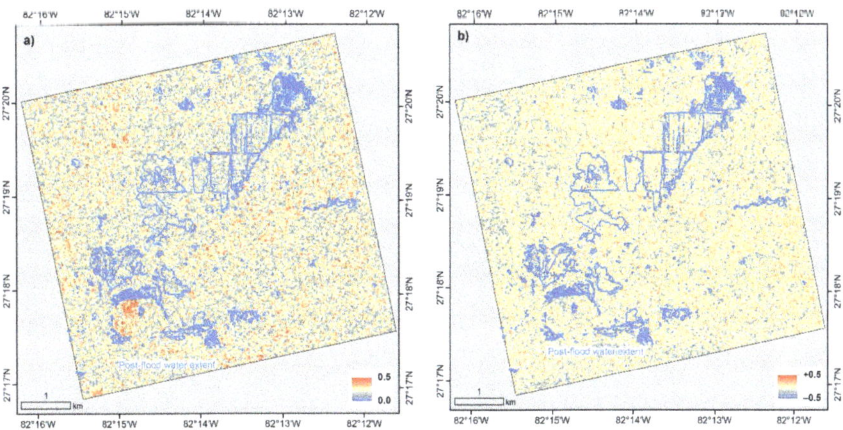

Figure 7. PALSAR-2 imagery of (**a**) coherence between post-flood (12 September 2017) and pre-flood (30 August 2016) scenes and (**b**) coherence decrease between the pre–post pair and two pre-flood scenes (30 August 2016 and 19 January 2016).

6. Discussion

6.1. Overview of NoBADI

In this study, a simple function of combining multiple SAR backscatter amplitude data, named NoBADI, was applied to hurricane-induced flood conditions in the study site where the water bodies potentially fluctuate. Single use of pre-flood data would derive different results according to the pre-flood conditions, whereas NoBADI has higher potential to exclude potential water bodies through the statistical algorithm. No significant difference was found between L-band and C-band observations. Indeed, hundreds of pre-flood data, as demonstrated with S1 data, support higher reliability of output information than small samples. According to the result of NoBADI calculated from PALSAR-2 data, however, four pre-flood data, at least, enable the correct output of NoBADI.

Detailed interpretation suggests slightly different characteristics of L-bands/C-bands in terms of volumetric scatter. As described in the Results section, the PALSAR-2 post-flood single image had higher contrast and flat surfaces that were easier to identify than the S1 image. On the other hand, the S1 image emphasizes water bodies, which have been identified by WV2 data. A major reason for the lower contrast for non-water flat domains in S1 data is possibly the difference of L-bands/C-bands. By comparing the two wavelengths, C-band is largely influenced by short grass and coarse vegetation (i.e., volumetric scattering).

6.2. SAR Backscatter Amplitude and Water Extent

To understand the sensitivity of backscatter amplitude relative to temporal changes of a water-covered area, we compared the relationship between the S1-derived backscatter amplitude (VV) and the S2-derived NDWI in temporal profiles (Figure 8). Each panel corresponds to the domains of representative post-flood water body from ID-1 to ID-6. The backscatter amplitude is negatively correlated with the water extent represented by the average NDWI value for the six domains. This means that backscatter amplitude is a highly sensitive proxy for water existence.

A detailed comparison shows that when the backscatter amplitude is below -16 dB, the NDWI is positive (most of the area is covered by water). The ID-2 domain has a significant drop of backscatter amplitude on September 2016. This means that the land surface turns into a water surface, possibly artificially. The ID-5 domain has mostly positive NDWIs, suggesting land cover in the long term. This domain was temporarily covered by water in response to Hurricane Irma during mid-September 2017, but it is not reflected in the profile possibly due to rapid recovery.

For all profiles, there is no uniform seasonal variation. It was not seasonal changes but transient and irregular events that were dominant for water extent changes. NDWI values have frequent and wide-range fluctuations compared to backscatter amplitude. This is because cloud mask is not applied on this analysis due to statistical limitations. However, we observe that the general trend is the same as in backscatter amplitude.

6.3. Optimization of Threshold Value for NoBADI Water Extraction

In the NoBADI analysis, one threshold value is set for binarization, then pixels with lower values are considered flood-derived water extent. Classification accuracy is evaluated by matching NoBADI-derived water pixels versus S2-derived water pixels above the WF of 0.2. In this accuracy validation, the Kappa coefficient is calculated by setting the NoBADI threshold (th.) to multiple values from -4 to 0.

Variations of the Kappa coefficient corresponding to multiple NoBADI threshold are obtained from PALSAR-2 HH, HV, S1 VV, and VH datasets (Figure 9). The highest coefficient value is 0.43 (th. = -1.6) on S1/VV scenes, followed by 0.39 (th. = -1.4) on S1/VH scenes. PALSAR-2 results derive lower maximum coefficients of 0.32 (th. = -0.9) on HH scenes and 0.27 (th. = -1.0) on HV scenes. Selecting these thresholds enables the most optimized extraction of flood-derived water extent. S1 will derive more than PALSAR-2, but PALSAR-2 has more flexible observation criteria for emergency observation under international partnerships. The choice of which one to use will be determined by the timing of observations after the disaster occurs.

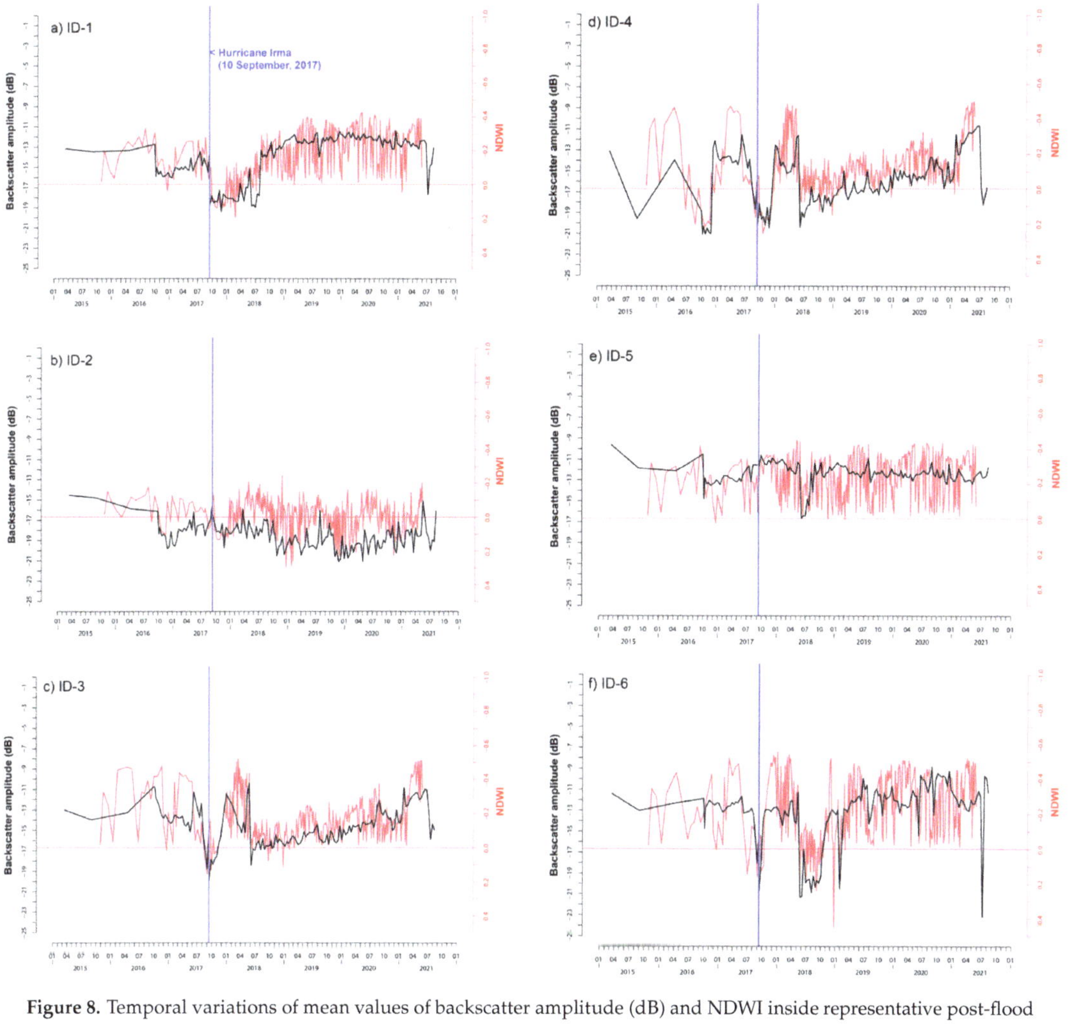

Figure 8. Temporal variations of mean values of backscatter amplitude (dB) and NDWI inside representative post-flood water bodies of ID-1-6 (**a**–**f**). A vertical blue line denotes the day Hurricane Irma affected the area (10 September 2017).

To assess sensitivity of the Kappa coefficient, the same calculation was performed for the validation site (see Figure 1). The highest coefficient value is 0.45 (th. = −1.3) on S1/VV scenes, followed by 0.44 (th. = −1.2) on S1/VH scenes. PALSAR-2 results derive lower maximum coefficients of 0.33 (th. = −1.4) on HH scenes and 0.28 (th. = −1.3) on HV scenes. The study site and the validation site, thus, have different values for optimized threshold, which have smaller differences in S1 than PALSAR-2. The S1/VH scenes have values of the Kappa coefficient above/below 0.4, but the VV scenes have similar values both above 0.4 between the two sites. The sensitivity of Kappa coefficient demonstrated in the two sites suggests that S1/VV scenes derive relatively robust outputs that are slightly influenced by selecting the area for analysis.

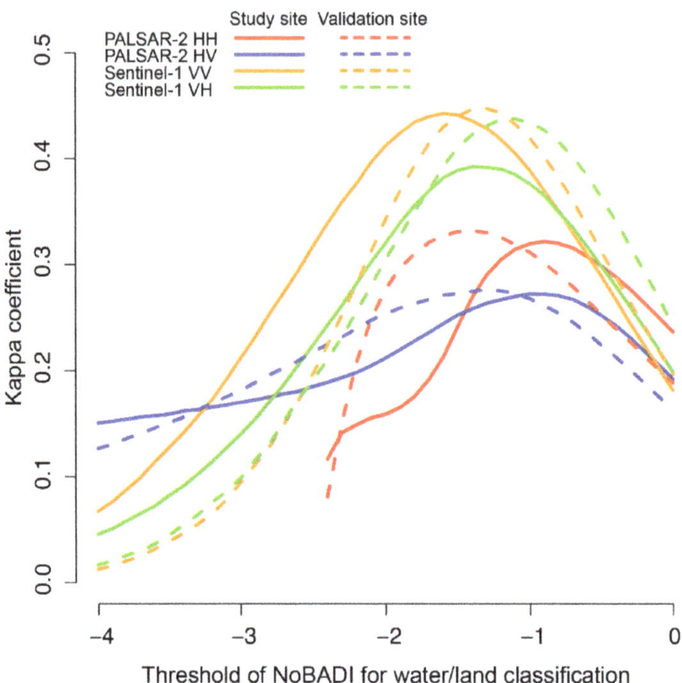

Figure 9. Variations of Kappa coefficients between NoBADI-/optically derived flood-derived water extent for the study site and the validation site. Values correspond to different settings of NoBADI threshold.

6.4. Advantages of NoBADI

Fundamentally, the NoBADI method, as we demonstrated in Florida by means of C-band and L-band SAR, is not affected by cloud cover and that supports urgent and promising observation just after or even during the effects of the hurricane. Figure 5 shows the comprehensively similar features of NoBADI's spatial distribution, although detailed differences are caused among C-bands/L-bands and HH/HV/VV/VH polarizations. Other SAR sensors such as Radarsat-2, NISAR, ALOS-4, and upcoming ones onboard commercial small satellites can potentially apply this method. Note that the X-band has not been verified at this time, so it is expected to be evaluated in the future.

Figure 6 shows that spatial distributions of backscatter amplitude decreasing over the flood event are different depending on the scene that is used as the pre-flood condition. These differences that happen to occur are especially significant within the post-flood water extents, except stable water extents (WF > 0.2) (i.e., HWE). Furthermore, Figure 8 shows that SAR backscatter amplitude and NDWI are irregularly fluctuated and not related to the seasonal cycle. It is possible that the dominant factor here is not the overall seasonality but the uneven cycle of precipitation and evaporation. This suggests that selecting a pre-flood scene in the same season as a post-flood scene does not necessarily improve the uncertainty.

The NoBADI value is not a physical value directly corresponding to backscatter amplitude (dB) but an index that expresses how rare the water extent is. Water extents in a location sometimes covered by water would denote moderate values near zero because standard deviation is relatively large due to frequent changes of land cover. On the other hand, water extents caused in surfaces with small opportunities of water cover would have significantly lower values, which is highlighted in the NoBADI map. For these reasons, spatial distributions of NoBADI (Figure 5) do not denote homogeneous values in the entire surfaces of the HWE. Instead, we can selectively recognize rare water extents as significantly lower NoBADI values.

6.5. Toward Operational Use of Crisis Response

Recent studies of SAR-based flood monitoring analyzed multi-temporal scenes acquired in their study sites [19,20]. Complicated and highly technical methods, however, need to be evaluated in terms of their applicability for each flood case in different topographies and the limitation of available SAR data archives. Our study proposes a simple function of NoBADI aiming to be a standard method for low and flat topography. We conclude that different wavelengths, different polarizations, and the number of pre-flood scenes have certain effects on the output results. These characteristics should be considered for practical applications in crisis response.

NoBADI results do not show neighboring and homogeneous distribution of pixel values for the HWE, but denser and lower pixel values with some deviations around expected places of the HWE (Figure 5). It is possibly the reason why values of the Kappa coefficient are relatively small values that are not larger than 0.5. This may have some difficulties for simple conversion to vector-format polygons. Therefore, further geoinformatics consideration would be needed, such as filter processing and segmentation. Thus, in the present status, NoBADI should be used for the purpose of narrowing down the target area in the first stage of operational use during crisis response. It would be followed by further rigorous analysis in order to collect more accurate information.

7. Conclusions

This study assessed Hurricane-induced Water Extent (HWE) by using SAR-based method of NoBADI and considering the potential fluctuation of water extent in Florida. Both of L-band and C-band SAR data extracted flood-derived water while excluding potential water distribution. These wavelengths can be used to obtain initial information about the flood distribution, and the classification accuracy is higher in the C-band. Future studies will be expected for the validation of NoBADI analysis in other locations that have different temporal patterns of water surface fluctuation.

Supplementary Materials: The following are available online at https://www.mdpi.com/article/10.3390/rs13204136/s1, Table S1: List of acquisition dates of Sentinel-2, Table S2: List of acquisition dates of Sentinel-1.

Author Contributions: H.N. designed the overall flow of the study and performed the main part of the methodology; T.A. and M.O. mainly performed the SAR analysis and improved the results and discussion. All authors have read and agreed to the published version of the manuscript.

Funding: The purchase of optical images for WorldView-2 was supported by the Waseda University Special Research Project Grant (Project No.: 2018S-059).

Institutional Review Board Statement: Not applicable.

Informed Consent Statement: Not applicable.

Acknowledgments: The initial study was conducted in the ALOS Research and Application Project, Earth Observation Research Center, JAXA. The authors thank the Japan Aerospace Exploration Agency (JAXA) for providing the PALSAR-2 data through ALOS-2 6th Research Announcement (PI-3263). We thank three anonymous reviewers for their valuable comments.

Conflicts of Interest: The authors declare no conflict of interest.

References

1. Agnihotri, A.K.; Ohri, A.; Gaur, S.; Shivam; Das, N.; Mishra, S. Flood inundation mapping and monitoring using SAR data and its impact on Ramganga River in Ganga basin. *Environ. Monit. Assess.* **2019**, *191*, 760. [CrossRef] [PubMed]
2. Bayik, C.; Abdikan, S.; Ozbulak, G.; Alasag, T.; Aydemir, S.; Sanli, F.B. Exploiting multi-temporal Sentinel-1 SAR data for flood extend mapping. In Proceedings of the International Archives of the Photogrammetry, Remote Sensing and Spatial Information Sciences—ISPRS Archives, Istanbul, Turkey, 18–21 March 2018; Volume 42.

3. Sanyal, J.; Lu, X.X. Application of Remote Sensing in Flood Management with Special Reference to Monsoon Asia: A Review. *Nat. Hazards* **2004**, *33*, 283–301. [CrossRef]
4. Long, S.; Fatoyinbo, T.E.; Policelli, F. Flood extent mapping for Namibia using change detection and thresholding with SAR. *Environ. Res. Lett.* **2014**, *9*, 035002. [CrossRef]
5. Meyer, F.J.; Ajadi, O.A.; Schultz, L.; Bell, J.; Arnoult, K.M.; Gens, R.; Nicoll, J.B. An automatic flood monitoring service from Sentinel-1 SAR: Products, delivery pipelines, and performance assessment. In Proceedings of the IGARSS 2018—2018 IEEE International Geoscience and Remote Sensing Symposium, Valencia, Spain, 22–27 July 2018; pp. 6576–6579. [CrossRef]
6. Meyer, F.J.; Ajadi, O.A.; Schultz, L.; Bell, J.; Arnoult, K.M.; Gens, R.; Molthan, A.L.; Nicoll, J.B.; Hogenson, K. Applications of a SAR-based flood monitoring service during disaster response and recovery. In Proceedings of the IGARSS 2019—2019 IEEE International Geoscience and Remote Sensing Symposium, Yokohama, Japan, 28 July–2 August 2019; pp. 4649–4652.
7. Watanabe, M.; Koyama, C.N.; Hayashi, M.; Nagatani, I.; Tadono, T.; Shimada, M. Refined algorithm for forest early warning system with ALOS-2/PALSAR-2 ScanSAR data in tropical forest regions. *Remote Sens. Environ.* **2021**, *265*, 112643. [CrossRef]
8. Florida State University Florida Climate Center. Available online: https://climatecenter.fsu.edu/ (accessed on 31 January 2021).
9. Cangialosi, J.P.; Latto, A.S.; Berg, R. *National Hurricane Center Tropical Cyclone Report: Hurricane Irma (30 August–12 September 2017)*; NOAA National Centers for Environmental Information: Asheville, NC, USA, 2018.
10. National Aeronautics and Oceanic Administration BULLETIN—Hurricane Irma Intermediate Advisory Number 45A. Available online: https://www.capeweather.com/weather-ftopicp-27027.html (accessed on 31 August 2021).
11. Chakalian, P.M.; Kurtz, L.C.; Hondula, D.M. After the Lights Go Out: Household Resilience to Electrical Grid Failure Following Hurricane Irma. *Nat. Hazards Rev.* **2019**, *20*, 05019001. [CrossRef]
12. Mitsova, D.; Esnard, A.M.; Sapat, A.; Lai, B.S. Socioeconomic vulnerability and electric power restoration timelines in Florida: The case of Hurricane Irma. *Nat. Hazards* **2018**, *94*, 689–709. [CrossRef]
13. Feng, K.; Lin, N. Reconstructing and analyzing the traffic flow during evacuation in Hurricane Irma (2017). *Transp. Res. Part D Transp. Environ.* **2021**, *94*, 102788. [CrossRef]
14. Issa, A.; Ramadugu, K.; Mulay, P.; Hamilton, J.; Siegel, V.; Harrison, C.; Campbell, C.M.; Blackmore, C.; Bayleyegn, T.; Boehmer, T. Deaths Related to Hurricane Irma—Florida, Georgia, and North Carolina, September 4–October 10, 2017. *MMWR. Morb. Mortal. Wkly. Rep.* **2018**, *67*, 829–832. [CrossRef] [PubMed]
15. MAXAR Technologies WorldView-2 Datasheet. Available online: https://resources.maxar.com/data-sheets/worldview-2 (accessed on 31 August 2021).
16. European Space Agency Sentinel-2 MSI User Guide. Available online: https://sentinel.esa.int/web/sentinel/user-guides/sentinel-2-msi (accessed on 31 August 2021).
17. Updike, T.; Comp, C. *Radiometric Use of WorldView-2 Imagery Technical Note*; DigitalGlobe: Westminster, CO, USA, 2010; pp. 1–17.
18. McFeeters, S.K. The use of the Normalized Difference Water Index (NDWI) in the delineation of open water features. *Int. J. Remote Sens.* **1996**, *17*, 1425–1432. [CrossRef]
19. Uddin, K.; Matin, M.A.; Meyer, F.J. Operational flood mapping using multi-temporal Sentinel-1 SAR images: A case study from Bangladesh. *Remote Sens.* **2019**, *11*, 1581. [CrossRef]
20. Martinis, S.; Rieke, C. Backscatter analysis using multi-temporal and multi-frequency SAR data in the context of flood mapping at River Saale, Germany. *Remote Sens.* **2015**, *7*, 7732–7752. [CrossRef]

MDPI

Article

Vegetation Effects on Soil Moisture Retrieval from Water Cloud Model Using PALSAR-2 for Oil Palm Trees

Veena Shashikant [1], Abdul Rashid Mohamed Shariff [1,2,3,*], Aimrun Wayayok [1,2,3], Md Rowshon Kamal [1], Yang Ping Lee [4] and Wataru Takeuchi [5]

1 Department of Biological and Agricultural Engineering, Faculty of Engineering, Universiti Putra Malaysia, Serdang 43400, Malaysia; gs52191@student.upm.edu.my (V.S.); aimrun@upm.edu.my (A.W.); rowshon@upm.edu.my (M.R.K.)
2 SMART Farming Technology Research Center, Faculty of Engineering, Universiti Putra Malaysia, Serdang 43400, Malaysia
3 Laboratory of Plantation System Technology and Mechanization (PSTM), Institute of Plantation Studies (IKP), Universiti Putra Malaysia, Serdang 43400, Malaysia
4 FGV R&D Sdn Bhd, Level 9, Wisma FGV, Jalan Raja Laut, Kuala Lumpur 50350, Malaysia; yangp.lee@fgvholdings.com
5 Department of Human and Social Systems, Institute of Industrial Science, The University of Tokyo, Tokyo 153-8505, Japan; wataru@iis.u-tokyo.ac.jp
* Correspondence: rashidpls@upm.edu.my; Tel.: +60-3-9769-6414

Abstract: In oil palm crop, soil fertility is less important than the physical soil characteristics. It is important to have a balance and sufficient soil moisture to sustain high yields in oil palm plantations. However, conventional methods of soil moisture determination are laborious and time-consuming with limited coverage and accuracy. In this research, we evaluated synthetic aperture radar (SAR) and in-situ observations at an oil palm plantation to determine SAR signal sensitivity to oil palm crop by means of water cloud model (WCM) inversion for retrieving soil moisture from L-band HH and HV polarized data. The effects of vegetation on backscattering coefficients were evaluated by comparing Leaf Area Index (LAI), Leaf Water Area Index (LWAI) and Normalized Plant Water Content (NPWC). The results showed that HV polarization effectively simulated backscatter coefficient as compared to HH polarization where the best fit was obtained by taking the LAI as a vegetation descriptor. The HV polarization with the LAI indicator was able to retrieve soil moisture content with an accuracy of at least 80%.

Keywords: SAR; backscattering; soil moisture content; LAI; HH and HV polarization

Citation: Shashikant, V.; Mohamed Shariff, A.R.; Wayayok, A.; Kamal, M.R.; Lee, Y.P.; Takeuchi, W. Vegetation Effects on Soil Moisture Retrieval from Water Cloud Model Using PALSAR-2 for Oil Palm Trees. *Remote Sens.* **2021**, *13*, 4023. https://doi.org/10.3390/rs13204023

Academic Editors: Takeo Tadono and Masato Ohki

Received: 31 August 2021
Accepted: 5 October 2021
Published: 9 October 2021

1. Introduction

Soil moisture content is a critical input variable in a wide variety of scientific studies in the field of agriculture. Soil moisture is an environmental element that connects the Earth's surface and the atmosphere. When soil moisture levels are balanced, agricultural yields improve, yield losses due to drought are reduced, and groundwater levels are recharged, ensuring the continuity of rivers and stream flows [1]. Oil palm trees require a reasonably steady high temperature, and continuous precipitation all year. Furthermore, prolonged dry periods of more than 2–3 months do not directly harm vegetative growth but have a significant impact on the yield and quality of fruit bunches [2]. The yield of oil palm is highly dependent on the availability of water during the sex differentiation of its inflorescences, which occurs approximately 28 months before bunch harvest. Soil moisture is therefore critical for optimal production in the oil palm crop [3]. Plants may be stressed if there is a water shortage. When water is not a constraint, potential evapotranspiration is the quantity of water that might be evaporated or transpired at a given temperature and humidity [4]. Water is a key component of plant tissue and is used to transport metabolites and minerals inside the plant. Water is also required for cell expansion, whereby it increases the turgor pressure [5]. Many of the physiological processes related to growth are harmed by water deficiency in the soil, and severe deficiency may even result in the death of the plant. The effect of water deficiency, on the other hand, varies depending on the degree and length of water stress as well as the oil palm's development stage [6]. Nitrogen, Potassium, Magnesium, Boron, Copper, and Zinc are said to be essential nutrient components in oil palm crop soil for maximum growth [7]. The soil system and its activities are governed by nutrient availability, which leads to greater drought, insect, pest, and disease resistance [8]. Moreover, oil palm fronds have a delicate water transport system, which may expose them to increasing drought stress when the environment heats and dries [9]. Soil fertility refers to the soil's ability to deliver nutrients to the palm, including water availability for nutrient absorption and yield where soil–water conservation is carried out commonly to prevent soil erosion beneath mature palms, build terracing and silt pits in steep areas, mulch with empty fruit bunches and trunk chips, and cultivate leguminous cover crops [10]. Therefore, effective water management is the key to achieving high oil palm yield. As a result, sufficient soil moisture in the root zones is required, as too little or too much of it would reduce oil palm yields [11]. Recently, it was found that oil palm crop demands range from 0.893 to 1.6 million cubic meters; where ultimately, the actual requirement is site-specific and varies based on the soil moisture deficit, root zone water availability, and rooting depth [12].

Soil moisture has traditionally been measured directly at sampling places using gravimetry, which is a highly reliable and therefore often preferred method [13]. The results from gravimetric methods, however, only reflect a very limited region that changes instantly as the sampled field changes around the sampling site [14]. Remote sensing, both active and passive, enables unique studies of soil moisture at multiple spatial scales; therefore addressing agricultural scientific and application demands [15–17]. In passive remote sensing, Google Earth-based imageries and Normalised Difference Vegetation Index (NDVI) were used to determine the effect of land use and change of water storage in an oil palm plantation by measuring variations in soil water content over time [18]. Plus, the use of the NDVI and the Soil Adjusted Vegetation Index (SAVI) in oil palm crop to determine plant health by adopting a regression model technique revealed a highly correlated relationship between plant health based on NDVI analysis and nitrogen content to SAVI [19]. Additionally, to overcome the lack of soil information for farmers, a real-time palm oil soil monitoring system was built for Palm Oil Soil Monitoring in a Smart Agriculture where it can process, transmit, display, and conclude the soil's state via smartphone [20].

On the other hand, active remote sensing or SAR sensors are mostly used to assess soil moisture and crop water usage over broad regions [21]. SAR sensors are capable of identifying the spatial pattern of volumetric soil moisture due to their ability to penetrate to a depth of approximately 5 cm below the surface [22]. The depth to which microwaves penetrate, on the other hand, is mostly controlled by the density of vegetation, the stage of crop development, and sensor-related factors such as incidence angle, polarization, and frequency [23]. Recently, SAR sensor usage has increased over time, especially when the soil dielectric constant shows a linear relationship to the backscatter coefficient in the unit of decibels [24]. The microwave signals from the HH polarization often penetrate efficiently in the vegetation while reducing the interactions resulting from trunk or branches when compared to VV polarization [25]. When the sensor incidence angle was studied, the higher incident angles provided a better penetration as well as higher accuracy for soil moisture retrieval for both HH and VV [26]. Additionally, longer wavelengths or L-bands provide an adequate level of soil moisture sensitivity beneath the majority of plant cover [27,28]. In oil palm crop, AIRSAR backscattering coefficients were found to initially increase with age and the oil palm biomass is mentioned to be highly correlated at r = 0.85; where a Gamma filter of 11 × 11 window discriminates oil palm age classes effectively [29]. Following that, using the backscatter coefficient of fully polarized ALOS PALSAR data, biomass estimations were generated for an oil palm plantation in Malaysia and higher correlations were achieved from VH polarization data [30]. Similarly, the biomass estimation of oil palm plantations using a regression analysis of HV polarized PALSAR was carried out [31].

A few SAR backscatter modelling techniques have been carried out to explore the influence of surface-related variables on the backscatter coefficient, such as topography and vegetation. These models are generally divided into theoretical, semi-empirical and empirical methods. Theoretical techniques such as the Integral Equation Model (IEM) and advanced IEM are complex, where the main challenge in assembling model parameters that accurately characterize the canopy is tough [32] as it requires a large number of parameters [33,34]. In addition to that, the empirical methods, for example the Dubois model [35], are simpler than the IEM models. However, the Dubois model has been reported to be dependent on experimental site and data conditions [36], where the saturation points were reported at lower NDVI values [37].

When the semi-empirical method was used, effects of vegetation cover using the WCM [38] were initially developed and later on were modified [39–41]. WCM is preferred for its simplicity [42] where it defines the overall backscatter coefficient obtained by the sensor over vegetated surfaces as the incoherent sum of the effects of vegetation and soil [43]. In this model, the canopy is often represented as a set of exact variables such as plant density per area, leaf size and orientation, which complicates and makes the model hard to interpret [44]. Many factors influence the backscatter coefficients from vegetation canopies, e.g., size of disperses in a canopy; the shape of scatters in a canopy; the orientation in a canopy; and the geometry of the canopy cover on ground [32]. To begin with, the WCM was modelled using plant height and water content of the vegetation layer to allow for the retrieval of soil moisture [38], and Leaf Area Index (LAI) was subsequently evaluated [45]. Other vegetation descriptors in the WCM were used from various combinations such as leaf water area index (LWAI), Normalized Plant Water Content (NPWC), vegetation water mass, and biomass [43,46–48].

The goal of this research is to optimize soil moisture retrieval using the WCM, thereby reducing the effect of vegetation on the crop-covered soil moisture backscatter coefficient. The WCM attempts to express vegetation cover scattering and attenuation terms in the concept through plain vegetation descriptors. Furthermore, a comparison analysis is performed on the use of the three primary vegetation descriptors, LAI, NPWC, and LWAI, singly or in combination, in the retrieval of soil moisture from PALSAR-2 data from Malaysian oil palm trees.

2. Materials and Methods

To recover soil moisture, a basic WCM was used in this investigation. Assuming that the effect of soil surface roughness on observed backscatter is consistent over a short time period for a given site, the temporal change in SAR backscattering only reflects changes in vegetation and soil moisture [49]. As a result, for this investigation, multi-temporal SAR data were used in the WCM. The vegetation descriptors and actual soil moisture were determined using field collected data. The Root Mean Square Error (RMSE) and mean absolute error coefficient were utilized to analyze the soil moisture retrieval accuracy. The next sections describe WCM, vegetation descriptors used in the model and evaluation of data processing methods.

2.1. Water Cloud Model (WCM)

WCM was developed assuming that the canopy "cloud" contains identical water droplets randomly distributed within the canopy [38]. In a water cloud model, the expression developed incorporated the soil moisture and the vegetation parameters in the equation. WCM for a given polarization (pp) is given as

$$\sigma^{\circ}{}_{tot,pp} = \sigma^{\circ}{}_{veg,pp} + \sigma^{\circ}{}_{soil+veg,\ pp} + \tau^{2}{}_{pp}\sigma^{\circ}{}_{soil,pp} \quad (1)$$

where $\sigma^{\circ}{}_{tot,pp}$ is the total backscatter coefficient, $\sigma^{\circ}{}_{veg,pp}$ is the backscatter contribution of the vegetation cover, $\sigma^{\circ}{}_{soil+veg,\ pp}$ is the multiple scattering involving vegetation elements and the soil surface, $\sigma^{\circ}{}_{soil,pp}$ is the backscatter contribution of the soil surface and $\tau^2{}_{pp}$ is the two-way vegetation attenuation. The second component in Equation (1) reflects the interaction of incident radiation with the underlying soil. Because the interaction is not a dominant factor in co-polarized radiation, it may be ignored [43,50]. Subsequent to the model development, there were studies that modified the mentioned model [45]. WCM, Equation (2) [43], with four empirical coefficients, A_{pp}, B_{pp}, C_{pp}, and D_{pp} where A_{pp} and B_{pp} are vegetative characteristics and C_{pp} and D_{pp} are soil parameters, was presented for the given polarization where pp is either HH or HV polarization. The parameter A_{pp} corresponds to the albedo of the vegetation, and B_{pp} is an attenuation factor seen in (3) and (4). The parameter D_{pp} indicates the sensitivity of the radar signal to soil moisture, and C_{pp} can be considered as a calibration constant in (5). Hence, the equation written for the given polarization where pp is HH and HV is modified to

$$\sigma^{\circ}{}_{tot,pp} = \sigma^{\circ}{}_{veg,pp} + \tau^2{}_{pp}\sigma^{\circ}{}_{soil,pp} \tag{2}$$

where

$$\sigma^{\circ}{}_{veg,pp} = A_{pp} \times V_1 \times \cos\theta\left(1 - \tau^2{}_{pp}\right) \tag{3}$$

$$\tau^2{}_{pp} = \left(\text{Exp}(-2 \times B_{pp} \times V_2 \times \sec\theta)\right) \tag{4}$$

V_1 and V_2 describe the effect of canopy water content and its geometry on backscatter coefficients. Vegetation descriptors have different effects on the WCM model. Therefore, several experimental studies on different combination of vegetation variables to quantify V_1 and V_2 in WCM were conducted by using plant height, LAI, LWAI and NPWC [51]. In Equations (3) and (4), θ is the incident angle of the image used, A_{pp} and B_{pp} are the vegetation parameter. M_v is the volumetric soil moisture and C_{pp} and D_{pp} are the soil parameter in Equation (5).

$$\sigma^{\circ}{}_{soil,pp} = D_{pp}M_v + C_{pp} \tag{5}$$

In this study, the LAI of palm fronds is defined by the amount of leaflet surface area per unit ground area [3]. LWAI is a product of LAI of palm fronds multiplied by the amount of water (W), expressed as the ratio of the difference between wet and dry mass to wet mass as in Equations (6) and (7).

$$LWAI = (LAI \times W) \tag{6}$$

$$W = \left(\frac{m_w - m_d}{m_w}\right) \tag{7}$$

where m_w and m_d are field records for freshly plucked and oven-dried mass of vegetation samples, respectively. Normalised plant water content, NPWC, also plays a dominant part in attenuating backscatter [38]. NPWC is like Equation (8) but divided with dry mass instead of wet mass.

$$NPWC = \left(\frac{m_w - m_d}{m_d}\right) \tag{8}$$

In this study, 5 combinations were used to understand the vegetation effects of WCM by using

Case 1 where V_1 = 1 and V_2 = LAI
Case 2 where V_1 = LAI and V_2 = 1
Case 3 where V_1 = LAI and V_2 = LAI
Case 4 where V_1 = LWAI and V_2 = LWAI
Case 5 where V_1 = NPWC and V_2 = NPWC

2.2. Estimating Parameters of A, B, C and D in the WCM

The estimation of parameters C_{pp} and D_{pp} are solved using a linear model fitting, following which, the values of C_{pp} and D_{pp} are substituted into Equation (2), which allows for parameter A_{pp} and B_{pp} to be solved using the Nonlinear Least Squares Method (NLSM) [51–54]. It was noted that using Levenberg–Marquardt (LM) optimization in NLSM, an estimation of A_{pp} and B_{pp} can be made [55]. According to the optimization, LM is a common approach for addressing nonlinear least square issues that emerge from fitting a parameterized function to a collection of observed data points by minimizing the sum of the squares of the errors between the observed data points and the function output [56]. The LM algorithm is an iterative process which combines the Gauss–Newton method and the gradient descent method [57].

2.3. Evaluating WCM

To validate the WCM prior to inversing the model for accuracy and estimation performance metrics which include the coefficient of determination (R^2), the RMSE and the mean absolute error (MAE) were examined. The RMSE derived using Equation (9) was tested in most studies [49,55,57,58]. RMSE is a frequently used measure of the difference between values predicted by a model and the values observed from the environment that is being modelled. RMSE evaluation is given in Equation (9); X_{sim} is the simulated $\sigma^{\circ}{}_{tot,pp}$, X_{obs} is the observed $\sigma^{\circ}{}_{tot,pp}$ and n is the sample count. These individual differences are also called residuals, and the RMSE serves to aggregate them into a single measure of predictive power [59].

$$RMSE = \left(\sqrt{\frac{\sum_{i=1}^{n} (X_{sim} - X_{obs})^2}{n}} \right) \tag{9}$$

Together with R^2 and RMSE, the MAE was explored to further evaluate the model [60]. MAE is the average magnitude of the errors in a set of predictions, without considering their direction [61]. It is the average over the test sample of the absolute differences between prediction and actual observation where all individual differences have equal weight, as seen in (10)

$$MAE = \frac{1}{n} \sum_{i=1}^{n} |X_{sim} - X_{obs}| \tag{10}$$

3. Study Area and Datasets

3.1. Study Area

The study area of this research was in Chuping district, Perlis state, which comprises a region of flat-terrain oil palm cultivation. The area mentioned is about 28 ha. It covers 4 years of old palm stands which have just started fruiting. The center point coordinates of the study area are 6°31′07.2″ N 100°19′07.7″ E. In Chuping district, there are seven sub-districts namely Panggas, Sungai Buloh, Kubang Perun, Guar Nangka, Felda Chuping, Sungai Buloh, and Kilang Gula Chuping. This research took place at the Kilang Gula Chuping subdistrict as shown in Figure 1. The field data collected in this research were soil moisture, leaf moisture content, and leaf area index from frond 17 of the palms. Field data collection was planned to match the acquisition date of PALSAR-2 as shown in Section 3.2.

In the research site, the soil type was identified as Chuping and Dampar soil series, where the Chuping soil type is categorized as sandy clay loam with a hue of 7.5–10 YR (Yellow to Red) following the Munsell color chart standards, and the latter soil type was identified as clay loam with a hue of 7.5 YR [62]. The site was considered flat topography with an elevation of 21.6 m while the slope class was identified as 4–12% [63]. As per the study site, the months of February to March of the calendar year are regarded as the driest season of the year; with a maximum of 28 days of dry spell every year and an average of 1362.38 mm of precipitation per year in the years of 2015–2017 [64]. In the PALSAR-2 image acquisition months, Table 1 shows the monthly meteorological data obtained from Malaysian Meteorological Department. The observation dates in Table 1 represent the annual weather of the study area where the month of January was observed to be drier than April and July; however, the Mean Evaporation and Mean Radiation were similarly recorded.

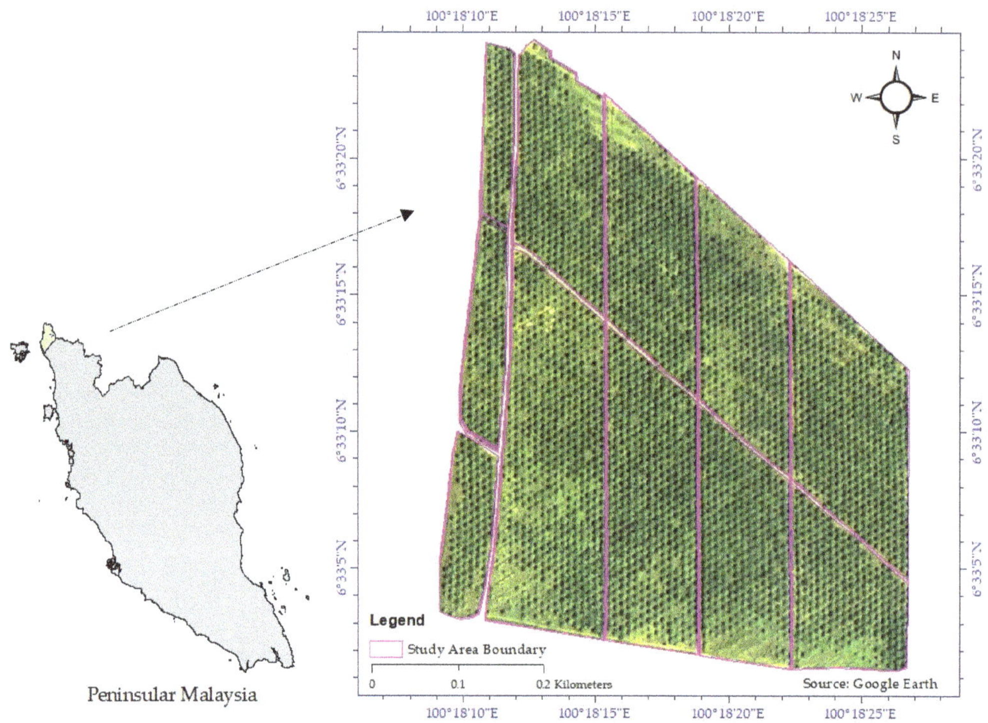

Figure 1. The oil palm cultivated area within the Chuping district in Perlis, Malaysia.

Table 1. Meteorological information of study area.

Month	Daily Meana Temperature (°C)	Total Precipitation for the Month (mm)	Daily Mean Radiation (MJm^{-2})	Daily Mean Evaporation (mm)
January	27.6	32.4	18.3	5.0
April	29.0	89.4	20.8	5.0
July	27.7	61.2	18.6	4.6

3.2. Data and Processing

PALSAR-2 signals were used for extracting the backscatter coefficient from the oil palm trees. High resolution scenes of PALSAR-2 were obtained under our collaboration with the Earth Observation Research Announcement 2 (EO-RA-2) of the Japanese Aerospace Exploration Agency (JAXA). We used a total of 3 PALSAR-2 scenes in this research, details of which are given in Table 2.

The acquired PALSAR-2 data were first converted from their Digital Number (DN) to backscatter coefficients in decibels, dB for both polarizations. Once the $\sigma^{\circ}{}_{HH}$ and $\sigma^{\circ}{}_{HV}$ for each of the field points was available, georeferencing of both images using SRTM (3 Sec) was done. The Lee filter was applied with a 5 × 5 window size to remove speckle and noise from the image. All the rectification process was completed in the SNAP open-source software. Additionally, Unmanned Aerial Vehicle (UAV) imagery was acquired on 17 January 2019 with the spatial resolution of 8 by 8 cm consisting of the red, blue, green, red edge and near infrared bands. The imagery from the UAV platform allowed us to compute the NDVI to identify bare soil areas in the field for the calibration process, with an NDVI value less than 0.2 taken to be bare soil. Ground truthing was done to validate the bare soil areas.

Table 2. Details of PALSAR-2 images.

Date of Acquisition	Flight Direction	Mode	Resolution	Polarization	Incident Angle
17 January 2019	Ascending	Strip Map 3	6.25 m × 6.25 m	HH + HV	30.4–42.4°
19 April 2019	Ascending	Strip Map 3	6.25 m × 6.25 m	HH + HV	41.2–53.3°
9 July 2019	Ascending	Strip Map 3	6.25 m × 6.25 m	HH + HV	30.4–42.4°

3.3. *In-Situ Data Collection*

The determination of soil moisture content at soil depth 0–5 cm was done using the soil gravimetric method. Fresh weight of soil moisture collected in field was recorded whereas the dry weight was determined after oven drying of the soil samples in the laboratory. A total of 32 points of soil sample were taken at each observation dated 17 January, 19 April and 9 July 2019, totalling to 96 soil samples altogether.

For the LAI for oil palm trees, a conventional method of destructive sampling was carried out using Equation (11). The frond to be chosen must not only be a good indicator of the nutrient status of the palm, but must also be one which is easy to identify and is consistent. By convention, frond 17 is used commonly in the oil palm fields. It is reported that frond 17 provides satisfactory results from its use as an input in the LAI formula [65]. According to the localised LAI formula, LAI was determined by the following equation:

$$LAI = \left(A_f \times F_n \times \frac{P_{DEN}}{10{,}000} \right) \mathrm{m}^2/\mathrm{m}^2 \tag{11}$$

where A_f is the leaf area per frond (m^2), F_n is the number of fronds per palm; P_{DEN} is the planting density where the number of palms per hectare is identified. Leaflets from the one-sided leaflet area were multiplied by two to obtain the total leaflet area of the frond in this research [66]. All leaflets were brought to the laboratory to determine the leaflet area using the LI-3100C, LI-COR Inc., which is USA-made equipment. Samples from the leaflets were taken and oven dried at 70 °C for 72 h until constant weight was achieved. An electronic balance was used to weigh the oven dried leaflets upon the completion of drying step. Similarly, the LWAI and NPWC were used from the same samples and methods as seen in Equations (6)–(8).

4. Results and Discussion

4.1. *In-Situ Results*

Field data collection for the three observation periods was based on the availability of PALSAR-2 orbit path data acquisition. In situ information was simultaneously collected for the soil moisture. LAI, LWAI and NPWC were calculated to understand the range values and average of each component in field. In Table 3, it can be noted that during the January and April data collection, the soil moisture average was 0.240 m^3/m^3 whereas the July data acquisition showed an increase in soil moisture recording at 0.273 m^3/m^3. Overall, for the palm trees cultivated areas, the soil moisture at the depths of 0–5 cm was averaged at 0.251 m^3/m^3. LAI for the palm fronds was averaged at 1.845 m^2/m^2 where it was seen to increase at every observation date, being 1.748, 1.784 and 2.005 m^2/m^2, respectively, during the January, April, and July observations dates. For the water content based on LAI in the palm fronds, an increase on each observation date was observed as well; the January observation date was at 0.205, 0.353 for the April observation date and 0.396 in the July observation date. This gave an average of 0.368 for all three observation dates. It can be noted that the study area faced drought stress in the earlier months of the calendar year where January to March recorded a lower rainfall [64]. Normalized plant water content was similar in all the observations at an average of 0.199%.

Table 3. Summary of in-situ data collection of Soil Moisture, LAI, LWAI and NPWC.

Observation Date	Soil Moisture (m^3/m^3)		LAI (m^2/m^2)		LWAI (% W in m^2/m^2)		NPWC (%)	
	Range	Mean	Range	Mean	Range	Mean	Range	Mean
17 January 2019	0.075–0.419	0.240	0.680–3.251	1.748	0.123–0.532	0.205	0.169–0.228	0.205
19 April 2019	0.170–0.316	0.240	0.662–3.174	1.784	0.125–0. 668	0.353	0.185–0.233	0.198
9 July 2019	0.119–0.454	0.273	1.214–3.078	2.005	0.076–0.661	0.396	0.052–0.221	0.195
Overall	0.075–0.454	0.251	0.661–3.251	1.845	0.076–0.668	0.368	0.052–0.233	0.199

4.2. Water Cloud Model Parameterization

The WCM model parameterization values were calibrated using bare soils and vegetation points from the NDVI values, derived from the RGB-NIR data obtained from UAV data acquisition, to fit the WCM model with the ground range values. Table 4 shows the vegetation parameters, A_{pp} and B_{pp} and the soil parameters C_{pp} and D_{pp} for the given polarization together with the combination indicator that represents the simulated scenarios of Cases 1–5. Using these vegetation and soil parameters together with LAI, LWAI, and NPWC as the input values, the WCM model was able to replicate the backscatter coefficients.

Table 4. WCM PALSAR-2 polarization calibration parameters for oil palm trees and combination vegetation descriptors used in this study.

Image Polarization	Combination Indicator	Vegetation Parameters		Soil Parameters	
		A_{pp}	B_{pp}	C_{pp}	D_{pp}
HH	Case 1	0.0118	0.0006	−26.0150	−2.8638
	Case 2	0.2188	0.0027	−26.0150	−2.8638
	Case 3	0.8467	0.0134	−26.0150	−2.8638
	Case 4	0.7122	0.0063	−26.0150	−2.8638
	Case 5	0.7457	0.0089	−26.0150	−2.8638
HV	Case 1	0.0850	0.0011	22.2070	−23.8660
	Case 2	0.1634	0.0047	22.2070	−23.8660
	Case 3	0.2530	0.0019	22.2070	−23.8660
	Case 4	0.4193	0.0321	22.2070	−23.8660
	Case 5	0.0182	0.0751	22.2070	−23.8660

4.3. Backscatter Simulation Based on the Proposed Vegetation Descriptors

Backscatter coefficients at frequency range of C and X bands are dominated by scattering activities in the crown layer of branches and foliage in the canopies, whereas scattering processes involving substantial trunks and branches would be dominated at lower frequencies like P and L bands [67]. In this study, the L band from PALSAR-2 was considered to give a better penetration in oil palm tree structure where backscatter coefficient characterizes the nature of oil palm structure. It was reported that oil palm trees using L band were capable to penetrate to the basal trunk in the oil palm plantation using both HH and HV polarizations [68]. Using the WCM model, the backscatter coefficients were simulated using a combination of vegetation descriptors. These simulated backscatter coefficients were then compared with the image-derived backscatter where the observed backscatter values were extracted from PALSAR-2. Table 5 shows the comparison of the backscatter coefficients observed and simulated for the study site. Both HH and HV polarizations showed a positive indication of a fitted WCM model using the respective vegetation descriptors where R^2 values ranged from 0.823 to 0.998, indicating the regression model fits the observed data well. In most HV polarization, the regression was found to be higher than HH polarization using the WCM model. On the evaluation of the RMSE and MAE, it was found that the errors were minimized in the HV polarization.

Table 5. Metrics comparison between observed and simulated backscatter coefficients from WCM for HH and HV polarization with respect to the vegetation descriptors.

Description (n = 96)	Case 1		Case 2		Case 3		Case 4		Case 5	
	HH	HV	HH	HV	HH	HV	HH	HV	HH	HV
R^2	0.962	0.997	0.956	0.965	0.823	0.998	0.919	0.995	0.969	0.951
RMSE (dB)	2.259	0.222	2.266	0.782	2.384	0.158	2.222	0.387	2.256	0.351
MAE	1.821	0.212	1.814	0.212	1.872	0.150	1.789	0.366	1.811	0.280

Using the LAI vegetation descriptor, the Case 3 where LAI was used in both V_1 and V_2, HV showed the highest R^2 value of 0.998 with a lowest RMSE of 0.158 dB, indicating an efficient estimation of simulated backscatter coefficient. In comparison, HH showed a lower R^2 value of 0.823. Based on the RMSE and MAE evaluation, HV recorded more accurate values of 0.158 dB and 0.150 for RMSE and MAE, respectively, when compared to HH polarization which had values of 2.384 dB and 1.872 for RMSE and MAE, respectively. In the Case 1 where $V_1 = 1$ and $V_2 = \text{LAI}$, the HV polarization was more efficient to the simulate the backscatter coefficient with RMSE 0.222 dB and $R^2 = 0.997$. However, the HH polarization, even though it shows good R^2 of 0.962, had higher RMSE with 2.259 dB. Using the LAI indicator as V_1 and V_2 as 1, both HH and HV polarizations showed a good R^2 with 0.956 and 0.965, respectively, but the RMSE was lower by 1.584 dB in the HV polarization. The MAE gave lower values for the HV polarization with 0.212. In Case 4 using LWAI, the model accurately predicted the simulated backscatter with $R^2 = 0.995$ in the HV with 0.387 dB RMSE and MAE of 0.366. When the NPWC vegetation descriptors was evaluated for HH polarization, its R^2 value of 0.969 was higher than the HV polarization R^2 of 0.951, but the MAE in HV polarization was computed to be lower than HH polarization by 1.531 and RMSE by 1.905 dB, which means the latter polarization is more accurately simulated. It is important to note that in the LAI combinations of the vegetation descriptors, dual use of LAI was found to be the most accurate parameter to simulate backscatter coefficient values. Furthermore, it is critical to produce an accurate simulation of the backscatter coefficient since it can convey the actual field range values obtained during the in-situ collection utilizing the model. This understanding is important to allow future work to be carried out with minimal calibration values and adopted into large areas of plantations where it is remotely challenging to access the collected field data and senses

4.4. Vegetation Effects on Soil Moisture Retrieval Based on Polarization

Vegetation descriptors play an equally important role in the estimation of the backscatter coefficient as well as providing reliable information to showcase the ground parameters in the WCM model. The backscattering coefficient mechanism interaction with soil moisture is complex [69]. Recently, in an attempt to reduce complexity in the modified IEM, the soil parameters were being simplified from three to two soil parameters [70]. Earlier studies have demonstrated the necessity to eliminate the effects of vegetation on soil moisture retrieval [71]. It was discovered that soil moisture retrieval is much more accurate when vegetation cover is considered [72]. To further explore the backscatter coefficients derived in Section 4.3 with different vegetation descriptors, the soil moisture was retrieved from the backscatter to allow comparison of observed soil moisture in the oil palm study area. It was seen that the simulated backscatter coefficient was predicted from the ground soil moisture values accurately given the observation period variability of lesser rainfall in the January observation when compared to the other observations. By making comparisons with the soil moisture retrieved from PALSAR-2 images, the model can be cross-evaluated to see the model fitting using the WCM. The vegetation descriptors evaluated as mentioned by Cases 1–5 can be seen in Table 6. In this comparison between the retrieved and observed soil moisture values, we have the best case with the highest R^2 value of 0.805 and the lowest RMSE of 0.046 m^3/m^3. When looking at Case 1 for HV polarization, the soil moisture retrieved was plotted against simulated backscatter coefficient where the R^2 was demonstrated at 0.916, as seen in Figure 2. Similar accuracy was achieved using Radarsat-2 utilizing the LAI descriptor, whereby RMSE of 0.069 m^3/m^3 was reported [73]. The initial combination introduced in the WCM [38], with V_1 as 1 and V_2 as LAI providing the most promising soil moisture retrieval in the oil palm trees in this study, as seen in Figure 2. LAI parameter was evaluated as one of the best parameters when VV polarization was compared to LWAI and NPWC, with 4.19–4.43% of the RMSE [51]. Signal backscattering is influenced by canopy structure where it is very sensitive to plant water content, a variable highly correlated with LAI during the vegetative phase [74].

Table 6. Comparison between statistical parameters of retrieved and observed soil moisture for HH and HV polarization in relation to the vegetation descriptors.

Image Polarization	Description (n = 96)	Case 1	Case 2	Case 3	Case 4	Case 5
HH	R^2	0.598	0.512	0.490	0.727	0.558
	RMSE (m^3/m^3)	0.088	0.091	0.101	0.085	0.089
	MAE	0.070	0.072	0.080	0.069	0.071
HV	R^2	0.805	0.609	0.675	0.459	0.301
	RMSE (m^3/m^3)	0.046	0.057	0.051	0.066	0.075
	MAE	0.043	0.047	0.044	0.050	0.058

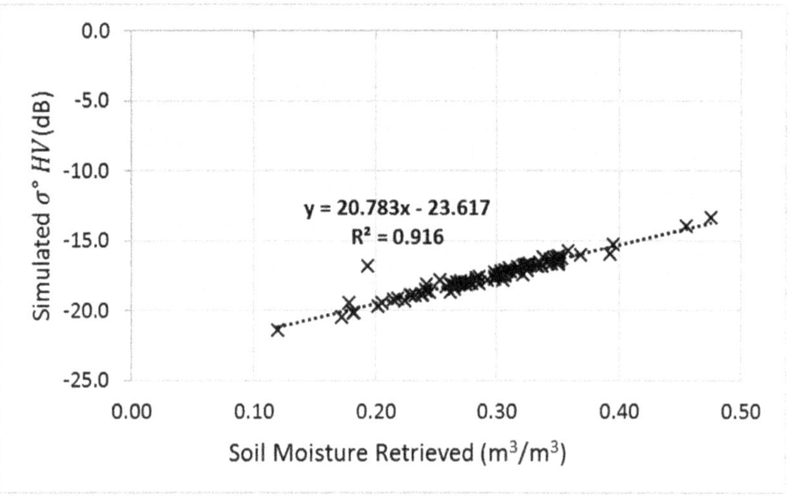

Figure 2. Soil moisture retrieved using vegetation descriptors of V_1 = 1 and V_2 = LAI, and simulated backscatter coefficient of HV polarization.

It can be observed that the Case 4 using LWAI as vegetation descriptors, using HH polarization gave an R^2 value of 0.727 where the RMSE was found to be 0.085 m^3/m^3, the HV polarization recorded a lower error with an RMSE value of 0.066 m^3/m^3 with R^2 value of 0.727. In contrast, the lowest R^2 value of 0.301 was from the NPWC case where RMSE was observed at 0.075 m^3/m^3. Figure 3 shows the observed soil moisture based on polarization of PALSAR images. It can be noted that the HV polarization gives better lower RMSE values as compared to HH polarization, considering all the vegetation descriptors [75]. It was reported that HH polarization could provide a low RMSE value of 0.049 m^3/m^3 [71] using remote sensing-based vegetation descriptors, in this case Normalized Difference Infrared Index (NDII). Similarly, when canopy water content based on LAI was applied as a vegetation descriptor, an RMSE of 0.039 was recorded [53]. Overall, when RMSE and

MAE are carefully evaluated, the cross-polarized HV backscatter coefficient is revealed to be more vulnerable than the co-polarized backscatter HH in terms of polarization response in all Cases 1–5 observed.

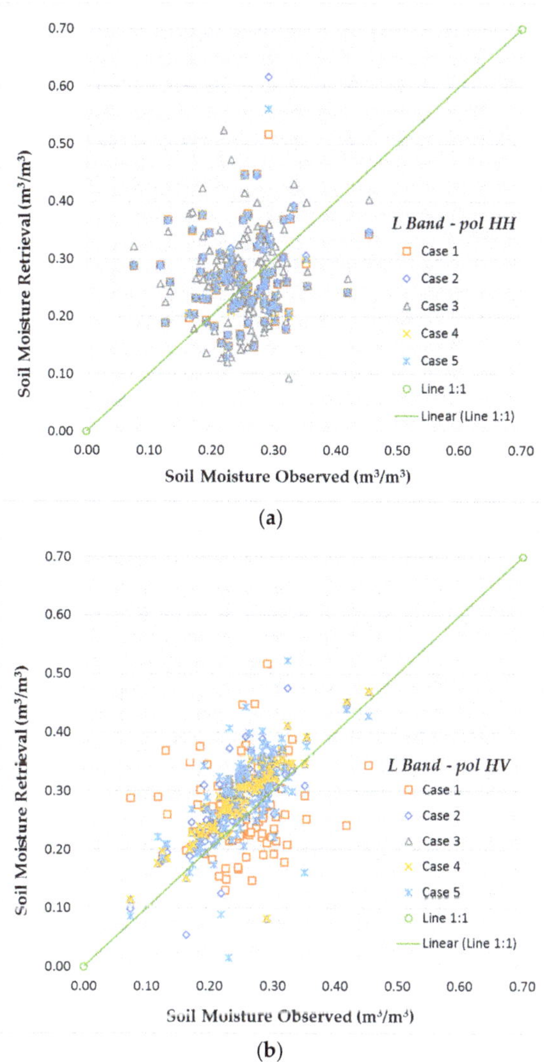

Figure 3. Observed soil moisture compared to retrieved soil moisture: Polarization in L band using (**a**) HH polarization and (**b**) HV polarization.

Our research results need to be interpreted with respect to the age of the oil palm stands which were 4 years old and our terrain type, whereby the terrain in our study was considered flat and the backscatter intensity represented in the acquired images was a composition of interactions with the crown, the trunk, and the ground surface of the current palm stands. However, it is known that oil palm crop growth or biomass increases over years, hence it is recommended that larger data sets need to be collected from the field to represent greater diversity of palm ages, possibly using fully polarimetric SAR data. Specifically, an understanding of the effects of backscattering on undulating grounds and peatlands in oil palm growing areas will be helpful in future. This could help in understanding the influence of SAR on other variables of importance to the oil palm industry, such as estimation of vegetation water content of oil palm canopy and yield. Benefits of this research can be useful for the plantation industry especially with the increase in available high-resolution multi-polarization satellite SAR sensors. This allows for the possibilities in exploiting the use of oil palm crop-related parameters supported with satellite data input to a decision-making platform for oil palm plantations.

5. Conclusions

In this study, L band from PALSAR-2 observations and in-situ data collected were used to evaluate a WCM model for soil moisture retrieval in oil palm cultivated regions. The WCM was calibrated using NDVI values, then fitted to a WCM model by different instances of vegetation descriptors to evaluate the best fit for the model. Using the WCM model and LAI specifically as V_1 and V_2, it was found to provide the best simulated backscatter coefficient in the HV polarization, secondly followed by V_1 which is set as constant with a value of 1 and V_2 as LAI. For HV polarization, LAI specifically as V_1 and V_2 showed an R^2 of 0.998 in simulating the backscatter coefficient accurately with RMSE of 0.158 dB, whereas HH polarization showed a lower R^2 of 0.823 with a higher RMSE of 2.384 dB. However, when LWAI and NPWC were considered, the model fitted better with co-polarized backscatter, where HH backscatter coefficient records R^2 of 0.727 and 0.558, respectively. It can be said that the co-polarized backscatters have shown lower RMSE values in the model fit for all cases where it can be predicted to be more useful in retrieving soil moisture in oil palm cultivated areas. In this respect, this research makes a useful and novel contribution on soil moisture for the benefit of the oil palm cultivation, and we found the PALSAR-2 sensor data beneficial for this purpose. The findings of this research will eventually help the oil palm growers to have systems in place to address the abrupt droughts brought about by climate change. Future work of this research can be explored with SAR-based vegetation descriptors, e.g., Radar Vegetation Index and optical derived indices such as NDVI, Normalised Difference Water Index and NDII.

Author Contributions: Conceptualization, V.S. and A.R.M.S.; Data curation, V.S.; Formal analysis, V.S.; Investigation, A.R.M.S.; Methodology, V.S., A.R.M.S., M.R.K., Y.P.L. and W.T.; Resources, A.R.M.S., A.W., M.R.K., Y.P.L. and W.T.; Software, V.S.; Supervision, A.R.M.S. and A.W.; Validation, V.S., A.R.M.S., M.R.K., Y.P.L. and W.T.; Visualization, A.R.M.S. and A.W.; Writing—original draft, V.S.; Writing—review and editing, V.S., A.R.M.S., A.W., M.R.K., Y.P.L. and W.T. All authors have read and agreed to the published version of the manuscript.

Funding: This research was funded by the Japanese Aerospace Agency (JAXA), using the Earth Observation-Research Announcement (EO-RA2) collaboration with PI number of ER2A2N180 to obtained PALSAR-2 data.

Data Availability Statement: PALSAR-2 data was requested via proposal where restrictions apply for high resolution data.

Acknowledgments: The authors would like to thank Rajah Selvarajah for providing financial support for purchasing UAV data and field work logistics to complete the research work. The authors would like to express their gratitude to the Soil and Water Conservation Laboratory, Department of Biological and Agricultural Engineering, Faculty of Engineering, University Putra Malaysia (UPM), particularly Mohamad Hafis bin Ramli, for assistance with soil-related matters, particularly providing equipment for extracting soil samples and oven drying lab equipment. The authors would also like to express their gratitude to Ghazali bin Kassim, who arranged for GPS equipment from the Spatial Information System Laboratory, Biological and Agricultural Engineering, Faculty of Engineering, UPM, to be utilized for locating soil samples points in the field. Authors would like to thank the FGV R&D Sdn. Bhd. to support in-field arrangement leading to a smooth data collection especially to Mohd Shahkhirat bin Norizan, Syahidah Abu Hassan, Mazuan, Haryati Abidin, Noorsusilawati bt. Mandangan, Mohd Mahfuz Roslan, Muhammad Farid Abdul Rahim and Mohd Na'aim bin Samad.

Conflicts of Interest: The authors declare no conflict of interest in the process of writing this paper locally and internationally.

References

1. Benites, J.; Castellanos, A. Improving soil moisture with conservation agriculture. *Leisa Mag.* **2003**, *19*, 6–7.
2. Verheye, W. Growth and production of oil palm. In *Encyclopedia of Life Support Systems (EOLSS)*; UNESCO-EOLSS Publishers: Oxford, UK, 2010.
3. Noor, M.R.M.; Harun, M.H. Water deficit and irrigation in oil palm: A review of recent studies and findings. *Oil Palm Bull.* **2004**, *49*, 1–6.
4. Abubaker, J. Irrigation scheduling for efficient water use in dry climates. Master's Thesis, Swedish University of Agricultural Sciences, Uppsala, Sweden, 2001; p. 47.
5. Fahad, S.; Bajwa, A.A.; Nazir, U.; Anjum, S.A.; Farooq, A.; Zohaib, A.; Sadia, S.; Nasim, W.; Adkins, S.; Saud, S.; et al. Crop Production under Drought and Heat Stress: Plant Responses and Management Options. *Front. Plant Sci.* **2017**, *8*, 1147. [CrossRef] [PubMed]
6. Cha-um, S.; Yamada, N.; Takabe, T.; Kirdmanee, C. Physiological features and growth characters of oil palm (*Elaeis guineensis* jacq.) in response to reduced water-deficit and rewatering. *Aust. J. Crop Sci.* **2013**, *7*, 432–439.
7. Ng, S.K. Nutrition and nutrient management of the oil palm. New thrust for the future perspective. In *Potassium for Sustainable Crop Production. International Symposium on Role of Potassium in India*; Potash Research Institute of India and International Potash Institute: New Delhi, India, 2002; pp. 415–429.
8. Sundram, S.; Angel, L.P.L.; Sirajuddin, S.A. Integrated balanced fertiliser management in soil health rejuvenation for a sustainable oil palm cultivation: A review. *J. Oil Palm Res.* **2019**, *31*, 348–363. [CrossRef]
9. Waite, P.-A.; Schuldt, B.; Link, R.M.; Breidenbach, N.; Triadiati, T.; Hennings, N.; Saad, A.; Leuschner, C. Soil moisture regime and palm height influence embolism resistance in oil palm. *Tree Physiol.* **2019**, *39*, 1696–1712. [CrossRef] [PubMed]
10. Haron, K.; Tarmizi, A.M. Techniques of Soil and Water Conservation and Nutrient Recycling in Oil Palm Plantations on Inland Soils. *Oil Palm Bull.* **2008**, *56*, 1–11.
11. May, C.Y. Malaysia: Economic transformation advances oil palm industry. *Am. Oil Chem. Soc.* **2012**, *23*, 536–542.
12. Adesiji, A.R.; Nik Daud, N.N.; Asogwa, E.O.; Jarumi, A.M.; Musa, H.H.; Adaudu, I.I. Irrigation and crop water requirement estimation for oil palms using soil moisture balance model in Peninsular Malaysia. *J. Agric. Food Eng.* **2020**, *3*, 1–8.
13. Teixeira, W.G.; Huwe, B.; Schroth, G.; Marques, J.D. Sampling and TDR probe insertion in the determination of the volumetric soil water content. *Revista Brasileira de Ciência do Solo* **2003**, *27*, 575–582. [CrossRef]
14. Cepuder, P.; Evett, S.; Heng, L.K.; Hignett, C.; Laurent, J.P.; Ruelle, P.; Moutonnet, P.; Nguyen, M.L. Field estimation of soil water content: A practical guide to methods, instrumentation, and sensor technology. *IAEA Vienna* **2008**, *30*, 24.
15. Sishodia, R.P.; Ray, R.L.; Singh, S.K. Applications of Remote Sensing in Precision Agriculture: A Review. *Remote Sens.* **2020**, *12*, 3136. [CrossRef]
16. Njoku, E.G.; Entekhabi, D. Passive microwave remote sensing of soil moisture. *J. Hydrol.* **1996**, *184*, 101–129. [CrossRef]
17. Schmugge, T.J.; Kustas, W.P.; Ritchie, J.C.; Jackson, T.J.; Rango, A. Remote sensing in hydrology. *Adv. Water Resour.* **2002**, *25*, 1367–1385. [CrossRef]
18. Rau, M.I. Remote Sensing Approaches to Observe the Significance of Landuse Change on Water Storage at Oil Palm Plantation in Kuala Langat South Forest Reserve, Malaysia. Master's Dissertation, University of Glasgow, Glasgow, UK, 2008.
19. Foong, A.; Sum, W.; Shukor, S.A.A. Oil Palm Plantation Monitoring from Satellite Image. *IOP Conf. Ser. Mate. Sci. Eng* **2019**, *705*, 1–8.
20. Rawi, R.; Miit, U.K.L.; Hasnan, M.S.I.; Sajak, A.A.B. Palm Oil Soil Monitoring System for Smart Agriculture. *Int. J. Integr. Eng.* **2020**, *12*, 189–199. [CrossRef]
21. Palazzi, V.; Bonafoni, S.; Alimenti, F.; Mezzanotte, P.; Roselli, L. Feeding the World with Microwaves: How Remote and Wireless Sensing Can Help Precision Agriculture. *IEEE Microw. Mag.* **2019**, *20*, 72–86. [CrossRef]
22. Zhuo, L.; Han, D. Science Direct the relevance of soil moisture by remote sensing and hydrological modelling. *Procedia Eng.* **2016**, *154*, 1368–1375. [CrossRef]
23. Lakshmi, V. Remote Sensing of Soil Moisture. *ISRN Soil Sci.* **2013**, *2013*, 33. [CrossRef]
24. Petropoulos, G.P.; Ireland, G.; Barrett, B. Surface soil moisture retrievals from remote sensing: Current status, products & future trends. *Phys. Chem. Earth Parts A/B/C* **2015**, *83–84*, 36–56.
25. ABalenzano, A.; Mattia, F.; Satalino, G.; Davidson, M.W.J. Dense Temporal Series of C- and L-band SAR Data for Soil Moisture Retrieval Over Agricultural Crops. *IEEE J. Sel. Top. Appl. Earth Obs. Remote Sens.* **2011**, *4*, 439–450. [CrossRef]
26. Joseph, A.T.; Van der Velde, R.; O'Neill, P.E.; Lang, R.; Gish, T. Effects of corn on C- and L-band radar backscatter: A correction method for soil moisture retrieval. *Remote Sens. Environ.* **2010**, *114*, 2417–2430. [CrossRef]
27. Barrett, B.; Dwyer, E.; Whelan, P. Soil Moisture Retrieval from Active Spaceborne Microwave Observations: An Evaluation of Current Techniques. *Remote Sens.* **2009**, *1*, 210–242. [CrossRef]
28. Ottinger, M.; Kuenzer, C. Spaceborne L-Band Synthetic Aperture Radar Data for Geoscientific Analyses in Coastal Land Applications: A Review. *Remote Sens.* **2020**, *12*, 2228. [CrossRef]

29. Nordin, L.; Shahruddin, A.; Mariamni, H. Application of AIRSAR data to oil palm tree characterization. In Proceedings of the Asian Conference on Remote Sensing, Bangkok, Thailand, 7–9 August 2002; pp. 1–7.
30. Shashikant, V.; Shariff, A.R.M.; Nordin, L.; Pradhan, B. Estimation of above ground biomass of oil palm trees by PALSAR. In Proceedings of the CHUSER 2012—2012 IEEE Colloquium on Humanities, Science and Engineering, Kota Kinabalu, Malaysia, 3–4 December 2012; pp. 838–841.
31. Morel, A.C.; Saatchi, S.S.; Malhi, Y.; Berry, N.J.; Banin, L.; Burslem, D.; Nilus, R.; Ong, R.C. Estimating aboveground biomass in forest and oil palm plantation in Sabah, Malaysian Borneo using ALOS PALSAR data. *For. Ecol. Manag.* **2011**, *262*, 1786–1798. [CrossRef]
32. Bindlish, R.; Barros, A.P. Parameterization of vegetation backscatter in radar-based, soil moisture estimation. *Remote Sens. Environ.* **2001**, *76*, 130–137. [CrossRef]
33. Chen, K.; Wu, T.-D.; Tsang, L.; Li, Q.; Shi, J.; Fung, A. Emission of rough surfaces calculated by the integral equation method with comparison to three-dimensional moment method simulations. *IEEE Trans. Geosci. Remote Sens.* **2003**, *41*, 90–101. [CrossRef]
34. Fung, A.K.; Chen, K.S. Dependence of the surface backscattering coefficients on roughness, frequency and polarization states. *Int. J. Remote Sens.* **2007**, *13*, 1663–1680. [CrossRef]
35. Dubois, P.; Van Zyl, J.; Engman, T. Measuring soil moisture with imaging radars. *IEEE Trans. Geosci. Remote Sens.* **1995**, *33*, 915–926. [CrossRef]
36. Gorrab, A.; Zribi, M.; Baghdadi, N.; Lili-Chabaane, Z.; Mougenot, B. Multi-frequency analysis of soil moisture vertical heterogeneity effect on radar backscatter. In Proceedings of the International Conference on Advanced Technologies for Signal and Image Processing (ATSIP), Sousse, Tunisia, 17–19 March 2014; pp. 379–384.
37. Dobson, M.C.; Ulaby, F.T.; Pierce, L.E.; Sharik, T.L.; Bergen, K.M.; Kellndorfer, J.; Kendra, J.R.; Li, E.; Lin, Y.C.; Nashashibi, A.; et al. Estimation of forest biophysical characteristics in Northern Michigan with SIR-C/X-SAR. *IEEE Trans. Geosci. Remote Sens.* **1995**, *33*, 877–895. [CrossRef]
38. Attema, E.P.W.; Ulaby, F.T. Vegetation modeled as a water cloud. *Radio Sci.* **1978**, *13*, 357–364. [CrossRef]
39. Ulaby, F.T.; Stiles, W.H.; Abdelrazik, M. Snowcover Influence on Backscattering from Terrain. *IEEE Trans. Geosci. Remote Sens.* **1984**, *GE-22*, 126–133. [CrossRef]
40. Baghdadi, N.; King, C.; Chanzy, A.; Wigneron, J.P. An empirical calibration of the integral equation model based on SAR data, soil moisture and surface roughness measurement over bare soils. *Int. J. Remote Sens.* **2010**, *23*, 4325–4340. [CrossRef]
41. Moran, M.S.; Peters-Lidard, C.D.; Watts, J.M.; McElroy, S. Estimating soil moisture at the watershed scale with satellite-based radar and land surface models. *Can. J. Remote Sens.* **2014**, *30*, 805–826. [CrossRef]
42. Gherboudj, I.; Magagi, R.; Berg, A.A.; Toth, B. Soil moisture retrieval over agricultural fields from multi-polarized and multi-angular RADARSAT-2 SAR data. *Remote Sens. Environ.* **2011**, *115*, 33–43. [CrossRef]
43. Prévot, L.; Champion, I.; Guyot, G. Estimating surface soil moisture and leaf area index of a wheat canopy using a dual-frequency (C and X bands) scatterometer. *Remote Sens. Environ.* **1993**, *46*, 331–339. [CrossRef]
44. Lang, R.H.; Saleh, H.A. Microwave Inversion of Leaf Area and Inclination Angle Distributions from Backscattered Data. *IEEE Trans. Geosci. Remote Sens.* **1985**, *GE-23*, 685–694. [CrossRef]
45. Ulaby, F.; Allen, C.; Eger, G.; Kanemasu, E. Relating the microwave backscattering coefficient to leaf area index. *Remote Sens. Environ.* **1984**, *14*, 113–133. [CrossRef]
46. Xu, H.; Steven, M.D.; Jaggard, K.W. Monitoring leaf area of sugar beet using ERS-1 SAR data. *Int. J. Remote Sens.* **2007**, *17*, 3401–3410. [CrossRef]
47. Dabrowska-Zielinska, K.; Inoue, Y.; Kowalik, W.; Gruszczynska, M. Inferring the effect of plant and soil variables on C- and L-band SAR backscatter over agricultural fields, based on model analysis. *Adv. Space Res.* **2007**, *39*, 139–148. [CrossRef]
48. Chauhan, S.; Srivastava, H.S.; Patel, P. Wheat crop biophysical parameters retrieval using hybrid-polarized RISAT-1 SAR data. *Remote Sens. Environ.* **2018**, *216*, 28–43. [CrossRef]
49. Li, J.; Wang, S. Using SAR-Derived Vegetation Descriptors in a Water Cloud Model to Improve Soil Moisture Retrieval. *Remote Sens.* **2018**, *10*, 1370. [CrossRef]
50. Ulaby, F.T.; Moore, R.K.; Fung, A.K. Microwave Remote Sensing. Active and Passive. From Theory to Applications. *Geol. Mag.* **1986**, *124*, 88.
51. Kumar, K.; Rao, H.P.S.; Arora, M.K. Study of water cloud model vegetation descriptors in estimating soil moisture in Solani catchment. *Hydrol. Process.* **2014**, *29*, 2137–2148. [CrossRef]
52. Khabazan, S.; Hosseini, M.; Saradjian, M.R.; Motagh, M.; Magagi, R. Accuracy Assessment of IWCM Soil Moisture Estimation Model in Different Frequency and Polarization Bands. *J. Indian Soc. Remote Sens.* **2015**, *43*, 859–865. [CrossRef]
53. Tao, L.; Wang, G.; Chen, X.; Li, J.; Cai, Q. Estimation of soil moisture using a vegetation scattering model in wheat fields. *J. Appl. Remote Sens.* **2019**, *13*, 044503. [CrossRef]
54. Zribi, M.; Baghdadi, N.; Bousbih, S.; El-Hajj, M.; Gao, Q.; Escorihuela, M.J.; Muddu, S. Soil Surface Moisture Estimation Using the Synergy S1/S2 Data. In Proceedings of the IGARSS 2018—2018 IEEE International Geoscience and Remote Sensing Symposium, Valencia, Spain, 22–27 July 2018; pp. 6119–6122.
55. Chauhan, S.; Srivastava, H.S.; Patel, P. Improved Parameterization of Water Cloud Model for Hybrid-Polarized Backscatter Simulation Using Interaction Factor. *ISPRS Int. Arch. Photogramm. Remote Sens. Spat. Inf. Sci.* **2017**, *XLII-4/W2*, 61–66. [CrossRef]

56. Gavin, H.P. *The Levenberg-Marquardt Algorithm for Nonlinear Least Squares Curve-Fitting Problems*; Department of Civil and Environmental Engineering, Duke University: Durham, NC, USA, 2020.
57. Mandal, D.; Kumar, V.; McNairn, H.; Bhattacharya, A.; Rao, Y. Joint estimation of Plant Area Index (PAI) and wet biomass in wheat and soybean from C-band polarimetric SAR data. *Int. J. Appl. Earth Obs. Geoinf.* **2019**, *79*, 24–34. [CrossRef]
58. Yadav, V.P.; Prasad, R.; Bala, R. Leaf area index estimation of wheat crop using modified water cloud model from the time-series SAR and optical satellite data. *Geocarto Int.* **2019**, *36*, 791–802. [CrossRef]
59. Irizarry, R.A. *Introduction to Data Science: Data Analysis and Prediction Algorithms with R*; CRC Press: Boca Raton, FL, USA, 2019.
60. Schönbrodt-Stitt, S.; Ahmadian, N.; Kurtenbach, M.; Conrad, C.; Romano, N.; Bogena, H.R.; Vereecken, H.; Nasta, P. Statistical Exploration of SENTINEL-1 Data, Terrain Parameters, and in-situ Data for Estimating the Near-Surface Soil Moisture in a Mediterranean Agroecosystem. *Front. Water* **2021**, *3*, 75. [CrossRef]
61. Hair, J.F.; Tomas, G.; Hult, M.; Ringle, C.M.; Sarstedt, M. *A Primer on Partial Least Squares Structural Equation Modeling (PLS-SEM)*, 3rd ed.; SAGE Publications: London, UK, 2021.
62. National Geospatial Centre. *Malaysian Standard Geographic Information/Geomatics Features and Attribute Codes*; Ministry of Energy and Natural Resources (KeTSA): Putrajaya, Malaysia, 2020.
63. Norizan, M.S.; Wayayok, A.; Karim, Y.A.; Abdullah, A.F.; Mahadi, M.R. Quantitative approach for irrigation requirement of oil palm: Case study in Chuping, Northern Malaysia. *J. Oil Palm Res.* **2021**, *33*, 278–288. [CrossRef]
64. Shashikant, V.; Mohamed Shariff, A.R.; Wayayok, A.; Kamal, M.R.; Lee, Y.P.; Takeuchi, W. Utilizing TVDI and NDWI to Classify Severity of Agricultural Drought in Chuping, Malaysia. *Agronomy* **2021**, *11*, 1243. [CrossRef]
65. Kok, Z.H.; Mohamed Shariff, A.R.; Khairunniza-Bejo, S.; Kim, H.-T.; Ahamed, T.; Cheah, S.S.; Wahid, S.A.A. Plot-Based Classification of Macronutrient Levels in Oil Palm Trees with Landsat-8 Images and Machine Learning. *Remote Sens.* **2021**, *13*, 2029. [CrossRef]
66. Awal, M.; Ishak, W.W. Measurement of Oil Palm LAI by Manual and LAI-2000 Method. *Asian J. Sci. Res.* **2007**, *1*, 49–56. [CrossRef]
67. Imhoff, M.L. Radar backscatter and biomass saturation: Ramifications for global biomass inventory. *IEEE Trans. Geosci. Remote Sens.* **1995**, *33*, 511–518. [CrossRef]
68. Hashim, I.; Mohamed Shariff, A.R.; Bejo, S.; Muharam, F.; Ahmad, K. Machine-Learning Approach Using SAR Data for the Classification of Oil Palm Trees That Are Non-Infected and Infected with the Basal Stem Rot Disease. *Agronomy* **2021**, *11*, 532. [CrossRef]
69. Jiang, N.; Xu, T.; Xu, Y.; Xu, G.; Schuh, H. Assessment of Different Stochastic Models for Inter-System Bias between GPS and BDS. *Remote Sens.* **2019**, *11*, 989. [CrossRef]
70. Baghdadi, N.; Zribi, M.; Paloscia, S.; Verhoest, N.E.C.; Lievens, H.; Baup, F.; Mattia, F. Semi-Empirical Calibration of the Integral Equation Model for Co-Polarized L-Band Backscattering. *Remote Sens.* **2015**, *7*, 13626–13640. [CrossRef]
71. Zhang, L.; Meng, Q.; Yao, S.; Wang, Q.; Zeng, J.; Zhao, S.; Ma, J. Soil Moisture Retrieval from the Chinese GF-3 Satellite and Optical Data over Agricultural Fields. *Sensors* **2018**, *18*, 2675. [CrossRef]
72. Kirimi, F.; Kuria, D.N.; Thonfeld, F.; Amler, E.; Mubea, K.; Misana, S.; Menz, G. Influence of Vegetation Cover on the Oh Soil Moisture Retrieval Model: A Case Study of the Malinda Wetland, Tanzania. *Adv. Remote Sens.* **2016**, *5*, 28–42. [CrossRef]
73. Yue, J.; Yang, G.; Qi, X.; Wang, Y. Soil moisture retrieval in well covered farmland by Radarsat-2 SAR data. *Int. Geosci. Remote Sens. Symp.* **2016**, *3*, 1699–1702.
74. Beriaux, E.; Lucau-Danila, C.; Auquiere, E.; Defourny, P. Multiyear independent validation of the water cloud model for retrieving maize leaf area index from SAR time series. *Int. J. Remote Sens.* **2013**, *34*, 4156–4181. [CrossRef]
75. Srivastava, H.S.; Patel, P.; Sharma, Y.; Navalgund, R.R. Multi-frequency and multi-polarized SAR response to thin vegetation and scattered trees. *Curr. Sci.* **2009**, *97*, 425–429.

Article

Soil Moisture Retrieval over a Vegetation-Covered Area Using ALOS-2 L-Band Synthetic Aperture Radar Data

Ya Gao [1], Maofang Gao [2,*], Liguo Wang [1,3] and Offer Rozenstein [4]

[1] College of Information and Communication Engineering, Harbin Engineering University, Harbin 150000, China; gaoya0001@hrbeu.edu.cn (Y.G.); wangliguo@hrbeu.edu.cn (L.W.)
[2] Key Laboratory of Agricultural Remote Sensing, Ministry of Agriculture and Rural Affairs, Institute of Agricultural Resources and Regional Planning, Chinese Academy of Agricultural Sciences, Beijing 100089, China
[3] College of Information and Communications Engineering, Dalian Minzu University, Dalian 116600, China
[4] Institute of Soil, Water and Environmental Sciences, Agricultural Research Organization—Volcani Institute, HaMaccabim Road 68, P.O. Box 15159, Rishon LeZion 7528809, Israel; offerr@volcani.agri.gov.il
* Correspondence: gaomaofang@caas.cn

Abstract: Soil moisture (SM) plays a significant part in regional hydrological and meteorological systems throughout Earth. It is considered an indispensable state variable in earth science. The high sensitivity of microwave remote sensing to soil moisture, and its ability to function under all weather conditions at all hours of the day, has led to its wide application in SM retrieval. The aim of this study is to evaluate the ability of ALOS-2 data to estimate SM in areas with high vegetation coverage. Through the water cloud model (WCM), the article simulates the scene coupling between active microwave images and optical data. Subsequently, we use a genetic algorithm to optimize back propagation (GA-BP) neural network technology to retrieve SM. The vegetation descriptors of the WCM, derived from optical images, were the normalized difference vegetation index (NDVI), the normalized difference water index (NDWI), and the normalized multi-band drought index (NMDI). In the vegetation-covered area, 240 field soil samples were collected simultaneously with the ALOS-2 SAR overpass. Soil samples at two depths (0–10 cm, 20–30 cm) were collected at each sampling site. The backscattering of the ALOS-2 with the copolarization was found to be more sensitive to SM than the crosspolarization. In addition, the sensitivity of the soil backscattering coefficient to SM at a depth of 0–10 cm was higher than at a depth of 20–30 cm. At a 0–10 cm depth, the best results were the mean square error (MAE) of 2.248 vol%, the root mean square error (RMSE) of 3.146 vol%, and the mean absolute percentage error (MAPE) of 0.056 vol%, when the vegetation is described as by the NDVI. At a 20–30 cm depth, the best results were an MAE of 2.333 vol%, an RMSE of 2.882 vol%, a MAPE of 0.067 vol%, with the NMDI as the vegetation description. The use of the GA-BP NNs method for SM inversion presented in this paper is novel. Moreover, the results revealed that ALOS-2 data is a valuable source for SM estimation, and ALOS-2 L-band data was sensitive to SM even under vegetation cover.

Keywords: soil moisture; ALOS-2, GA-BP; water cloud model; L-band

Citation: Gao, Y.; Gao, M.; Wang, L.; Rozenstein, O. Soil Moisture Retrieval over a Vegetation-Covered Area Using ALOS-2 L-Band Synthetic Aperture Radar Data. *Remote Sens.* **2021**, *13*, 3894. https://doi.org/10.3390/rs13193894

Academic Editors: Takeo Tadono and Masato Ohki

Received: 19 August 2021
Accepted: 24 September 2021
Published: 29 September 2021

1. Introduction

Soil moisture (SM) is an important state variable that significantly affects the water cycle, ecosystem, and energy exchange between the land and the atmosphere. SM information is important in different fields, such as agriculture, meteorology, hydrology, weather, and evapotranspiration forecasting [1–7]. In recent decades, in the large-scale domain, the development of remote sensing technology has provided more opportunities. Specifically, surface SM inversion based on remote sensing technology has become a hotspot of research [8–11]. Optical remote sensing cannot penetrate clouds and rain and is easily restricted by weather and solar illumination conditions. Therefore, it is impossible to perform an all-weather observation of the Earth in the optical and thermal spectral regions. However, microwave remote sensing is not subject to weather and light conditions and can monitor surface SM under all weather conditions at all hours of the day [12]. Moreover, it has a certain penetrating ability for vegetation and, therefore, it has the potential for the continuous monitoring of surface SM over a large area [13–20].

In the past few decades, synthetic aperture radar (SAR) has been the primary active microwave remote sensing means of monitoring SM [6,21]. However, in vegetation-covered areas, the signal includes direct scattering from plant cover and attenuated backscatter from the ground. As a result, the observed backscattering signal includes the vegetation, the surface, and the interaction between vegetation and the surface simultaneously [22], making it extremely difficult to retrieve SM under the vegetation coverage. Therefore, the main challenge to estimating SM vegetation coverage is the elimination of the impact of surface roughness and vegetation [1,23–25]. Many scholars have proposed different solutions to the impact of vegetation on radar backscattering in vegetation-covered areas. In many studies, the WCM is applied in an inversion method to estimate SM in vegetation-covered areas. In the WCM, the total reflected radar signal is modeled as a function of soil and vegetation contributions. The direct contribution of the vegetation's scattering and attenuation is mainly calculated by using biophysical parameters representing vegetation. Optical data can be used to estimate biophysical parameters [12,26]. Therefore, combining optical and SAR data is beneficial for SM retrieval in vegetation-covered areas [27–38].

However, due to the topography, the actual measurement (SM, soil roughness, etc.), and other factors, the lack of soil roughness and other relevant information will affect the inversion accuracy. In order to solve this problem, the purpose of this study is to evaluate the potential of combining L-band SAR data and optical data to estimate SM under vegetation. Research on the basics of the inversion of the WCM using GA-BP neural networks was developed to solve the problem of the lack of soil roughness and other factors. This study consists of four main parts: (1) the parameterization of WCM; (2) a learning simulation of synthetic SAR data; (3) training of a GA-BP neural network; (4) applied training and the verification of the results of the inversion method on real datasets. Section 2 of the paper describes the study area and in situ measurements. Section 3 investigates the methods. Section 4 features the results. Section 5 presents the discussion. Finally, Section 6 outlines the main conclusions.

2. Study Area

2.1. Study Area

The area of interest, situated in the Liuzhi Special District of Guizhou province (centered at 105.159°E, 26.541°N), was selected for SM estimation research (Figure 1). The Liuzhi Special Zone is marked by a warm and humid subtropical monsoon climate with abundant rainfall. It is located between mountains with steep terrain and a high altitude. In addition, it is characterized by high vegetation cover, with a forest coverage rate of 51.05%, rich crop species, and a wide grassland range. Because of the influence of topography and vegetation, natural disasters, such as landslides and debris flow, are prone to occur in this area. Therefore, the motivation to estimate SM in this region is very high since disasters are driven by soil water content changes that significantly affect the economy, the environment, and people's lives.

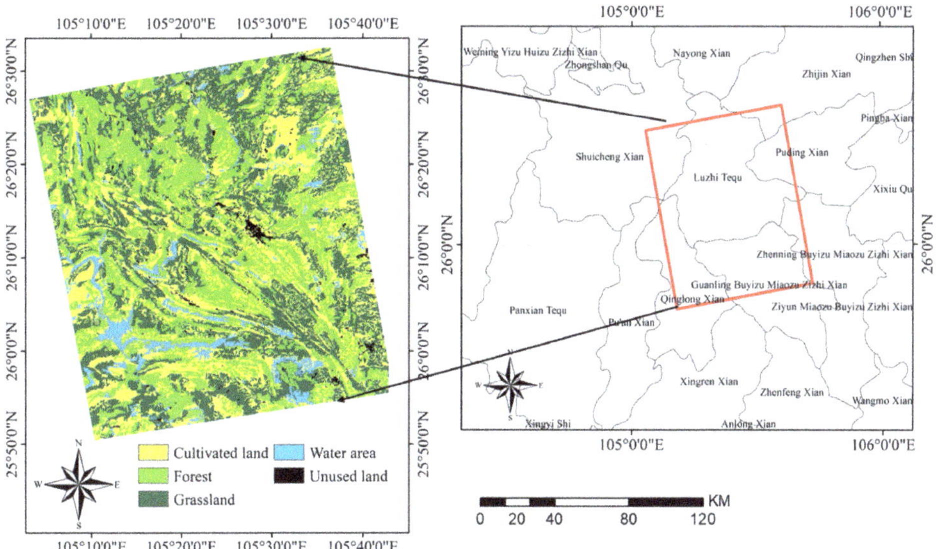

Figure 1. The study region of the land use patterns in Liuzhi (**left**), and a district map (**right**).

2.2. Radar Data

In this research, the ALOS-2 SAR data was utilized for SM inversion research. The ALOS-2 is the only L-band SAR satellite currently in orbit. Its frequency is 1.2 GHz. This study uses fine dual-polarization modes (HH and HV) with a nominal spatial coverage resolution of 3 m. Using an algorithm to calibrate SAR images by the Japan Aerospace Exploration Agency (JAXA), the digital number (*DNi*) of each pixel was converted to the radar backscatter coefficient (σ^0_{pq}) by radiation calibration. The calculation formula for radiation calibration of the ALOS-2 data is:

$$\sigma^0_{pq} = K \cdot DN_i^2 \tag{1}$$

The absolute calibration constant is represented by *K*. Following radiometric calibration, the backscattering coefficient had linear units, which was then converted to dB units for convenience [35] using Equation (2):

$$\sigma^0_{dB} = 10 \cdot \log_{10}\left(\sigma^0_{pq}\right) \tag{2}$$

SARscape, an advanced radar image processing software developed by SARmap (Purasca, Switzerland), was used to process the ALOS-2 data by performing multilooking, filtering, geocoding, and radiometric calibration.

2.3. Optical Data

The operational land imager (OLI) sensor, and the thermal infrared sensor (TIRS), are two instruments on NASA's Landsat-8 satellite [39]. Landsat-8 and Landsat 4–7 products have similar spatial resolution and spectroscopic characteristics. There is a total of 11 bands in Landsat-8 imagery. The spatial resolution of bands 1–7 and 9–11 is 30 m. The spatial resolution of band 8 is 15 m. The satellite achieves global coverage every 16 days (https://earthexplorer.usgs.gov/, accessed on 19 August 2021).

The NDVI is a significant measure reflecting crop growth and nutrition information [40]. The NDWI is an important index for evaluating vegetation water status [41]. The NMDI was calculated based on a near-infrared and two short wave infrareds [42,43].

2.4. In Situ Measurements

While acquiring ALOS-2 SAR data, the field measurements were performed in the study area from 26 September 2020 to 27 September 2020. During this period, there was no precipitation or significant temperature changes in the experimental area. In the research field, the appropriate sampling points were selected to collect soil samples. The terrain of the Liuzhi Special Zone in Guizhou Province is complex and there are many mountains, which increases the difficulty of sampling. Therefore, the areas we sampled were all flat areas, such as farmland and grassland. At the same time, in order to make the samples diverse, we tried to distribute them as widely as possible. A handheld GPS was used to record the longitude and latitude of each sample point, take photos of the sampling point (one photo), and take pictures of the surrounding environment (four photos from the front, back, left, and right). We finally selected 120 sampling points. 119 samples were taken at depths of 0–10 cm, and 120 samples were taken at depths of 20–30 cm. Figure 2 shows the distribution of sampling points.

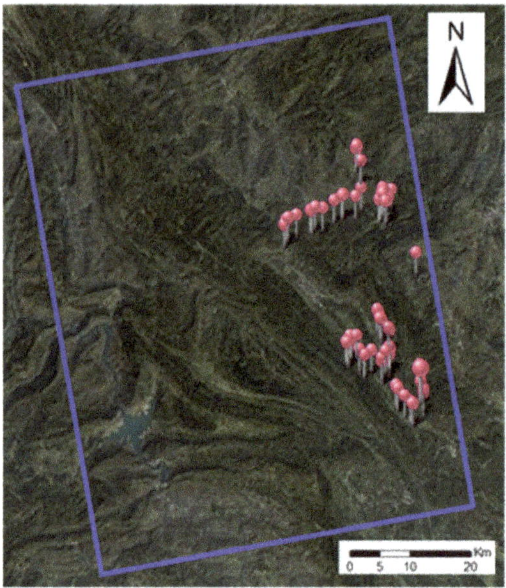

Figure 2. Distribution map of sampling points in the study area.

The determination of the soil moisture content was obtained by the laboratory weighing and drying of soil samples. Water content can be expressed as gravimetric water content and volumetric water content. The ratio of the quantity of water in the soil and the dry soil is gravimetric water content (M_g). It is represented by Formula (3).

$$M_g = \frac{M_w}{M_s} \tag{3}$$

where M_w represents the quality of water in the soil, and M_s represents the mass of dry soil.

The volumetric water content (of M_v) is the ratio of the volume of the soil water in soil.

$$M_v = \frac{V_w}{V} \tag{4}$$

where V_w is the volume occupied by water in the soil, and V is the total volume of the soil.

The relationship between soil volumetric water content and mass water content can be expressed as:

$$M_v = M_g \rho \tag{5}$$

where ρ represents the bulk density.

119 samples from a 0–10 cm depth were collected, and 120 samples were collected at a depth of 20–30 cm. The values of soil volumetric water content less than 20% and greater than 62% were removed. A total of 116 sample points remained. Following outlier removal, the measured range of SM at a 0–10 cm depth was between 22.90 and 60.82 vol%, and the average SM was 39.49 vol%. The measured range of SM at a 20–30 cm depth was between 21.17 and 61.92 vol%, and the average SM was 37.28 vol%.

3. Methods

The research is divided into two parts, the WCM and GA-BP, which are used for SM estimation. The GA-BP algorithm is used to analyze the empirical model, 3.1. radar signal modeling.

In 1978, Attema and Ulaby took crops as the research object and proposed a semi-empirical vegetation backscattering model ground on the first-order solution of the radiation transfer equation, namely, the WCM [44]. The intensity of radar backscattering is easily affected by the surface roughness and vegetation [22]. This is accounted for in the WCM, due to the joint scattering contribution of the vegetation and the underlying surface scattering to determine the total backward canopy scattering coefficient. The vegetation layer reduces the contribution of the underlying surface scattering to a certain extent. When the influence of the radar shadow and terrain undulation is neglected, the WCM can be expressed as Equation (6):

$$\sigma^0 = \sigma^0_{veg} + \tau^2 \sigma_{soil} \tag{6}$$

The total backscattering coefficient is represented by σ^0 in the vegetation coverage area; σ^0_{soil} represents the backscattering coefficient for the soil surface; σ^0_{veg} shows the backscattering coefficient produced by the surface plants; τ^2 is the two-way attenuation coefficient; σ^0_{veg} and τ^2 are expressed in Equations (7) and (8):

$$\sigma^0_{veg} = AV_1 \cos\theta \left(1 - \tau^2\right) \tag{7}$$

$$\tau^2 = \exp(-2BV_2 / \cos\theta) \tag{8}$$

$$\sigma^0_{soil} = \frac{\sigma^0 - AV_1 * \cos\theta[1 - \exp(-2 * B * V_2 * \sec\theta)]}{\exp(-2 * B * V_2 * \sec\theta)} \tag{9}$$

where θ is the angle of radar incidence; V_1 represents the direct scattering of vegetation; and V_2 represents the attenuation of vegetation. The common feature is that optical vegetation parameters are needed to parameterize the scattering component of the vegetation. In this study, V_1 and V_2 were replaced by the NDWI, the NDVI, and the NMDI, calculated from Landsat-8 imagery; A and B are empirical constants. The values of A and B are obtained by nonlinear least-square fitting.

3.1. GA-BP Neural Network

3.1.1. Genetic Algorithm (GA)

GA is a computational model simulating the natural selection and genetic mechanisms of Darwinian biological evolution. This is a method used to find the optimal solution according to the natural evolutionary process. The core content of the genetic algorithm is divided into five steps: parameter coding, initial population setting, fitness function design, genetic operation design, and control parameter setting.

The basic operation of the genetic algorithm is split into three steps: selection, crossover, and mutation operation [45–47]. Selection is the operation of selecting superior individuals from the group and eliminating inferior individuals. Crossover refers to replacing and recombining the partial structures of two parent individuals to generate new individuals. The basic content of the mutation operator is to change the gene value on some loci of the individual string in the population.

3.1.2. Back Propagation (BP) Neural Network

The BP neural network is divided into an input layer, a hidden layer, and an output layer. It is a multilayer feedforward neural network.

The learning process of the algorithm was divided into two stages: the first stage was the forward propagation process, in which the actual output values of nodes in each layer were calculated layer by layer from the input layer through the hidden layer. The nodes in each layer only accepted the input from the nodes in the previous layer and influenced only the state of nodes in the next layer. The second stage was the process of back propagation. If the output layer failed to get the expected output value, the error between the actual output and the expected output was calculated recursively layer by layer. The weight of the previous layer was corrected according to the error to minimize the error signal. In the direction of the decline of the error function slope, the network weight and threshold changes were constantly adjusted to gradually approach the objective function. Each time, the weight and error changes were proportional to the influence of the network error [48,49].

3.2. Soil Moisture Retrieval

SM was estimated using a GA-BP algorithm. In order to study the performance of the inversion method, the BP neural network was trained and verified on the synthetic dataset. The specific steps were as follows:

(1) The BP neural network consists of three layers. The layers are completely interconnected, with each layer having layers of simple processing units (neurons). The input data information is assigned to the input layer, multiplied, and forwarded through a weighting factor, and a deviation is added to the hidden layer. The output layer neurons obtained by the control are considered the input values of the output layer [27]. In this study, based on the data, we will set two inputs and one output. The soil backscattering coefficient under different polarizations (HH, HV), excluding the influence of vegetation, was used as input. These synthetic SAR backscatter datasets are obtained from the WCM. The parameterization uses soil volumetric moisture, vegetation descriptors, and incident angle values as input variables to simulate the backscatter coefficient of HH and HV polarization. Only parameters that can be easily estimated from optical images, such as the NDVI, the NDWI, and the NMDI, were considered in the generation of the synthetic dataset. When the WCM was coupled with the surface scattering model used to retrieve SM under vegetation cover, the separation of the vegetation-scattering contribution was mainly through synchronous optical data or auxiliary data measured on the ground. However, there is no unified standard for vegetation parameterization at present, and there is no theoretical basis to support which vegetation parameter can effectively and accurately represent vegetation scattering. Therefore, different vegetation parameters are used to characterize the contribution of vegetation scattering. This paper aims at the estimation of SM under vegetation cover. Therefore, before the active microwave method is used to retrieve SM, the data should be firstly downscaled. According to the resampling method, the radar backscattering coefficient, with a resolution of 3 m, is downscaled to the backscattering coefficient with a resolution of 30 m, and the total backscattering coefficient δ_{tot} of the vegetation-covered surfaces under HH and HV polarization is obtained, respectively. The WCM model is parameterized. Firstly, the least-square method was used to estimate parameters A and B by fitting the model based on the ground-truthed measurements (Equations (7)–(9)). Among them, the parameters of V_1 and V_2 were described by the NDVI, the NDWI, and the NMDI, and the incident angle was obtained from the radar image. With parameters A and B, it becomes possi-

ble to predict the WCM components (δ^0_{veg}, τ^2, and δ^0_{soil}) and, consequently, the total backscattering coefficient (δ_{tot}) using one vegetation descriptor and the SM values as inputs in the WCM.

(2) GA was used to optimize the weight and threshold of the BP neural network. Each individual in the population contained a network ownership value and threshold. The individual calculated the individual fitness value through a fitness function, and the genetic algorithm found the corresponding individual with the optimal fitness value through selection, crossover, and mutation operations.

(3) The set-up BP neural network topology: the BP neural network was optimized using a genetic algorithm to get the optimal individual to assign the initial weight and threshold of the network. The prediction function was output after the network was trained. The GA model was used to optimize the BP neural network and improve inversion accuracy. In the GA module, iterations, population, crossover probability, mutation probability, and BP network evolution are important input parameters. The nonlinear function to be fitted in this paper has two input parameters and one output parameter, so the BP neural network structure set was 2-5-1, that is, the input layer had two nodes, the hidden layer had five nodes, and the output layer had one node, with a total of 15 weights and six thresholds. Hence, the individual code length of the genetic algorithm was 21. The two polarized backscattering coefficients of HH and HV were taken as the input, and the measured SM corresponding to longitude and latitude was the output. The parameters of the genetic algorithm were set as follows: population size = 70; evolution times = 300; crossover probability = 0.6; and mutation probability = 0.2. Figure 3 presents the flow chart of SM inversion based on the GA-BP neural network.

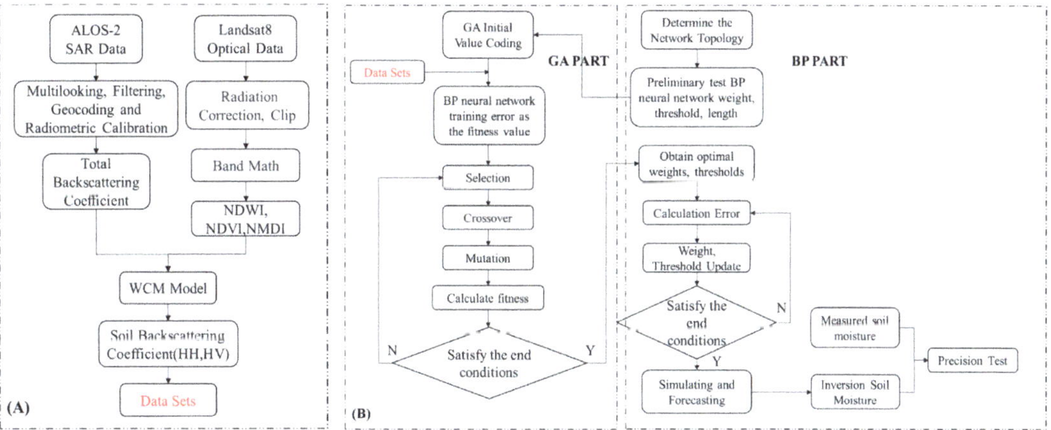

Figure 3. Soil moisture inversion method flow. (**A**) is a flow chart for establishing a data set based on WCM, and (**B**) shows the detailed process of the GA-BP algorithm. (**B, left**) is the GA algorithm part, and (**B, right**) is the BP algorithm part.

4. Results

4.1. Sensitivity Analysis of the Radar Signal

4.1.1. Water Cloud Model Parameterization

The WCM parameterization results were calculated and analyzed according to Section 3.2. Table 1 details the WCM input parameters.

Table 1. The input parameters of the WCM.

Parameter	Min Value	Max Value	Mean	Unit
NDVI	0	0.76	0.08	-
NDWI	0.3	0.94	0.72	-
NMDI	0.58	0.97	0.85	-
Incidence Angle	-	-	39.663	°

The backscatter coefficient of soil without vegetation influence was calculated from the WCM. The HH polarization and HV polarization backscatter values were lower than the total backscatter.

By comparison, because of the higher sensitivity of the HH polarization to the dihedral angular reflection effect, the backscattering coefficient was higher than that of the HV polarization. This is because the branches and shapes of the surface vegetation have more influence on crosspolarization than on copolarization. The inclination angle of branches and leaves affected the degree of response of different polarizations. Thus, the HH polarization and HV polarization of the radar scattering were affected differently.

4.1.2. The Sensitivity of ALOS-2 Data to SM under Vegetation Cover

The relationship with the soil backscattering coefficient and SM, presented in Table 2, was obtained by replacing the vegetation water content with different vegetation indices. At depths of 0–10 cm, the quality of the fit was approximately the same for all the vegetation descriptors used, with an MAE of the predicted backscattering coefficients between 2.792 and 3.142 dB in HH, and between 3.083 and 3.469 dB in HV polarization. The RMSE of the predicted backscattering coefficients was between 3.606 and 4.053 dB in HH, and between 3.755 and 4.226 dB in HV polarization. The MAPE of the predicted backscattering coefficients was −0.25 dB in HH, and −0.16 dB in HV polarization. At depths of 20–30 cm, the quality of the fit was approximately the same for all the used vegetation descriptors, with an MAE of the predicted backscattering coefficients between 2.843 and 3.199 dB in HH, and between 3.085 and 3.472 dB in HV polarization. The RMSE on the predicted backscattering coefficients was between 3.67 and 4.13 dB in HH, and between 3.743 and 4.13 dB in HV polarization. The MAPE of the predicted backscattering coefficients was −0.25 dB in HH, and −0.16 dB in HV polarization. Therefore, the copolarizations of the ALOS-2 were found to be more sensitive to SM than the crosspolarizations. In addition, the sensitivity to SM at a depth of 0–10 cm was higher than at a depth of 20–30 cm.

Table 2. The results of the relationship between the backscattering coefficient and SM.

	(dB)	0–10 cm (vol%)			20–30 cm (vol%)		
		MAE	RMSE	MAPE	MAE	RMSE	MAPE
WCM ($V_1 = V_2$ = NDVI)	HH	2.792	3.606	−0.25	2.843	3.67	−0.26
	HV	3.083	3.755	−0.16	3.085	3.743	−0.16
WCM ($V_1 = V_2$ = NDWI)	HH	3.006	3.882	−0.25	3.06	3.951	−0.26
	HV	3.319	4.043	−0.16	3.321	4.03	−0.16
WCM ($V_1 = V_2$ = NMDI)	HH	3.142	4.058	−0.25	3.199	4.13	−0.26
	HV	3.469	4.226	−0.16	3.472	4.212	−0.16

4.2. *Modeling Results*

4.2.1. GA-BP Results Analysis

In order to improve the sensitivity of the radar signal, HH and HV were used as inputs in the BP neural network model. The GA-BP parameter settings are displayed in Table 3.

Table 3. GA-BP preferences.

GA Preferences	Value
Iterations	300
Population	70
Crossover probability	0.6
Mutation probability	0.2
BP Preferences	**Value**
Maximum number of training	100
The training accuracy	0.00001
Learning rate	0.1

Figure 4 shows the training results obtained by using the soil backscattering coefficients (HH, HV) as inputs in the WCM model (V_1 = V_2 = NDVI in Equation (9)) at a depth of 0–10 cm. The GA-BP network became stable, as seen in the fitness curve region, after 160 generations, and the GA algorithm can search the appropriate weight and threshold at this time (Figure 4A). The error percentage of the BP neural network ranges from −0.4 to 0.6. (Figure 4B). The prediction of BP is between −0.25 and 0.2 (Figure 4C). The BEST dotted line indicates that the BP training result is ideal when the BP network is trained to the sixth generation (Figure 4D).

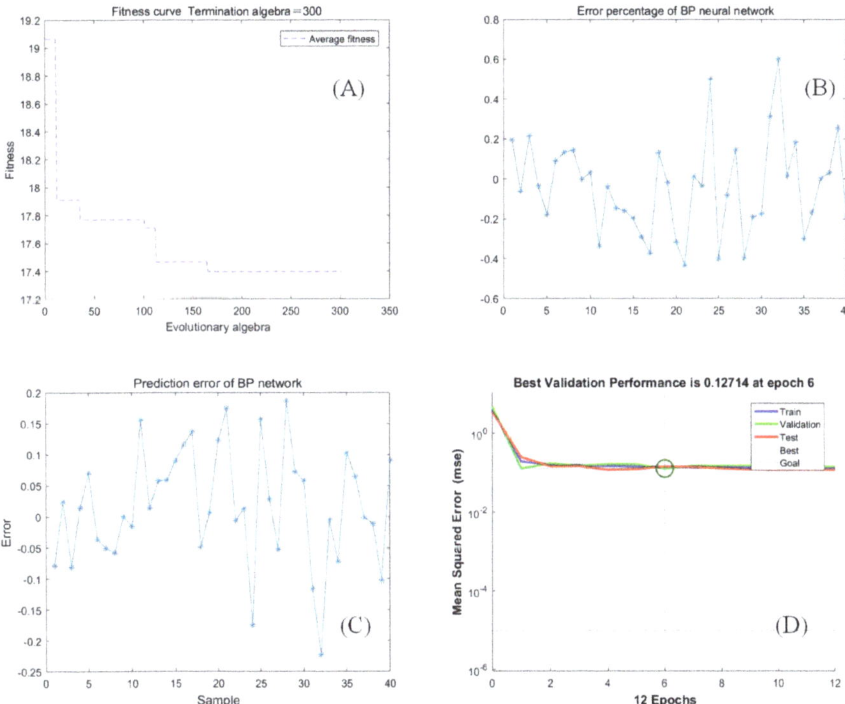

Figure 4. GA-BP training results of the vegetation description in the WCM (V_1 = V_2 = NDVI in Equation (9)), sampling depth is 0–10 cm. (**A**) shows the fitness curve, (**B**) shows error percentage of BP neural network, (**C**) is prediction error of BP network, and (**D**) is BP training process. The three solid colored lines in the figure: the blue line represents the performance of the MSE index in the BP training process in each generation; the green line shows the performance of the MSE index in the BP crossvalidation process in each generation; and the red line represents the performance of the MSE index in the BP testing process in each generation. The red line represents the test condition, which is the result of BP calculation and training (**D**).

Figure 5 shows the training results obtained by using the soil backscattering coefficients (HH, HV) as inputs in the WCM model (V_1 = V_2 = NDVI in Equation (9)) at a depth of 20–30 cm. The GA-BP network is stable in the fitness curve region after 60 generations, and the GA can search the appropriate weight and threshold at this time (Figure 5A). The error percentage of the BP neural network ranges from −0.4 to 0.5 (Figure 5B). The prediction of BP is −0.15 to 0.2 (Figure 5C). The BEST dotted line indicates that the BP training result is ideal when the BP network is trained to the second generation (Figure 5D).

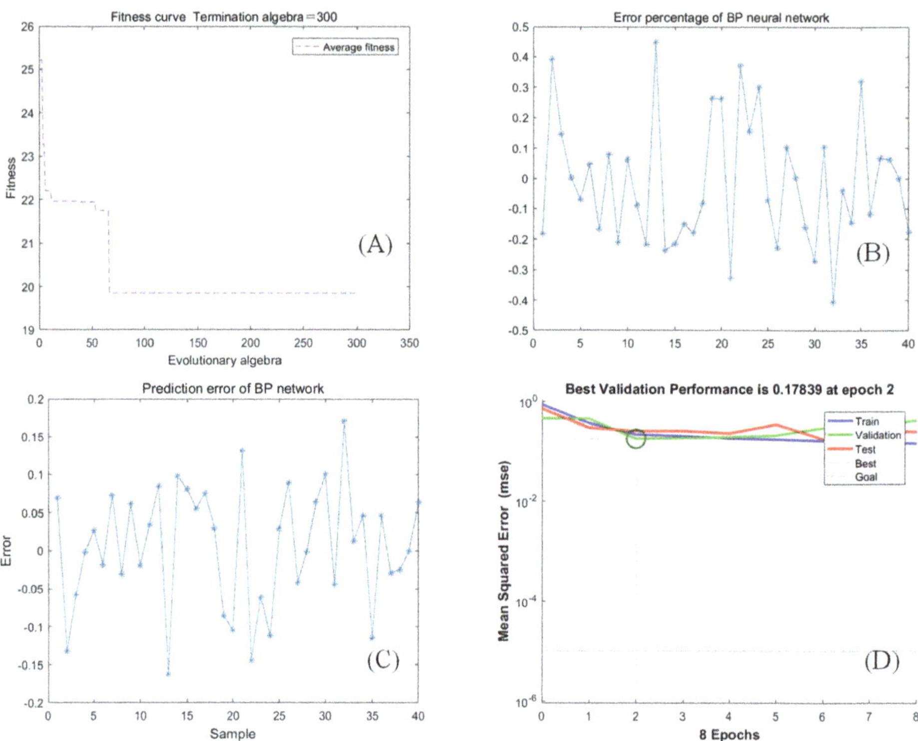

Figure 5. GA-BP training results of the vegetation description in the WCM (V_1 = V_2 = NDVI in Equation (9)), sampling depth is 20–30 cm. (**A**) shows the fitness curve, (**B**) shows error percentage of BP neural network, (**C**) is prediction error of BP network, and (**D**) is BP training process. The three solid colored lines in the figure: the blue line represents the performance of the MSE index in the BP training process in each generation; the green line shows the performance of the MSE index in the BP crossvalidation process in each generation; and the red line represents the performance of the MSE index in the BP testing process in each generation. The red line represents the test condition, which is the result of BP calculation and training (**D**).

Figure 6 shows the training results obtained by using the soil backscattering coefficients (HH, HV) as inputs in the WCM model (V_1 = V_2 = NDWI in Equation (9)) at a depth of 0–10 cm. The GA-BP network was stable in the fitness curve region after 105 generations, and the GA can search for the appropriate weight and threshold at this time (Figure 6A). The error percentage of the BP neural network ranged from −0.5 to 0.5 (Figure 6B). The prediction of BP was −0.25 to 0.25 (Figure 6C). The BEST line indicates that an ideal BP training result was reached at the 11th generation (Figure 6D).

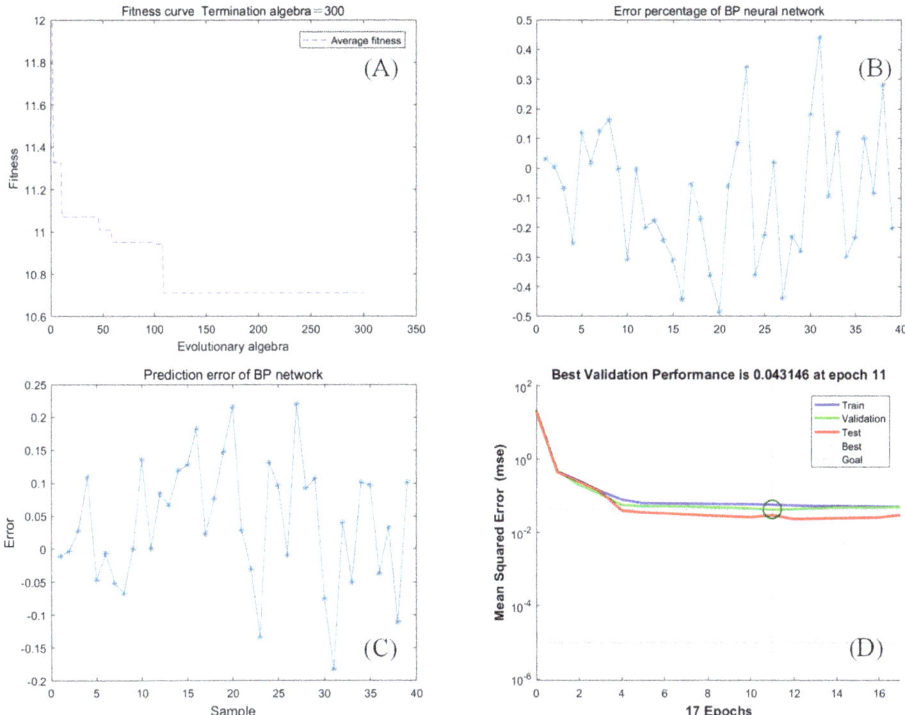

Figure 6. GA-BP training results of the vegetation description in the WCM ($V_1 = V_2$ = NDWI in Equation (9)), sampling depth is 0–10 cm. (**A**) shows the fitness curve, (**B**) shows error percentage of BP neural network, (**C**) is prediction error of BP network, and (**D**) is BP training process.The three solid colored lines in the figure: the blue line represents the performance of the MSE index in the BP training process in each generation; the green line shows the performance of the MSE index in the BP crossvalidation process in each generation; and the red line represents the performance of the MSE index in the BP testing process in each generation. The red line represents the test condition, which is the result of BP calculation and training (**D**).

Figure 7 shows the training results obtained by using the soil backscattering coefficients (HH, HV) as inputs in the WCM model ($V_1 = V_2$ = NDWI in Equation (9)) at a depth of 20–30 cm. The GA-BP network was stable in the fitness curve region after 75 generations, and the GA can search for the appropriate weight and threshold at this time (Figure 7A). The error percentage of the BP neural network ranges from −0.5 to 0.5 (Figure 7B). The prediction of BP was −0.2 to 0.2 (Figure 7C). The BEST dotted line indicates that the BP training result was ideal in first generation (Figure 7D).

Figure 8 shows the training results obtained by using the soil backscattering coefficients (HH, HV) as inputs in the WCM model ($V_1 = V_2$ = NMDI in Equation (9)) at a depth of 0–10 cm. The GA-BP network was stable in the fitness curve region after 120 generations, and the GA can search for the appropriate weight and threshold at this time (Figure 8A). The error percentage of the BP neural network ranges from −0.5 to 0.5 (Figure 8B). The prediction of BP was −0.2 to 0.2 (Figure 8C). The BEST dotted line indicates that the BP training result was ideal in the third generation (Figure 8D).

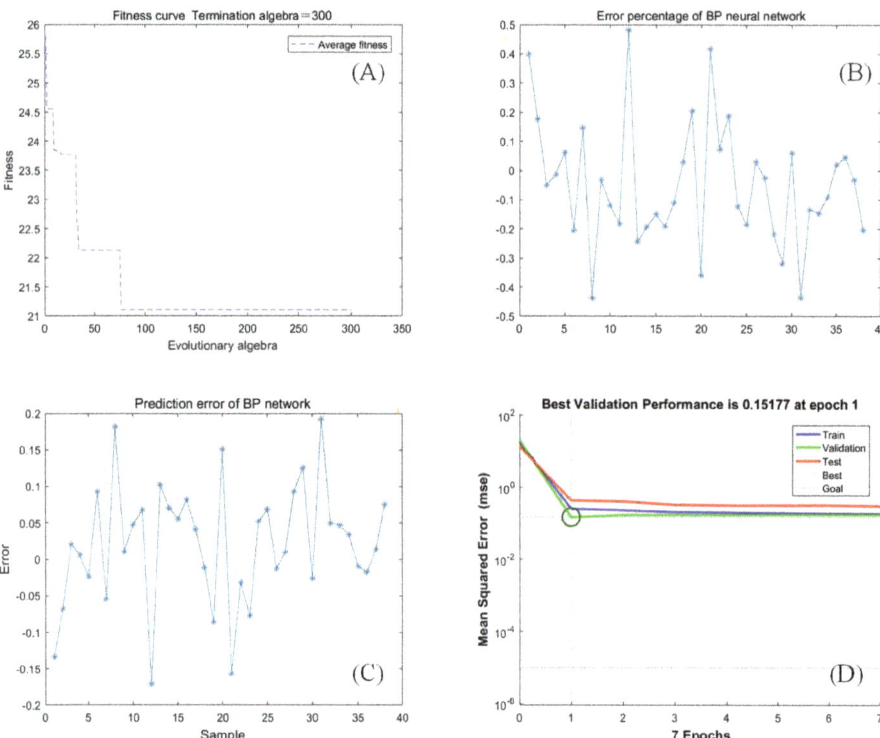

Figure 7. GA-BP training results of the vegetation description in the WCM ($V_1 = V_2$ = NDWI in Equation (9)), sampling depth is 20–30 cm. (**A**) shows the fitness curve, (**B**) shows error percentage of BP neural network, (**C**) is prediction error of BP network, and (**D**) is BP training process.The three solid colored lines in the figure: the blue line represents the performance of the MSE index in the BP training process in each generation; the green line shows the performance of the MSE index in the BP crossvalidation process in each generation; and the red line represents the performance of the MSE index in the BP testing process in each generation. The red line represents the test condition, which is the result of BP calculation and training (**D**).

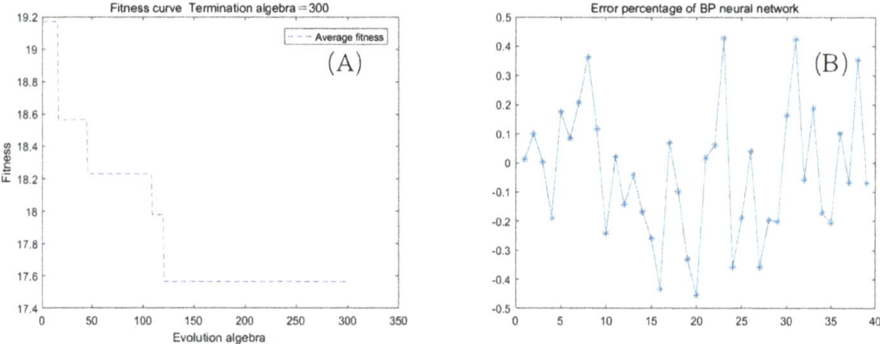

Figure 8. *Cont.*

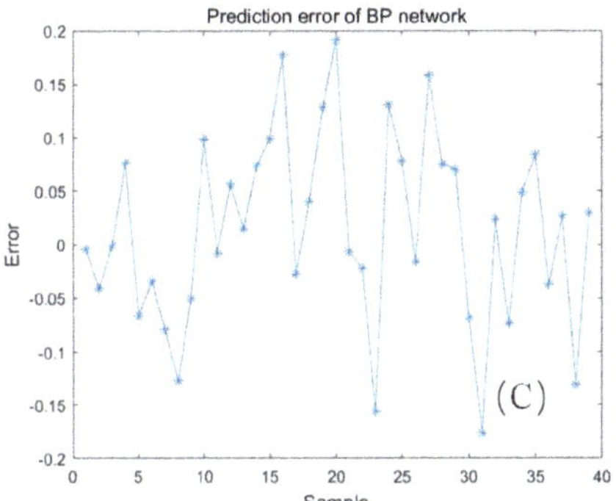

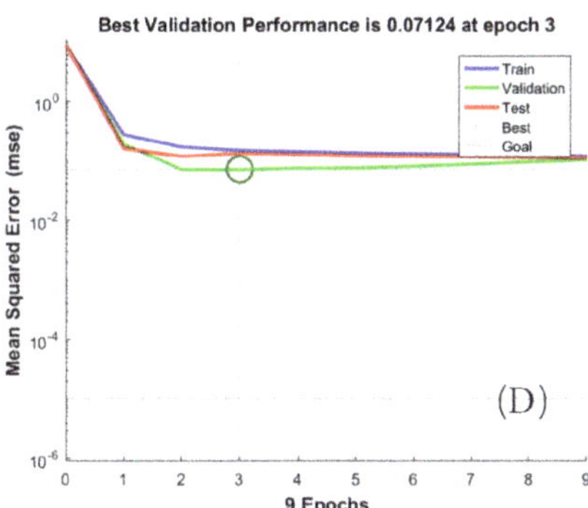

Figure 8. GA-BP training results of the vegetation description in the WCM ($V_1 = V_2$ = NMDI in Equation (9)), sampling depth is 0–10 cm. (**A**) shows the fitness curve, (**B**) shows error percentage of BP neural network, (**C**) is prediction error of BP network, and (**D**) is BP training process.The three solid colored lines in the figure: the blue line represents the performance of the MSE index in the BP training process in each generation; the green line shows the performance of the MSE index in the BP crossvalidation process in each generation; and the red line represents the performance of the MSE index in the BP testing process in each generation. The red line represents the test condition, which is the result of BP calculation and training (**D**).

Figure 9 shows the training results obtained by using the soil backscattering coefficients (HH, HV) as inputs in the WCM model ($V_1 = V_2$ = NMDI in Equation (9)) at a depth of 20–30 cm. The GA-BP network was stable in the fitness curve region after 75 generations, and the GA can search for the appropriate weight and threshold at this time (Figure 9A). The error percentage of the BP neural network ranges from −0.4 to 0.6 (Figure 9B). The prediction of BP was −0.2 to 0.15 (Figure 9C). The BEST dotted line indicates that the BP training result was ideal at the third generation (Figure 9D).

4.2.2. Soil Moisture Retrieval

As stated in Section 3.2, different datasets were used to retrieve SM using the GA-BP neural network: (1) using the radar signal in both HH and HV ($V_1 - V_2$ = NDVI in Equation (9)); (2) using the radar signal in both HH and HV, which come from the WCM ($V_1 = V_2$ = NDWI in Equation (9)); (3) the radar signal in both HH and HV ($V_1 = V_2$ = NMDI in Equation (9)). Estimates of SM and SM reference scores were compared to assess the accuracy of the SM inversion.

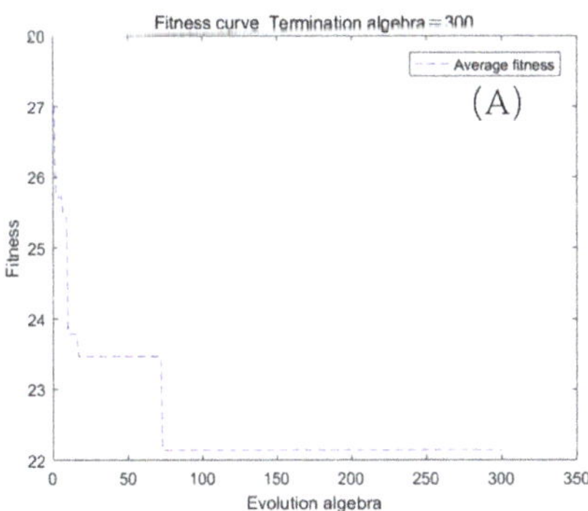

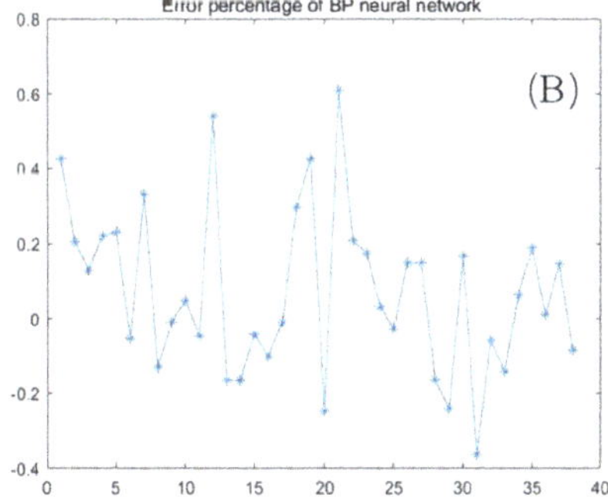

Figure 9. *Cont.*

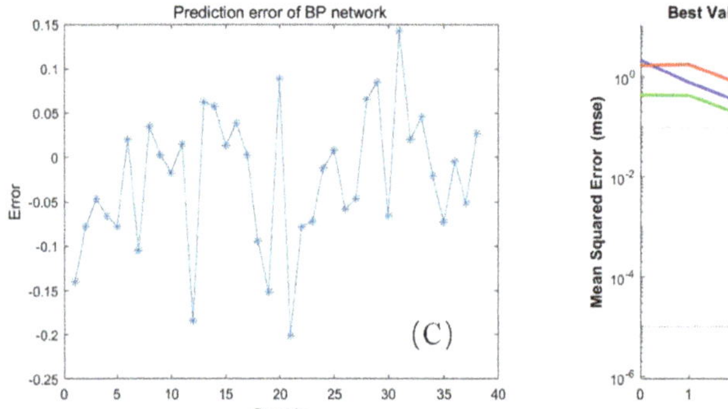

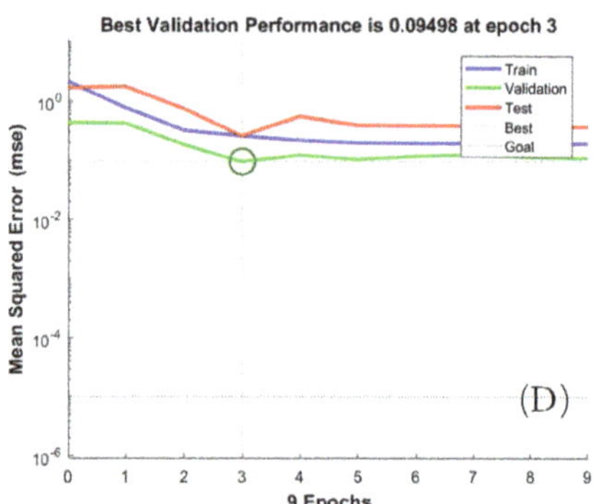

Figure 9. GA-BP training results of the vegetation description in the WCM (V_1 = V_2 = NMDI in Equation (9)), sampling depth is 20–30 cm. (**A**) shows the fitness curve, (**B**) shows error percentage of BP neural network, (**C**) is prediction error of BP network, and (**D**) is BP training process. The three solid colored lines in the figure: the blue line represents the performance of the MSE index in the BP training process in each generation; the green line shows the performance of the MSE index in the BP crossvalidation process in each generation; and the red line represents the performance of the MSE index in the BP testing process in each generation. The red line represents the test condition, which is the result of BP calculation and training (**D**).

To calculate the SM, use two-thirds of the data as the training set, and the rest as the validation set.

Figures 10–12 show the results of the GA-BP method.

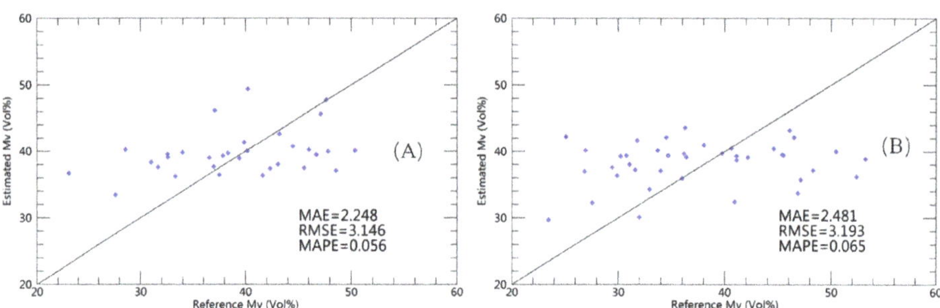

Figure 10. The relationship between the estimated Mv of GA-BP and reference Mv, in which the vegetation description in the WCM model is the NDVI (0–10 cm (**A**), 20–30 cm (**B**)).

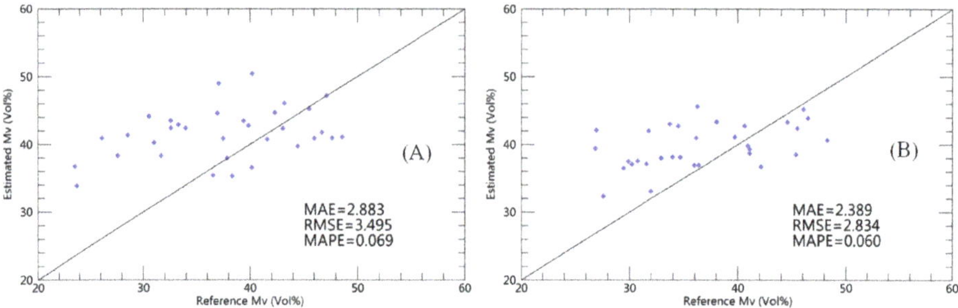

Figure 11. The relationship between the estimated Mv of GA-BP and reference Mv, in which the vegetation description in the WCM model is the NDWI (0–10 cm (**A**), 20–30 cm (**B**)).

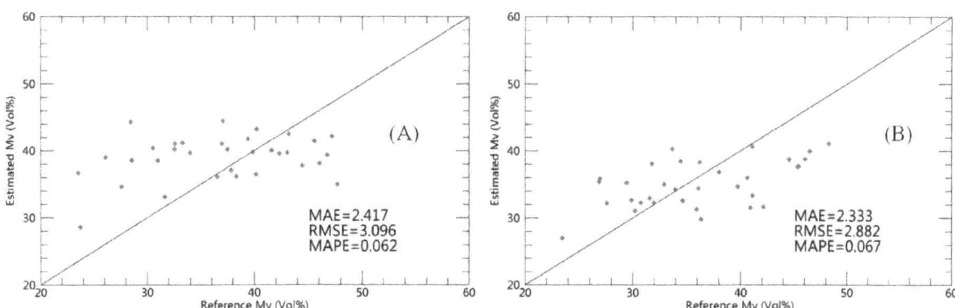

Figure 12. The relationship between the estimated Mv of GA-BP and reference Mv, in which the vegetation description in the WCM model is the NMDI (0–10 cm (**A**), 20–30 cm (**B**)).

These two backscattering coefficients (HH and HV) were used in GA-BP (WCM, $V_1 = V_2$ = NDVI in Equation (9)) analysis of the ALOS-2. According to the GA-BP results, the MAE, RMSE, and MAPE values were calculated as 2.248 vol%, 3.146 vol%, and 0.056 vol% at the depth of 0–10 cm, respectively (Figure 10A). At a 20–30 cm depth, the MAE, RMSE and MAPE values were 2.481 vol%, 3.196 vol%, and 0.065 vol%, respectively (Figure 10B).

The results of the backscattering coefficients were used in GA-BP (WCM, $V_1 = V_2$ = NDWI in Equation (9)). Correspondingly, at the depth of 0–10 cm, based on the GA-BP results, the MAE, RMSE and MAPE values were calculated as 2.883 vol%, 3.495 vol%, and 0.069 vol%, respectively (Figure 11A). At a 20–30 cm depth, the MAE, RMSE and MAPE values were calculated as 2.389 vol%, 2.834 vol%, and 0.06 vol%, respectively (Figure 11B).

It follows that two backscattering coefficients were used in GA-BP (WCM, $V_1 = V_2$ = NMDI in Equation (9)). Underlying the GA-BP results, the MAE, RMSE, and MAPE values were calculated as 2.417 vol%, 3.096 vol%, and 0.062 vol%, respectively, at the depth of 0–10 cm (Figure 12A). At a 20–30 cm depth, the MAE, RMSE, and MAPE values' calculated results were 2.333 vol%, 2.882 vol%, and 0.067 vol%, respectively (Figure 12B).

Combined with the data analysis, the experiments show that, at a depth of 0–10 cm, the accuracy of the results obtained by the NDVI as a description of vegetation is higher than that obtained by other parameters.. The findings were better when vegetation was described as the NDWI at depths of 20–30 cm. Table 4 presents the MAE, RMSE, and MAPE analyses between the GA-BP-based Mv and the in situ Mv for ALOS-2 data in detail.

Table 4. The results of the MAE, RMSE, and MAPE between the GA-BP-based Mv and in situ Mv for ALOS-2 data.

	0–10 cm (vol%)			20–30 cm (vol%)		
	MAE	RMSE	MAPE	MAE	RMSE	MAPE
$V_1 = V_2$ = NDVI	2.248	3.146	0.056	2.481	3.195	0.065
$V_1 = V_2$ = NDWI	2.883	3.495	0.069	2.389	2.834	0.06
$V_1 = V_2$ = NMDI	2.417	3.096	0.062	2.333	2.883	0.067

Figures 10–12 show the scatter plots between the measured and inverted soil moisture. It can be seen from the scatter plots that the retrieved soil moisture was underestimated in relatively humid regions because of the sensitivity of radar signals to soil moisture. In drier areas, the retrieved soil moisture was overestimated because the soil moisture content in the surface layer was lower than that in the deep layer. At the same time, the consistency of the soil moisture gradient also led to an overestimation of moisture. Because the soil moisture range in this study area was too extensive, the estimated value was underestimated or overestimated to a certain extent because of the actual local conditions. A large number of experimental analyses have verified this result.

Considering the GA-BP inversion analysis of the ALOS-2, the MAE, RMSE, and MAPE values were computed with different inputs (WCM, Equation (9)): Figures 10–12, respectively. The results show that the WCM (V_1 = V_2 = NDVI, NDWI, NMDI) can effectively eliminate the backscattering effect of vegetation, and the results of the GA-BP inversion using ALOS-2 data to estimate SM are satisfactory.

5. Discussion

This novel study shows the sensitivity of ALOS-2 radar data to SM. Only a few studies have performed SM inversion using ALOS-2 data [7,35,50,51]. Sekertekin et al. compared the potentials of ALOS-2 L-band and Sentinel-1 C-band data for SM estimation in bare and vegetation-covered agricultural fields using the WCM. The NDVI and plant coverage (PC) were considered the vegetation description, and the NDVI performed better than PC. The results show that the WCM can effectively eliminate the backscattering effect of vegetation, and the inversion of WCM presented satisfactory results in estimating SM with the ALOS-2 and Sentinel-1 data. Table 5 is a summary of the above methods. These findings show that the WCM method could effectively remove the influence of vegetation backscattering, which is consistent with our conclusion.

Table 5. The methods summary.

Author	Data	Method
Skkertekin et al.	ALOS-2 Sentinel-1	WCM Dubois MLR
El Hajj et al.	TerraSAR-X COSMO-SkyMed SPOT4/5 Landdat 7/8	WCM Multi-layer perceptron neural networks (NNs)
Zribi et al.	ALOS-2	WCM Dubois Baghdadi et al.

The potential of the C-band and L-band in SM retrieval is compared by El Hajj et al. However, for both frequencies, they only examine the potential of a copolarized HH HV C-band and L-band to estimate the SM because they believe that previous studies show that the use of crosspolarized (HV or VH) and copolarized SM data does not improve the estimation accuracy. Nevertheless, they did not consider crosspolarization [52,53]. This study, however, used ALOS-2 dual-polarization radar data and considered both polarization modes (HH and HV) while overcoming the lack of soil roughness.

The method used by Zribi et al. for estimating SM using ALOS-2 L-band radar data was compared for different types of crops (turmeric, marigold, and sorghum). In areas covered by vegetation, soil roughness measurements are rare. Accordingly, only the WCM has been considered to simulate the relationship between SM and the radar backscattering coefficient to retrieve the SM. Zribi et al. obtained moderately accurate estimates of SM for turmeric and marigold fields, with errors equal to 6.7 vol% and 7.9 vol% for HH and HV polarization, respectively, for turmeric, and 8.7 vol% and 11 vol%, in the HH and HV polarizations, respectively, for marigold. This result can be explained by the fact that the multiscattering effect is not considered in the simplified first-order radiation transfer equation of the WCM. The method presented in this paper also encountered challenges, such as the lack of measured soil roughness data. Therefore, we propose a GA-BP neural network algorithm, which overcomes the above problems to a certain extent while considering all polarization (HH and HV) modes. Even if there is a lack of measured data, such as soil roughness, an error analysis is carried out each time according to the results obtained from the training, and the expected results according to the nature of the BP neural network. Then, the weights and thresholds are modified step by step to get the model that can output the same as the expected result.

However, in this study, with the WCM model, different vegetation descriptions were established. When the vegetation index was the NDVI, the accuracy analysis results of the SM estimated by the GA-BP method, and the SM measured, were the best. The MAE, RESE, and MAPE results were 2.248 vol%, 3.146 vol%, 0.056 vol%, respectively, at a 0–10 cm depth.

When the vegetation index was the NDWI, the accuracy analysis results of SM estimated by the GA-BP method, and the SM measured, were the best at a 20–30 cm depth. The MAE, RESE, and MAPE results were 2.389 vol%, 2.834 vol%, and 0.06 vol%, respectively. We can also see from the results that when the SM is less than about 30 vol%, the inversion results are slightly higher than the measured values. When the SM was higher than approximately 45 vol%, the inversion result was slightly lower than the measured value. The reason is that the sensitivity of the backscatter coefficient to the SM increases in relatively low and dry areas. The sensitivity of the backscattering coefficient to the SM decreases in the relatively humid area. In the following research, the method will be analyzed and verified according to different humidity gradients to discuss the applicability of the method.

By using the three vegetation indices as the vegetation input in the algorithm, we can observe the following conclusions. When the soil moisture is less than about 34%, we find that the slope of the scatter plot of the measured and inverted soil moisture is higher. At this time, it was in a relatively dry area, and the retrieved soil moisture was overestimated because the soil moisture at this time was lower than the deeper part. When the soil moisture is greater than 44%, the slope of the scatter plot of the measured and inverted soil moisture is low. At this time, it is in a relatively humid area. This is because the sensitivity of the radar signal to soil moisture is reduced in the humid area. A large number of experimental analyses have verified this result.

This study has some limitations. Because the sampling area has many hillsides, sampling is more difficult. Because of the influence of the topography of the sampling points, the distribution of the sampling points should choose flat areas as much as possible. However, the distribution should be as uniform as possible throughout the study area. At the same time, only soil samples were collected. Soil roughness, soil type, etc., can be considered in future research. In addition to these limitations, further research is needed to obtain better results for SM estimation using SAR data. We believe that the results of this study provide a new idea for future research.

Future studies should look at the following:

1. The addition of different radar backscatter models to find out which model can improve the estimation accuracy of SM.
2. In the WCM model, more vegetation descriptions can be added.
3. More intelligent optimization algorithms and machine-learning algorithms can be applied to radar SM inversion.
4. More soil parameters can be added to increase the accuracy of SM inversion.
5. In the follow-up research, the different ranges of soil moisture will be studied and discussed separately.

6. Conclusions

According to this study, the potential of ALOS-2 L-band radar data for SM calculation was investigated over vegetation-covered fields.

(1) The results revealed that ALOS-2 L-band data was sensitive to SM in vegetation-covered surfaces.
(2) The backscattering of ALOS-2 with the copolarization was more sensitive to SM than the crosspolarization. In addition, at a depth of 0–10 cm, the sensitivity was higher than at a depth of 20–30 cm. It can be shown that radar penetration decreases with increasing depth.
(3) The NDVI was more sensitive than the NDWI and the NMDI as a vegetation description in the WCM model for estimating SM based on the ALOS-2 radar backscatter.
(4) The WCM can effectively eliminate the vegetation's backscattering effect, and the WCM shows satisfactory results in SM estimation using ALOS-2 data.
(5) Combining the two polarization modes of ALOS-2 using the novel GA-BP neural network method improved the estimation of SM in the absence of soil roughness and soil type. This might be the key component in future attempts to overcome SM retrieval by microwave remote sensing.

Author Contributions: Conceptualization, Y.G., M.G.; Methodology, Y.G., M.G.; Formal analysis, M.G., L.W.; Data curation, Y.G.; Writing—original draft, Y.G.; Writing—review & editing, M.G., L.W., O.R.; Valuable advice, M.G., L.W., O.R.; Funding acquisition, M.G. All authors have read and agreed to the published version of the manuscript.

Funding: This research was funded by the National Natural Science Foundation of China (grant number 41871282, 62071084, 62001434).

Conflicts of Interest: The authors declare no conflict of interest.

References

1. Liu, Y.; Qian, J.; Yue, H. Combined Sentinel-1A With Sentinel-2A to Estimate Soil Moisture in Farmland. *IEEE J. Sel. Top. Appl. Earth Obs. Remote Sens.* **2021**, *14*, 1292–1310. [CrossRef]
2. Du, W.; Chen, N.; Yan, S. Online soil moisture retrieval and sharing using geospatial web-enabled BDS-R service. *Comput. Electron. Agric.* **2016**, *121*, 354–367. [CrossRef]
3. Wang, Z.; Zhao, T.; Qiu, J.; Zhao, X.; Li, R.; Wang, S. Microwave-based vegetation descriptors in the parameterization of water cloud model at L-band for soil moisture retrieval over croplands. *GISci. Remote Sens.* **2021**, *58*, 48–67. [CrossRef]
4. Petropoulos, G.; Srivastava, P.; Piles, M.; Pearson, S. Earth Observation-Based Operational Estimation of Soil Moisture and Evapotranspiration for Agricultural Crops in Support of Sustainable Water Management. *Sustainability* **2018**, *10*, 181. [CrossRef]
5. Torres-Rua, A.; Ticlavilca, A.; Bachour, R.; McKee, M. Estimation of Surface Soil Moisture in Irrigated Lands by Assimilation of Landsat Vegetation Indices, Surface Energy Balance Products, and Relevance Vector Machines. *Water* **2016**, *8*, 167. [CrossRef]
6. Guo, S.; Bai, X.; Chen, Y.; Zhang, S.; Hou, H.; Zhu, Q.; Du, P. An Improved Approach for Soil Moisture Estimation in Gully Fields of the Loess Plateau Using Sentinel-1A Radar Images. *Remote Sens.* **2019**, *11*, 349. [CrossRef]
7. Gururaj, P.; Umesh, P.; Shetty, A. Assessment of surface soil moisture from ALOS PALSAR-2 in small-scale maize fields using polarimetric decomposition technique. *Acta Geophys.* **2021**, *69*, 579–588. [CrossRef]
8. Tian, J.; Philpot, W.D. Relationship between surface soil water content, evaporation rate, and water absorption band depths in SWIR reflectance spectra. *Remote Sens. Environ.* **2015**, *169*, 280–289. [CrossRef]
9. Entekhabi, D.; Njoku, E.G.; O'Neill, P.E.; Kellogg, K.H.; Crow, W.T.; Edelstein, W.N.; Entin, J.K.; Goodman, S.D.; Jackson, T.J.; Johnson, J.; et al. The Soil Moisture Active Passive (SMAP) Mission. *Proc. IEEE* **2010**, *98*, 704–716. [CrossRef]
10. Sadeghi, M.; Babaeian, E.; Tuller, M.; Jones, S.B. The optical trapezoid model: A novel approach to remote sensing of soil moisture applied to Sentinel-2 and Landsat-8 observations. *Remote Sens. Environ.* **2017**, *198*, 52–68. [CrossRef]
11. Zhang, L.; Meng, Q.; Yao, S.; Wang, Q.; Zeng, J.; Zhao, S.; Ma, J. Soil Moisture Retrieval from the Chinese GF-3 Satellite and Optical Data over Agricultural Fields. *Sensors* **2018**, *18*, 2675. [CrossRef]
12. Kaplan, G.; Fine, L.; Lukyanov, V.; Manivasagam, V.S.; Tanny, J.; Rozenstein, O. Normalizing the Local Incidence Angle in Sentinel-1 Imagery to Improve Leaf Area Index, Vegetation Height, and Crop Coefficient Estimations. *Land* **2021**, *10*, 680. [CrossRef]
13. Miernecki, M.; Wigneron, J.; Lopez-Baeza, E.; Kerr, Y.; De Jeu, R.; De Lannoy, G.J.M.; Jackson, T.J.; O'Neill, P.E.; Schwank, M.; Moran, R.F.; et al. Comparison of SMOS and SMAP soil moisture retrieval approaches using tower-based radiometer data over a vineyard field. *Remote Sens. Environ.* **2014**, *154*, 89–101. [CrossRef]
14. Han, L.; Wang, C.; Yu, T.; Gu, X.; Liu, Q. High-Precision Soil Moisture Mapping Based on Multi-Model Coupling and Background Knowledge, Over Vegetated Areas Using Chinese GF-3 and GF-1 Satellite Data. *Remote Sens.* **2020**, *12*, 2123. [CrossRef]
15. Ma, C.; Li, X.; Notarnicola, C.; Wang, S.; Wang, W. Uncertainty Quantification of Soil Moisture Estimations Based on a Bayesian Probabilistic Inversion. *IEEE Trans. Geosci. Remote* **2017**, *55*, 3194–3207. [CrossRef]
16. Mattar, C.; Wigneron, J.; Sobrino, J.A.; Novello, N.; Calvet, J.C.; Albergel, C.; Richaume, P.; Mialon, A.; Guyon, D.; Jimenez-Munoz, J.C.; et al. A Combined Optical—Microwave Method to Retrieve Soil Moisture Over Vegetated Areas. *IEEE Trans. Geosci. Remote* **2012**, *50*, 1404–1413. [CrossRef]
17. Mishra, V.N.; Prasad, R.; Kumar, P.; Gupta, D.K.; Srivastava, P.K. Dual-polarimetric C-band SAR data for land use/land cover classification by incorporating textural information. *Environ. Earth Sci.* **2017**, *76*, 1–16. [CrossRef]
18. Yang, L.; Li, Y.; Li, Q.; Sun, X.; Kong, J.; Wang, L. Implementation of a multiangle soil moisture retrieval model using RADARSAT-2 imagery over arid Juyanze, northwest China. *J. Appl. Remote Sens.* **2017**, *11*, 1. [CrossRef]
19. Zhang, X.; Chen, B.; Fan, H.; Huang, J.; Zhao, H. The Potential Use of Multi-Band SAR Data for Soil Moisture Retrieval over Bare Agricultural Areas: Hebei, China. *Remote Sens.* **2016**, *8*, 7. [CrossRef]
20. Zhu, B.; Song, X.; Leng, P.; Sun, C.; Wang, R.; Jiang, X. A Novel Simplified Algorithm for Bare Surface Soil Moisture Retrieval Using L-Band Radiometer. *ISPRS Int. J. Geo-Inf.* **2016**, *5*, 143. [CrossRef]
21. Aubert, M.; Baghdadi, N.N.; Zribi, M.; Ose, K.; El Hajj, M.; Vaudour, E.; Gonzalez-Sosa, E. Toward an Operational Bare Soil Moisture Mapping Using TerraSAR-X Data Acquired Over Agricultural Areas. *IEEE J. Sel. Top. Appl. Earth Obs. Remote Sens.* **2013**, *6*, 900–916. [CrossRef]

22. Rozenstein, O.; Siegal, Z.; Blumberg, D.G.; Adamowski, J. Investigating the backscatter contrast anomaly in synthetic aperture radar (SAR) imagery of the dunes along the Israel—Egypt border. *Int. J. Appl. Earth Obs.* **2016**, *46*, 13–21. [CrossRef]
23. Bao, Y.; Lin, L.; Wu, S.; Kwal Deng, K.A.; Petropoulos, G.P. Surface soil moisture retrievals over partially vegetated areas from the synergy of Sentinel-1 and Landsat 8 data using a modified water-cloud model. *Int. J. Appl. Earth Obs.* **2018**, *72*, 76–85. [CrossRef]
24. De Roo, R.D.; Du, Y.; Ulaby, F.T.; Dobson, M.C. A semi-empirical backscattering model at L-band and C-band for a soybean canopy with soil moisture inversion. *IEEE T. Geosci. Remote* **2001**, *39*, 864–872. [CrossRef]
25. Qiu, J.; Crow, W.T.; Wagner, W.; Zhao, T. Effect of vegetation index choice on soil moisture retrievals via the synergistic use of synthetic aperture radar and optical remote sensing. *Int. J. Appl. Earth Obs.* **2019**, *80*, 47–57. [CrossRef]
26. Kaplan, G.; Fine, L.; Lukyanov, V.; Manivasagam, V.S.; Malachy, N.; Tanny, J.; Rozenstein, O. Estimating Processing Tomato Water Consumption, Leaf Area Index, and Height Using Sentinel-2 and VENµS Imagery. *Remote Sens.* **2021**, *13*, 1046. [CrossRef]
27. Kumar, P.; Prasad, R.; Choudhary, A.; Gupta, D.K.; Mishra, V.N.; Vishwakarma, A.K.; Singh, A.K.; Srivastava, P.K. Comprehensive evaluation of soil moisture retrieval models under different crop cover types using C-band synthetic aperture radar data. *Geocarto Int.* **2019**, *34*, 1022–1041. [CrossRef]
28. Huang, S.; Ding, J.; Liu, B.; Ge, X.; Wang, J.; Zou, J.; Zhang, J. The Capability of Integrating Optical and Microwave Data for Detecting Soil Moisture in an Oasis Region. *Remote Sens.* **2020**, *12*, 1358. [CrossRef]
29. Hosseini, M.; McNairn, H. Using multi-polarization C- and L-band synthetic aperture radar to estimate biomass and soil moisture of wheat fields. *Int. J. Appl. Earth Obs.* **2017**, *58*, 50–64. [CrossRef]
30. Jain, A.; Singh, D. An Information fusion approach for PALSAR data to retrieve soil moisture. *Geocarto Int.* **2017**, *32*, 1017–1033. [CrossRef]
31. Kong, J.; Yang, J.; Zhen, P.; Li, J.; Yang, L. A Coupling Model for Soil Moisture Retrieval in Sparse Vegetation Covered Areas Based on Microwave and Optical Remote Sensing Data. *IEEE Trans. Geosci. Remote* **2018**, *56*, 7162–7173. [CrossRef]
32. Mandal, D.; Hosseini, M.; McNairn, H.; Kumar, V.; Bhattacharya, A.; Rao, Y.S.; Mitchell, S.; Robertson, L.D.; Davidson, A.; Dabrowska-Zielinska, K. An investigation of inversion methodologies to retrieve the leaf area index of corn from C-band SAR data. *Int. J. Appl. Earth Obs.* **2019**, *82*, 101893. [CrossRef]
33. Meng, Q.; Zhang, L.; Xie, Q.; Yao, S.; Chen, X.; Zhang, Y. Combined Use of GF-3 and Landsat-8 Satellite Data for Soil Moisture Retrieval over Agricultural Areas Using Artificial Neural Network. *Adv. Meteorol.* **2018**, *2018*, 9315132. [CrossRef]
34. Prakash, R.; Singh, D.; Pathak, N.P. A Fusion Approach to Retrieve Soil Moisture With SAR and Optical Data. *IEEE J. Sel. Top. Appl. Earth Obs. Remote Sens.* **2012**, *5*, 196–206. [CrossRef]
35. Sekertekin, A.; Marangoz, A.M.; Abdikan, S. ALOS-2 and Sentinel-1 SAR data sensitivity analysis to surface soil moisture over bare and vegetated agricultural fields. *Comput. Electron. Agric.* **2020**, *171*, 105303. [CrossRef]
36. Tao, L.; Li, J.; Jiang, J.; Chen, X. Leaf Area Index Inversion of Winter Wheat Using Modified Water-Cloud Model. *IEEE Geosci. Remote Sens.* **2016**, *13*, 816–820. [CrossRef]
37. Wang, L.; He, B.; Bai, X.; Xing, M. Assessment of Different Vegetation Parameters for Parameterizing the Coupled Water Cloud Model and Advanced Integral Equation Model for Soil Moisture Retrieval Using Time Series Sentinel-1A Data. *Photogramm. Eng. Remote Sens.* **2019**, *85*, 43–54. [CrossRef]
38. Mandal, D.; Kumar, V.; Lopez-Sanchez, J.M.; Bhattacharya, A.; McNairn, H.; Rao, Y.S. Crop biophysical parameter retrieval from Sentinel-1 SAR data with a multi-target inversion of Water Cloud Model. *Int. J. Remote Sens.* **2020**, *41*, 5503–5524. [CrossRef]
39. Rozenstein, O.; Qin, Z.; Derimian, Y.; Karnieli, A. Correction: Rozenstein, O.; et al. Derivation of Land Surface Temperature for Landsat-8 TIRS Using a Split Window Algorithm. *Sensors* **2014**, *14*, 5768–5780. *Sensors* **2014**, *14*, 11277. [CrossRef]
40. Serrano, L.; Ustin, S.L.; Roberts, D.A.; Gamon, J.A.; Peñuelas, J. Deriving Water Content of Chaparral Vegetation from AVIRIS Data. *Remote Sens. Environ.* **2000**, *74*, 570–581. [CrossRef]
41. Chen, D.; Huang, J.; Jackson, T.J. Vegetation water content estimation for corn and soybeans using spectral indices derived from MODIS near- and short-wave infrared bands. *Remote Sens. Environ.* **2005**, *98*, 225–236. [CrossRef]
42. Wang, L.; Qu, J.J. NMDI: A normalized multi-band drought index for monitoring soil and vegetation moisture with satellite remote sensing. *Geophys. Res. Lett.* **2007**, *34*. [CrossRef]
43. Wang, L.; Qu, J.J.; Hao, X. Forest fire detection using the normalized multi-band drought index (NMDI) with satellite measurements. *Agric. For. Meteorol.* **2008**, *148*, 1767–1776. [CrossRef]
44. Attema, E.P.W.; Ulaby, F.T. Vegetation modeled as a water cloud. *Radio Sci.* **1978**, *13*, 357–364. [CrossRef]
45. Carter, J.N. Chapter 3 Introduction to using genetic algorithms. In *Developments in Petroleum Science*; Nikravesh, M., Aminzadeh, F., Zadeh, L.A., Eds.; Elsevier: Amsterdam, The Netherlands, 2003; Volume 51, pp. 51–76.
46. Juang, C.F. A Hybrid of Genetic Algorithm and Particle Swarm Optimization for Recurrent Network Design. *IEEE Trans. Syst. Man Cybern. Part B* **2004**, *34*, 997–1006. [CrossRef] [PubMed]
47. Morris, G.M.; Goodsell, D.S.; Halliday, R.S.; Huey, R.; Hart, W.E.; Belew, R.K.; Olson, A.J. Automated docking using a Lamarckian genetic algorithm and an empirical binding free energy function. *J. Comput. Chem.* **1998**, *19*, 1639–1662. [CrossRef]
48. Rumelhart, D.E. Learning internal representations by error back propagation. *Nature* **1986**, *323*, 533–536. [CrossRef]
49. Syarif, J.; Detak, Y.P.; Ramli, R. Modeling of Correlation between Heat Treatment and Mechanical Properties of Ti-6Al-4V Alloy Using Feed Forward Back Propagation Neural Network. *ISIJ Int.* **2010**, *50*, 1689–1694. [CrossRef]

50. El Hajj, M.; Baghdadi, N.; Zribi, M. Comparative analysis of the accuracy of surface soil moisture estimation from the C- and L-bands. *Int. J. Appl. Earth Obs.* **2019**, *82*, 101888. [CrossRef]
51. Zribi, M.; Muddu, S.; Bousbih, S.; Al Bitar, A.; Tomer, S.K.; Baghdadi, N.; Bandyopadhyay, S. Analysis of L-Band SAR Data for Soil Moisture Estimations over Agricultural Areas in the Tropics. *Remote Sens.* **2019**, *11*, 1122. [CrossRef]
52. El Hajj, M.; Baghdadi, N.; Zribi, M.; Bazzi, H. Synergic Use of Sentinel-1 and Sentinel-2 Images for Operational Soil Moisture Mapping at High Spatial Resolution over Agricultural Areas. *Remote Sens.* **2017**, *9*, 1292. [CrossRef]
53. El Hajj, M.; Baghdadi, N.; Zribi, M.; Belaud, G.; Cheviron, B.; Courault, D.; Charron, F. Soil moisture retrieval over irrigated grassland using X-band SAR data. *Remote Sens. Environ.* **2016**, *176*, 202–218. [CrossRef]

 MDPI

Article

Combination of Sentinel-2 and PALSAR-2 for Local Climate Zone Classification: A Case Study of Nanchang, China

Chaomin Chen [1], Hasi Bagan [1,2,*], Xuan Xie [1], Yune La [3,4] and Yoshiki Yamagata [2]

1 School of Environmental and Geographical Sciences, Shanghai Normal University, Shanghai 200234, China; 1000480060@smail.shnu.edu.cn (C.C.); 1000480061@smail.shnu.edu.cn (X.X.)
2 Center for Global Environmental Research, National Institute for Environmental Studies, Ibaraki 305-8506, Japan; yamagata@nies.go.jp
3 Cryosphere Research Station on the Qinghai-Tibetan Plateau, State Key Laboratory of Cryospheric Science, Northwest Institute of Eco-Environment and Resource, Chinese Academy of Sciences, Lanzhou 730000, China; layune20@mails.ucas.ac.cn
4 University of Chinese Academy Sciences, Beijing 100049, China
* Correspondence: hasi.bagan@nies.go.jp

Abstract: Local climate zone (LCZ) maps have been used widely to study urban structures and urban heat islands. Because remote sensing data enable automated LCZ mapping on a large scale, there is a need to evaluate how well remote sensing resources can produce fine LCZ maps to assess urban thermal environments. In this study, we combined Sentinel-2 multispectral imagery and dual-polarized (HH + HV) PALSAR-2 data to generate LCZ maps of Nanchang, China using a random forest classifier and a grid-cell-based method. We then used the classifier to evaluate the importance scores of different input features (Sentinel-2 bands, PALSAR-2 channels, and textural features) for the classification model and their contribution to each LCZ class. Finally, we investigated the relationship between LCZs and land surface temperatures (LSTs) derived from summer nighttime ASTER thermal imagery by spatial statistical analysis. The highest classification accuracy was 89.96% when all features were used, which highlighted the potential of Sentinel-2 and dual-polarized PALSAR-2 data. The most important input feature was the short-wave infrared-2 band of Sentinel-2. The spectral reflectance was more important than polarimetric and textural features in LCZ classification. PALSAR-2 data were beneficial for several land cover LCZ types when Sentinel-2 and PALSAR-2 were combined. Summer nighttime LSTs in most LCZs differed significantly from each other. Results also demonstrated that grid-cell processing provided more homogeneous LCZ maps than the usual resampling methods. This study provided a promising reference to further improve LCZ classification and quantitative analysis of local climate.

Keywords: local climate zone; random forest; feature importance; land surface temperature; grid cells; Sentinel-2; PALSAR-2; ASTER

Citation: Chen, C.; Bagan, H.; Xie, X.; La, Y.; Yamagata, Y. Combination of Sentinel-2 and PALSAR-2 for Local Climate Zone Classification: A Case Study of Nanchang, China. *Remote Sens.* **2021**, *13*, 1902. https://doi.org/10.3390/rs13101902

Academic Editor: Takeo Tadono

Received: 16 March 2021
Accepted: 5 May 2021
Published: 13 May 2021

1. Introduction

With continuous urbanization and the increasing settlement in global cities, natural landscapes are constantly converted to impervious surfaces in urban areas, altering the natural surface energy and water balances, which often results in altered climatic conditions in urban areas and the formation of the urban heat island (UHI) phenomenon [1–3]. As a key topic in urban climate studies, the concept of a "local climate zone" (LCZ) was introduced in 2012 by Stewart and Oke [4] to quantify the relationship between urban morphology and the UHI phenomenon. LCZs provide a standardized framework to link land cover types and urban morphology with corresponding thermal properties, so LCZs have been the systematic criteria for UHI comparisons [5]. Notably, the World Urban Database and Access Portal Tools (WUDAPT) project was developed as a new global initiative to produce standardized LCZ maps [6–8]. Because remote sensing data are widely available, they have been routinely used for LCZ mapping and have shown great potential for that purpose [9–12]. It is necessary to explore the combination of multi-source remote sensing data to generate LCZ mapping.

Because of the heterogeneity and complexity of the composition and configuration of urban pixels in remote sensing images, urban land cover maps based on remote sensing data are characterized by inherent uncertainties [13]. Unlike optical sensors that capture the spectral characteristics of objects on the ground, synthetic aperture radar (SAR) sensors can record the characteristics of light scattered by objects on the ground. Previous studies have demonstrated that the synergistic use of optical imagery and SAR data can facilitate urban land cover classification [14–16]. The cost-free, high-spatial-resolution imagery from the Sentinel-2 multispectral instrument (MSI) has been found to be suitable for large-scale LCZ mapping [17–20]. In addition, high-spatial-resolution phased array L-band SAR-2 (PALSAR-2) data have been used for large-scale land use and land cover mapping [21]. The use of a combination of Sentinel-2 imagery and PALSAR-2 data, therefore, has the potential to produce large-scale LCZ classification maps.

Random forest (RF) models [22] have become popular in the classification of land cover using remote sensing data because their classifications are highly accurate, their computational costs are low, and they can handle high-dimensional datasets [23,24]. Various studies have examined the importance of input features for the classifier [15,25–29] and for each class [30–33] in the context of RF classification. However, the contributions of the different bands and features of remote sensing data to the classification model and its classes have not been systematically studied in the case of LCZ classification. Only a few studies have examined the importance of features for LCZ mapping [12,17,34,35]. The feature contribution method based on decision paths [36,37] must be further investigated to take advantage of the RF model in LCZ land cover classification.

The land surface temperature (LST) observed by satellites is widely used for urban climate research, where pixel values are time-synchronized and spatially continuous [38–40]. Medium-resolution thermal satellite imagery is readily available and can provide a better alternative for urban land surface thermal analysis (e.g., surface UHI) than in situ thermal data [41]. Many studies have recently applied the LCZ classification scheme to understand the thermal characteristics of cities based on LSTs retrieved from thermal remote-sensing data [38,42–45]. Previous studies have indicated that nighttime LST could observe climatic conditions more accurately than daytime LST [46,47]. Given that summer nighttime is a crucial temporal period for surface UHI [48], it is important to explore the relationship between summer nighttime LST and LCZs.

Typically, the scales of LCZs vary from about a hundred meters to several kilometers that represent relatively homogeneous urban surfaces that share a similar energy budget [4,49]. To achieve a suitable resolution for LCZ classification, the common approach to generate LCZ maps is to preprocess the remote sensing images by resampling or to post-process the classified LCZ maps by resampling. Considering the different spatial resolutions of LCZ maps and LST data, as an alternative, the grid-cell-based method has been found to be a powerful tool for linking data at different spatial resolutions; it enables the user to fine-tune an analysis of data from multiple sources and to strike a compromise between the need for details and the feasibility of computations [50,51].

In this study, we combined the Sentinel-2 MSI imagery and PALSAR-2 data to generate LCZ maps of Nanchang City, Jiangxi Province, China, based on the RF classifier. The main objectives of this study were (i) to classify different combinations of spectral, backscattering, and textural features in Sentinel-2 and PALSAR-2, (ii) to assess the importance and contribution of the input features from Sentinel-2 MSI imagery and PALSAR-2 data to LCZ classification, and (iii) to compare the advantages and disadvantages of the resampling method and the grid-cell-based method in the process of LCZ mapping, and then to perform spatial statistical analysis of the best LCZs map and LST derived from summer nighttime Advanced Spaceborne Thermal Emission and Reflection Radiometer (ASTER) thermal imagery.

2. Materials

2.1. Study Area

Nanchang City, which is located between 115°27′–116°35′ E and 28°09′–29°11′ N (Figure 1), was selected as our study area to fill the research gap for LCZ maps in this region. Nanchang is the capital city of Jiangxi Province in southeastern China. It is one of the central cities in the middle reaches of the Yangtze River and covers about 7402 km^2. Since the 1980s, Nanchang has experienced rapid economic development, industrialization, and urbanization [52]. The Gan River runs through Nanchang from south to north and divides it into two parts. The eastern bank of the Gan River is the old urban district, while the western bank of the Gan River is the emerging urban district. As of the end of 2019, the permanent population of Nanchang was 5.6 million. The area on the eastern bank of the Gan River in Nanchang has a higher population density than other areas. In addition, Nanchang is one of the hottest cities in China, with a strong urban heat island effect [53].

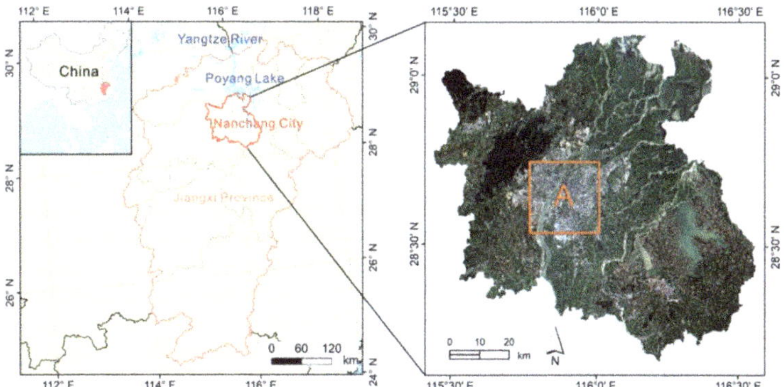

Figure 1. Left: Location of the study area (Nanchang City, China). Right: Sentinel-2 MSI image of the study area (R/G/B = bands 4/3/2). The square labeled "A" indicates a subregion shown in Figure 7.

Nanchang is located on the southwest shore of Poyang Lake, China's largest freshwater lake and the link between the Gan River and the Yangtze River. Nanchang lies within the Poyang Plain, which is rich in vegetation, rivers, and lakes. The city has rolling hills to the northwest and relatively flat terrain to the southeast. Nanchang has a subtropical, humid monsoon climate, with annual precipitation of 1613.3 mm, an average annual temperature of 19.1 °C, the highest temperature of 37.5 °C, and the lowest temperature of 0 °C, based on the meteorological statistics of 2019 [54]. Nanchang has a large diversity of land use and land cover types, which mainly includes urban and industrial land, rural settlements, paved land, rivers and bottomlands, ponds and reservoirs, cultivated land, forests, bush, grassland, and bare land [52]. The main types of buildings in Nanchang are residential buildings (e.g., elevator buildings, walk-up buildings, townhouses, bungalows, and villas), public buildings, industrial buildings, and agricultural buildings.

2.2. Remote Sensing Data

We chose Sentinel-2 MSI imagery and PALSAR-2 data to generate LCZ maps in the study area. To minimize classification errors due to different acquisition dates, we chose PALSAR-2 data with the acquisition date closest to that of Sentinel-2. The coverage of the PALSAR-2 scene is not the same as that of Sentinel-2. Therefore, we selected PALSAR-2 acquired in summer and late spring, which are closest to the acquisition date of Sentinel-2, to cover the whole study area. ASTER land surface temperature products (AST_08) were selected to investigate the relationship between the LCZs and the LST. Table 1 provides the details of the remote sensing images used in this study.

Table 1. Summary of remote-sensing data used in this study.

Remote Sensing Data	Date	Local Time at the Start of the Observation	Location in the Study Area	Spatial Resolution (m)
Sentinel-2B MSI L2A	17 September 2019 17 September 2019	10:55:49 10:55:49	Northwest Southwest	10, 20, 60
Sentinel-2A MSI L2A	19 September 2019 19 September 2019	10:45:51 10:45:51	Northeast Southeast	10, 20, 60
PALSAR-2 L3.1	19 May 2019 19 May 2019 19 May 2019 28 July 2019 28 July 2019 28 July 2019	00:12:54 00:13:02 00:13:10 00:12:54 00:13:02 00:13:10	Southwest West Northwest Southeast East Northeast	6.25
ASTER L2 AST_08	29 July 2019 29 July 2019 29 July 2019 23 August 2019 23 August 2019 23 August 2019	22:31:08 22:31:17 22:31:26 22:25:01 22:25:10 22:25:18	Southeast East Northeast Southwest West Northwest	90

All remote sensing data were acquired in 2019 and were projected to the same coordinate system by transforming projection (universal transverse mercator (UTM) zone 50 north map projection, World Geodetic System 84 (WGS-84) datum). For each source of remote sensing data, multiple scenes were mosaicked using a histogram-matching method.

2.2.1. Sentinel-2 MSI Imagery

Four Sentinel-2 MSI level-2A images (bottom-of-atmosphere reflectance) acquired in September 2019 were selected to generate a cloud-free image of the study area (https://scihub.copernicus.eu/dhus/#/home, (last accessed on 8 May 2021)). Sentinel-2 data are acquired in 13 spectral bands ranging from the visible and near-infrared (VNIR) to the short wave infrared (SWIR) at spatial resolutions of either 10 m, 20 m, or 60 m [55]. Band 10 (SWIR/cirrus) was excluded because it does not contain information about the land surface. To maintain consistency and facilitate calculations, we resampled bands with 20 m and 60 m resolutions to 10 m using a bilinear interpolation method based on Sentinel application platform (SNAP) 7.0 software (https://step.esa.int/main/download/snap-download/, (last accessed on 8 May 2021)).

2.2.2. PALSAR-2 Data

The L-band PALSAR-2 level 3.1 products were produced by the Japan Aerospace Exploration Agency (JAXA) (https://auig2.jaxa.jp/ips/homepalsar, (last accessed on 8 May 2021)) [56,57]. The data were acquired in stripmap fine beam dual (FBD) mode (HH and HV) during an ascending orbit with a right-looking observation direction, a pixel spacing of 6.25 m, and off-nadir angles of 28.6° (for 19 May 2019) and 32.9° (for 28 July 2019). To combine the PALSAR-2 data with the Sentinel-2 imagery at the pixel level, we transformed the PALSAR-2 data into the same coordinate system as the Sentinel-2 imagery and resampled it to a spatial resolution of 10 m using a bilinear interpolation method. The PALSAR-2 data were coregistered by using dispersed ground control points selected from Sentinel-2 imagery and applying a quadratic polynomial transformation and bilinear interpolation. The root-mean-square error of the ground control points was less than 0.5 pixels.

2.2.3. ASTER Land Surface Temperature Products

ASTER level-2 AST_08 (surface kinetic temperature) products are generated from ASTER's five thermal infrared bands at 90 m resolution and produced by the temperature and emissivity separation (TES) algorithm [58]. The AST_08 products were downloaded from the Land Processes Distributed Active Archive Center (https://lpdaac.usgs.gov/products/ast_08v003/, (last accessed on 8 May 2021)) and processed by the science scalable scripts-based science processor for missions (S4PM) algorithm (Version 3.4) [59]. Because a very small amount of data covering the study area were missing in 2019, we used the values of their nearest neighbors according to Euclidean distance to substitute for the missing data based on the nibble tool in ArcGIS 10.8 software.

3. Methods

3.1. Local Climate Zones Scheme

LCZs are climate-related regions that span hundreds of meters to several kilometers on a horizontal scale and are functions of surface cover, structures, construction material, and human activity [4]. As depicted in Figure 2a, the standard LCZ scheme comprises two major types: built types (LCZ classes 1–10) and land cover types (LCZ classes A–G). The 17 standard classes of LCZs are determined by surface characteristics; each provides a unique thermal environment that is most apparent in areas of simple relief, over dry surfaces, and on calm nights [4].

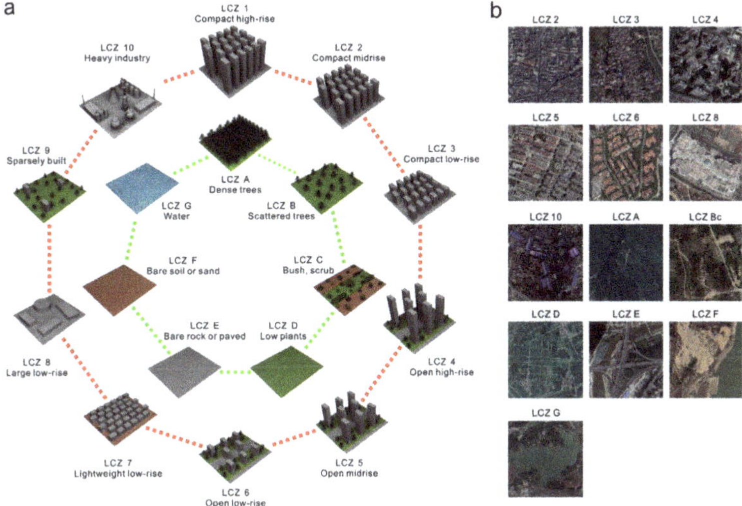

Figure 2. (**a**) Standard local climate zone (LCZ) scheme modified from Stewart and Oke [4]. (**b**) Google Earth images of typical samples of the LCZs in Nanchang.

3.2. Training and Test Datasets

To collect field-based land cover observations, we conducted field surveys in Nanchang from May to September 2019. To reduce the effects caused by the imbalance of classes [60], roughly balanced ground reference samples of 13 LCZ classes were randomly collected throughout the study area based on this field investigation and visual interpretation of high-spatial-resolution Google Earth imagery from May to September 2019. The reference samples were then randomly split into two sets of disjoint training and test pixels to ensure spatial separation of training and test sites [61] (Table 2). Figure 2b shows Google Earth images of typical samples of the LCZ classes in Nanchang. It should be pointed out that LCZ 1 (compact high-rise) was not included because there was almost no LCZ 1 in our study area. Furthermore, we merged LCZ B (scattered trees) and LCZ C (bush, scrub) into a new class LCZ B_C (scattered trees with bush and scrub) because in most cases, shrubs, short trees, and scattered trees were mixed.

Table 2. Description of LCZ classes and the number of training and test pixels in the classification.

Class	Description	Training Pixels	Test Pixels
LCZ 2	Compact mid-rise	4078	1211
LCZ 3	Compact low-rise	4377	1336
LCZ 4	Open high-rise	4500	1297
LCZ 5	Open mid-rise	4843	1343
LCZ 6	Open low-rise	4226	1420
LCZ 8	Large low-rise	4134	1303
LCZ 10	Heavy industry	4046	1310
LCZ A	Dense trees	4616	1448
LCZ B_C	Scattered trees with bush and scrub	4063	1214
LCZ D	Low plants	4283	1387
LCZ E	Bare rock or paved	4723	1430
LCZ F	Bare soil or sand	4654	1289
LCZ G	Water	4727	1381
Total		57,272	17,369

3.3. Input Features

The 12 spectral bands of Sentinel-2 MSI imagery (bottom-of-atmosphere reflectance), four backscattering intensity features obtained from dual-polarized PALSAR-2 (HH and HV backscattering coefficients, and the difference and ratio between the two polarization bands), and 24 textural features were used for the LCZ classification (Table 3). To explore the effects of different combinations of input features on classification accuracy, we set up six datasets designated as D1–D6 using these 40 features (Table 3). The textural features were extracted by using ENVI 5.5 software as follows: First, we performed a minimum noise fraction (MNF) transformation [62] on four bands at 10 m (bands 2, 3, 4, and 8) in the Sentinel-2 image. Second, the gray-level co-occurrence matrix (GLCM) [63] was computed considering a processing window of 3 $\times$ 3, the grayscale quantization level of 64, and the distance of 1. For the Sentinel-2 based GLCM, we selected the first MNF component (MNF 1) as the input. For the PALSAR-2 based GLCM, we selected the two polarization bands (HH and HV) as the input, respectively. Third, based on the obtained GLCM, we averaged eight textural features (contrast, correlation, dissimilarity, entropy, homogeneity, mean, angular second moment, and variance) in four directions (0°, 45°, 90°, and 135°) to achieve rotational invariance.

3.4. Random Forest Classification

The RF [22] is a parallel ensemble based on a classification and regression tree and can be generated simultaneously without strong dependencies between individual learners [64]. We implemented the RF classifier by using the scikit-learn library [65] and the Geospatial Data Abstraction Library (GDAL, https://gdal.org/, (last accessed on 8 May 2021)) in Python.

We used out-of-bag (OOB) samples for selecting the hyperparameters of the model. Before launching the RF classifier, two important hyperparameters that determine the randomness of the RF model had to be set: the number of trees (T) and the number of features (as listed in Table 3) randomly selected at each node (nr). We kept the other hyperparameters of the RF classifier as defaults and performed a grid search. The searching range of T was between 100 and 2000 using intervals of 100, whereas the searching range of nr was between the total number of features in intervals of 1. Based on the OOB scores of different RF models using various combinations of hyperparameters, we selected the optimal combination of hyperparameters (Table 3).

Table 3. Six datasets of different input features for LCZ classification and hyperparameters used for RF classifiers (*T*: the number of trees; *nr*: the number of features randomly selected at each node).

Dataset	Features	Number of Features	Source	Hyperparameters (*T* and *nr*)
D1	Sentinel-2 bands (1–8, 8a, 9, 11–12)	12	Sentinel-2	$T = 2000$, $nr = 4$
D2	Sentinel-2 bands (1–8, 8a, 9, 11–12) + MNF 1_GLCM (contrast, correlation, dissimilarity, entropy, homogeneity, mean, angular second moment (ASM), variance)	20	Sentinel-2	$T = 2000$, $nr = 12$
D3	Backscattering intensity (HH, HV, HH–HV, HH/HV)	4	PALSAR-2	$T = 2000$, $nr = 2$
D4	Backscattering intensity (HH, HV, HH–HV, HH/HV) + HH_GLCM (contrast, correlation, dissimilarity, entropy, homogeneity, mean, ASM, variance) + HV_GLCM (contrast, correlation, dissimilarity, entropy, homogeneity, mean, ASM, Variance)	20	PALSAR-2	$T = 2000$, $nr = 12$
D5	Sentinel-2 bands (1–8, 8a, 9, 11–12) + backscattering intensity (HH, HV, HH–HV, HH/HV)	16	Sentinel-2 + PALSAR-2	$T = 2000$, $nr = 6$
D6	D2 + D4	40	Sentinel-2 + PALSAR-2	$T = 2000$, $nr = 18$

3.5. Grid-Cell Processing and Postprocessing

Because LCZs are defined at the local scale (10^2–10^4) [4,49], we used a grid-cell (100 m × 100 m) process for pixel aggregation. First, we used ArcGIS 10.8 software to create nets of grid cells with sizes of 100 m × 100 m covering the entire study area. The 100 m × 100 m grid cells were intersected with the LST data (90 m spatial resolution), and the area of each intersected portion was calculated. The LST attribute of a grid cell was then obtained by the weighted average of the LST values of the intersected portion according to the area percentage. Next, for each grid cell, the area of each LCZ class within a grid cell was calculated and stored in the attribute table. To calculate the percentage of each LCZ class within each grid cell, we divided the area of each LCZ class by the area of the grid cell. For a single grid cell, we assigned the dominant LCZ class that accounted for the largest area to the corresponding grid cell. Finally, we used a 3 × 3 majority filter for LCZ classification maps to include more contextual information.

3.6. Usual Resampling Methods

To explore the differences between the grid-cell-based method and the usual resampling methods, we used the D6 dataset to generate 100 m LCZ maps based on ArcGIS 10.8 software. We performed majority resampling and nearest neighbor resampling on the classified LCZ map. For the classified LCZ map (categorical data), we did not include bilinear interpolation or cubic convolution in the comparison because they alter the pixel values so that the original categories are not maintained. In addition, we applied nearest neighbor resampling, bilinear interpolation resampling, and cubic convolution resampling to the original images before executing the classification. To ensure the consistency of the comparison, the 100 m LCZ classification results obtained by each resampling method were subjected to the same 3 × 3 majority filter as those obtained by the grid-cell-based method. Finally, we used the grid-cell-based method as the baseline to compare the differences between the other five resampling methods and the grid-cell-based method.

3.7. Feature Importance for the RF Model and Feature Contributions for Each Class

To understand how each feature affected the RF classification model, we used the mean decrease in Gini/Gini importance and the mean decrease in accuracy/permutation importance [22] based on the training set. The Gini importance of a feature was obtained by averaging the decrease of the Gini impurity at all nodes where this feature was used in all trees. The permutation importance was expressed as the value of the change in the accuracy of a trained model when the values of a feature in the dataset were randomly permuted. For the second of these calculations, we performed 100 repeated shuffles for all features separately and averaged the decrease of accuracy to reduce randomness.

To explore the impact of each feature on each class, we employed a feature contribution method using the tree-interpreter package. The feature contribution method is based on decision paths through each tree in a forest and can reveal the relationship between features and predictions [36,37].

For classification tasks, consider a dataset of m samples $D = \{(x_1, y_1), (x_2, y_2), \ldots, (x_m, y_m)\}$ consisting of n input features and one label y_i, where $x_{i,j}$ $(1 \le i \le m, 1 \le j \le n)$ is the value of the j-th feature at the i-th sample. Denote classes by k $(1 \le k \le K)$, where K is the total number of classes. Let t $(1 \le t \le T)$ be the t-th tree in a forest, where T is the total number of trees. For a single input x_i, there is a decision path from the root node to the leaf node in each tree. At a node in the decision path, if this node (parent node) is split into child nodes by feature j, then the contribution of feature j to class k is defined as:

$$FC_{j,k} = \begin{cases} p_{j,k}^{\text{child}} - p_{j,k}^{\text{parent}}, & \text{if the split in a parent node is performed over the feature } j; \\ 0, & \text{otherwise,} \end{cases} \tag{1}$$

where $p_{j,k}^{\text{child}}$ is the fraction of samples that belong to class k at the child node, corresponding to the feature j, and $p_{j,k}^{\text{parent}}$ is the fraction of samples that belong to class k at the parent node, corresponding to the feature j.

The predicted probability P_k that x_i belongs to class k can be written as:

$$P_k = \frac{1}{T}\sum_{t=1}^{T} p_k^{(t,\text{root})} + \sum_{j=1}^{n}\left(\frac{1}{T}\sum_{t=1}^{T} FC_{j,k}^{(t)}\right), \tag{2}$$

where $p_k^{(t,\text{root})}$ is the fraction of samples that belong to class k at the root node in the t-th tree and $FC_{j,k}^{(t)}$ is the sum of $FC_{j,k}$ over all nodes on the decision path in the t-th tree.

To obtain the feature contributions of each class, we averaged the results computed from all training samples belonging to the same class.

3.8. Statistical Analysis for Nighttime LST within LCZs

To examine the spatial autocorrelation of nighttime LST, we used the global Moran's I statistic and the Anselin local Moran's I statistic based on ArcGIS 10.8 software. For grid cells, we used an inverse distance conceptualization to generate a spatial weight matrix with a default threshold value of 270 m. Subsequently, the global Moran's I index for all grid cells was computed based on the spatial weight matrix. The Anselin local Moran's I analysis for all grid cells was also based on the spatial weight matrix. In addition, to explore the differences in LST among LCZ classes, we carried out statistical analysis using SPSS Statistics 26 software. First, we examined the normality of the LST in each LCZ class by using histogram comparisons, Q–Q plots, and Kolmogorov–Smirnov tests. We then performed Levene's test to examine the homogeneity of variances. Based on the results of these two tests, to estimate the statistical significance of the LST differences between LCZ classes, we finally chose nonparametric tests, including the Kruskal–Wallis one-way analysis of variance (ANOVA) test followed by all pairwise multiple comparisons and a median test followed by all pairwise multiple comparisons.

4. Results

4.1. Accuracy Assessment of LCZ Maps

Figure 3a shows the LCZ maps obtained with different datasets (D1–D6), respectively. The percentages of the area occupied by each LCZ class in different datasets (D1–D6) are shown in Figure 3b. The accuracies of the classification were evaluated in terms of user's accuracy (UA), producer's accuracy (PA), and overall accuracy (OA), which were derived from the confusion matrix based on test pixels [61]. The confusion matrices of LCZ maps obtained using different datasets (D1–D6) are shown in Figure 4. Figure 5 also shows the differences in the PAs and UAs of each LCZ class for the different LCZ maps.

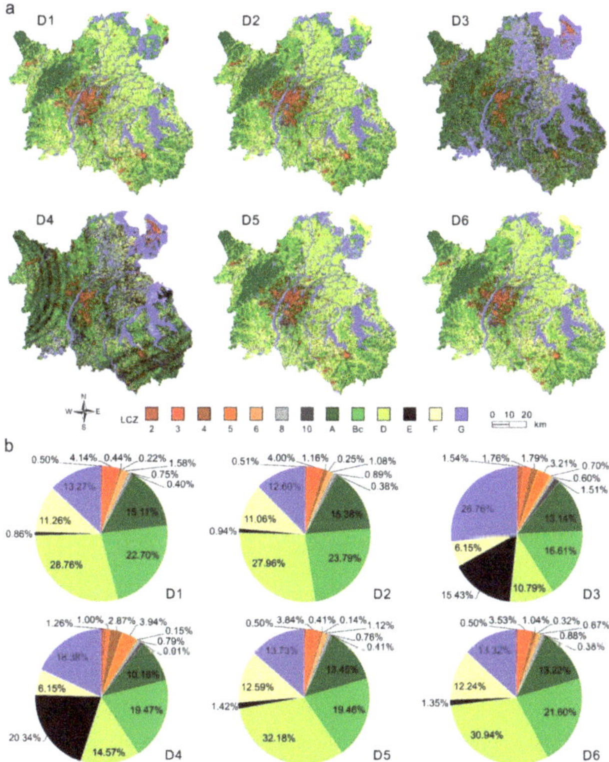

Figure 3. (**a**) LCZ maps obtained with each of the six datasets (**D1–D6**); (**b**) percentages of LCZ classes with each of the six datasets (**D1–D6**).

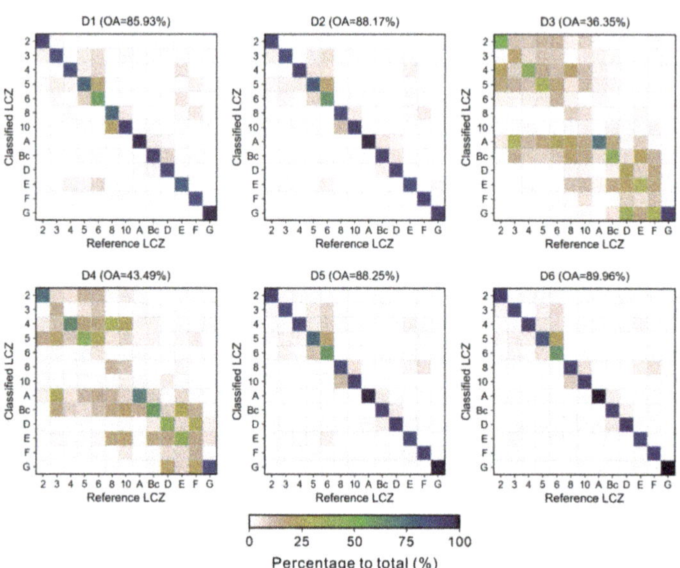

Figure 4. Confusion matrices and overall accuracies (OAs) of LCZ maps obtained by RF classification using different datasets (**D1–D6**). The confusion matrices are expressed as percentages to the total number of test pixels.

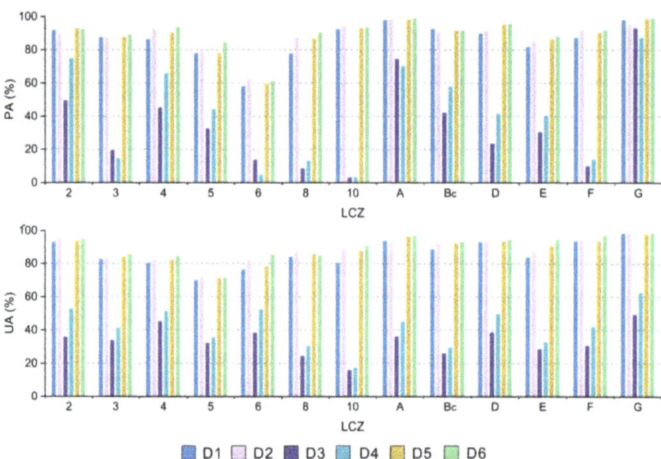

Figure 5. Producer's accuracies (PAs) and user's accuracies (UAs) of different LCZ maps obtained by RF classification using different datasets (**D1–D6**).

Compared to using the D1 dataset, the OAs were improved by 2.24% using D2, 2.32% using D5, and 4.03% using D6. There was a small improvement in the OA after using textural features. For example, the D2 dataset improved 2.24% over the D1 dataset, the D4 dataset improved 7.14% over the D3 dataset, and the D6 dataset improved 1.71% over the D5 dataset. When using the D5 dataset, the OAs were improved by 51.9% over the D3 dataset and 44.76% over the D4 dataset. The LCZ map derived using only the dual-polarized PALSAR-2 data (the D3 dataset) had the lowest OA. Using the D3 or D4 dataset, either the OAs were relatively low, or the land cover was not satisfactorily categorized. The highest OA was 89.96%, obtained from the D6 dataset by using all 40 input features.

For the D6 dataset, land cover LCZ types were generally classified with higher accuracy than built LCZ types (except for LCZs E and F). The confusion was manifested mainly among the built LCZ types. For the land cover LCZ types, LCZs E (bare rock or paved) and F (bare soil or sand) tended to be confused with built LCZ types. For the D6 dataset, LCZs A (dense tree), G (water), and D (low plants) had relatively high PA and UA among the land cover types. Among the built types, LCZs 2 (compact mid-rise) and 4 (open high-rise) had relatively high PA and UA. For the D6 dataset, open buildings (LCZs 4–6) were generally more difficult to distinguish than compact buildings (LCZs 2 and 3). This difficulty reflects that compact buildings are clustered in high-to-medium-spatial-resolution (10-m to 100-m) satellite imagery, whereas open buildings are scattered and occupy small pixels.

To measure the compliance or divergence of individual LCZ classifications, we computed the number of the same classes for a given location (individual cells of the grid) (Figure 6). The most obvious differences among the six LCZ maps were located in the northeastern part (close to Poyang Lake), the eastern part, and the urban district. A total of 86.2% of the grid cells showed good compliance for all datasets (Figure 6b).

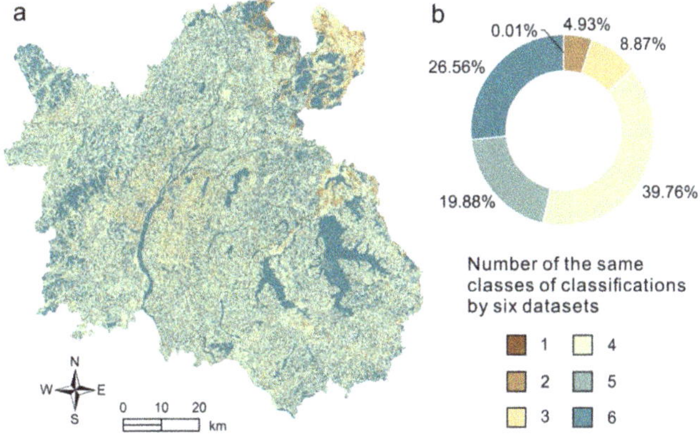

Figure 6. Difference (**a**) and its percentages (**b**) between six LCZ maps using different datasets (**D1–D6**). The difference is presented as the number of the same classes for individual cells of the grid.

To visualize the discrimination of LCZ classes using the datasets D1–D6, we extracted a subregion A in the urban district of Nanchang (Figure 7). This subregion is a typical urban region consisting of different types of buildings and land cover. It could be visually observed that the classification using PALSAR-2 polarimetric data alone did not yield a satisfactory result. Using the D3 dataset, most LCZ classes were under-represented. When using D4 by adding textural features to D3, there was a slight improvement in the classification of built LCZ types. Nevertheless, worse performance on LCZ classification was obtained using the D3 or D4 dataset. When using D2 by adding textural features to D1, the discrimination among built LCZ types was notably improved, especially for LCZ 4 (open high-rise). Compared to the classification results obtained from datasets D5 and D6, LCZ E (bare rock or paved) was under-represented using the D1 or D2 dataset. Compared to the D6 dataset, LCZ 4 was under-represented, while LCZ 6 (open low-rise) was over-represented using the D5 dataset. The most desirable result was produced when all 40 input features (the D6 dataset) were used because the confusion among LCZ classes was markedly reduced.

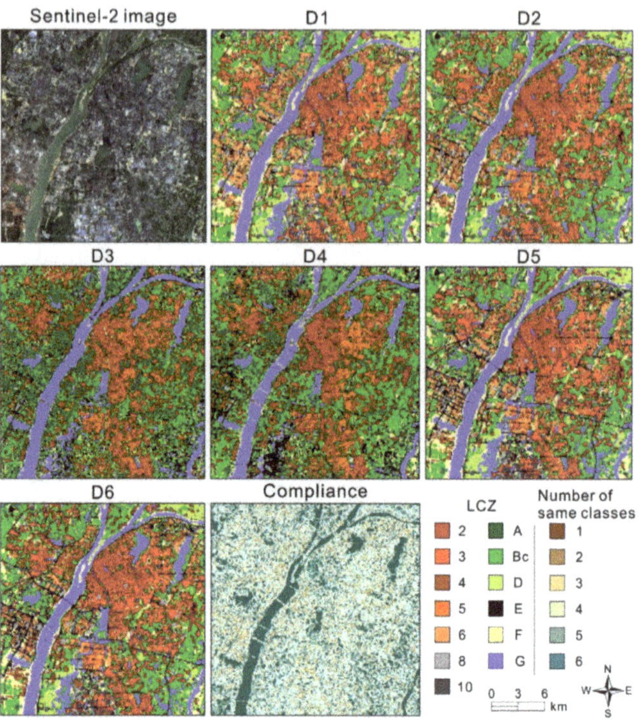

Figure 7. Sentinel-2 MSI image (RGB = bands 4, 3, 2) and LCZ classification maps using six datasets (**D1–D6**) of a subregion A in the urban district of Nanchang.

4.2. Comparison of the Grid-Cell-Based Method and Resampling Methods

As shown in Figure 8, using the nearest neighbor resampling, bilinear interpolation resampling, and cubic convolution resampling produced salt-and-pepper noise. Visually, the grid-cell-based method generated a more homogeneous result. The result of the majority resampling lay between those of the grid-cell-based method and the other resampling methods. Compared with other resampling methods, the difference between the majority resampling after classification and the grid-cell-based method was relatively small. As shown in Figure 8b–e, the basic patterns of these maps were relatively similar.

4.3. Importance and Contributions of Features for LCZ Classification

As mentioned above, the best LCZ classification was obtained using the D6 dataset. Therefore, we analyzed the feature importance of the RF model trained by all features (the D6 dataset) (Figure 9). The patterns of these two importance measures differed slightly from each other. In general, spectral features showed greater importance than polarimetric features and textural features. For both measures of importance, the most beneficial feature in the LCZ classification was S2_B12. Polarimetric features were also helpful for LCZ classification, especially the backscattering intensity at the HV polarization. In the eight textural features, GLCM_Mean was found to be the most useful feature. In addition, GLCM_Mean at the HV polarization of PALSAR-2 was more important than those extracted by Sentinel-2.

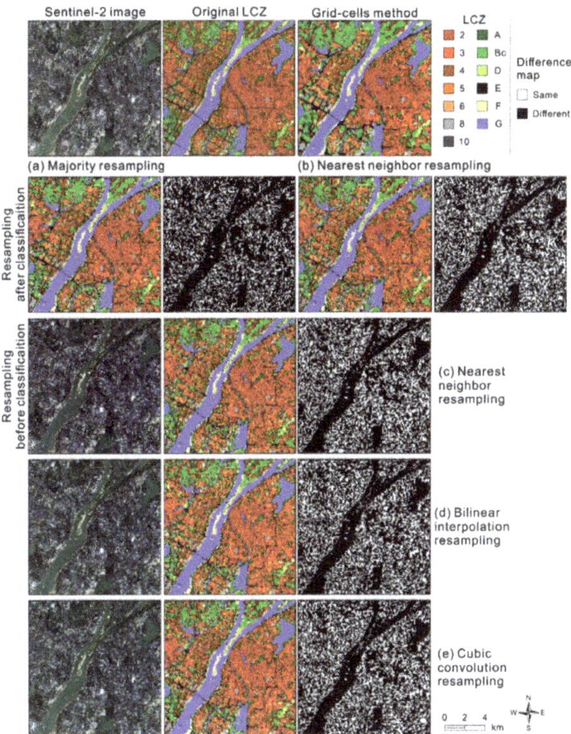

Figure 8. Detail of the comparison between the grid-cell-based method and the resampling methods using the D6 dataset. (**a**) Majority resampling after classification; (**b**) nearest neighbor resampling after classification; (**c**) nearest neighbor resampling before classification; (**d**) bilinear interpolation resampling before classification; (**e**) cubic convolution resampling before classification.

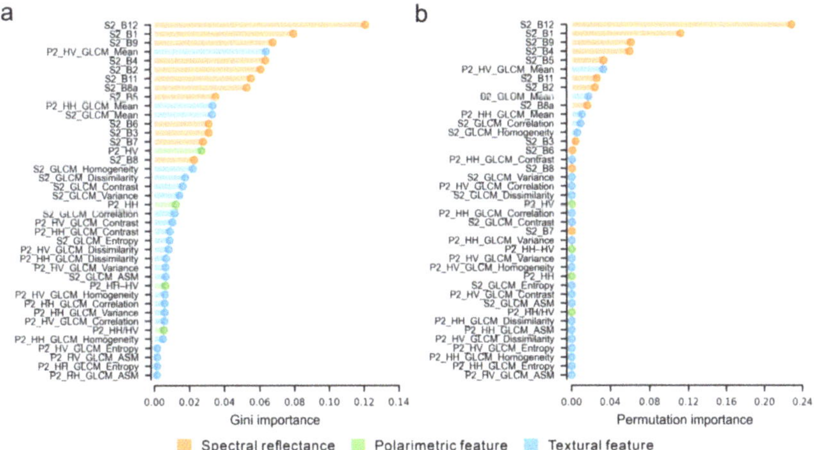

Figure 9. (**a**) Gini importance and (**b**) permutation importance of 40 input features for the RF model using the training set (D6). (S2: Sentinel-2. P2: PALSAR-2).

Figure 10 shows the feature contributions for each LCZ class for the RF model trained by all features (the D6 dataset). Land cover LCZ types exhibited more variability across input features than built LCZ types. In general, the same trends appeared in the feature importance for the RF model (Figure 9) and in the feature contributions for each LCZ class (Figure 10). For instance, S2_B12 was a beneficial feature for most of the LCZ classes. However, the contributions of a feature to each LCZ class differed to varying degrees. For example, compared with built LCZ types, the HV polarization band made a higher contribution to land cover LCZ types, especially LCZs A (dense trees), G (water), and E (bare rock or paved). As shown in Figure 10, there was no significant difference between each feature for LCZs 4 (open high-rise), 5 (open mid-rise), and 6 (open low-rise).

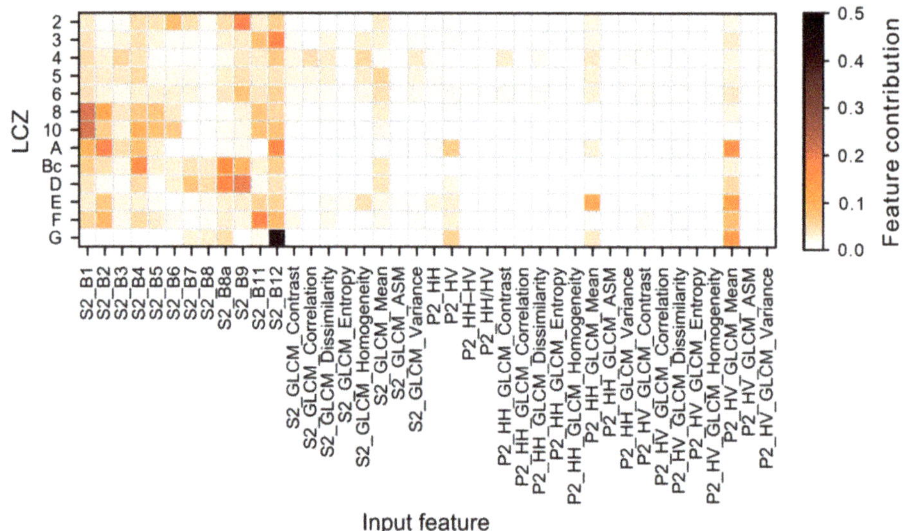

Figure 10. Mean of feature contributions for each LCZ class for the RF model using the training set (D6).

As shown in Figures 9 and 10, the combinations of features with low importance and contributions for LCZ classification did not have high classification accuracies. For the D3 and D4 datasets, except for P2_HV_GLCM_Mean and P2_HH_GLCM_Mean, the importance and contributions of the remaining features were not high, and therefore, their classification accuracies were not high. The importance and contributions of the 12 features of Sentinel-2 were relatively high, explaining the good classification accuracy achieved by using only Sentinel-2 imagery (the D1 dataset).

4.4. Relationships between LCZs and Nighttime LST

As shown in Figure 11a–b, high nighttime LSTs were dominant mainly in urban areas and water bodies. The fact that the global Moran's I index for all grid cells was 0.78 ($p < 0.001$) indicated a strong positive spatial autocorrelation for LST. The LST of both LCZ A (dense trees) and LCZ D (low plants) was clustered mostly as low-low, whereas the LST of LCZs G (water), E (bare rock or paved), and built LCZ types (LCZs 2–6, 8, and 10) was clustered mostly as high-high (Figure 11c).

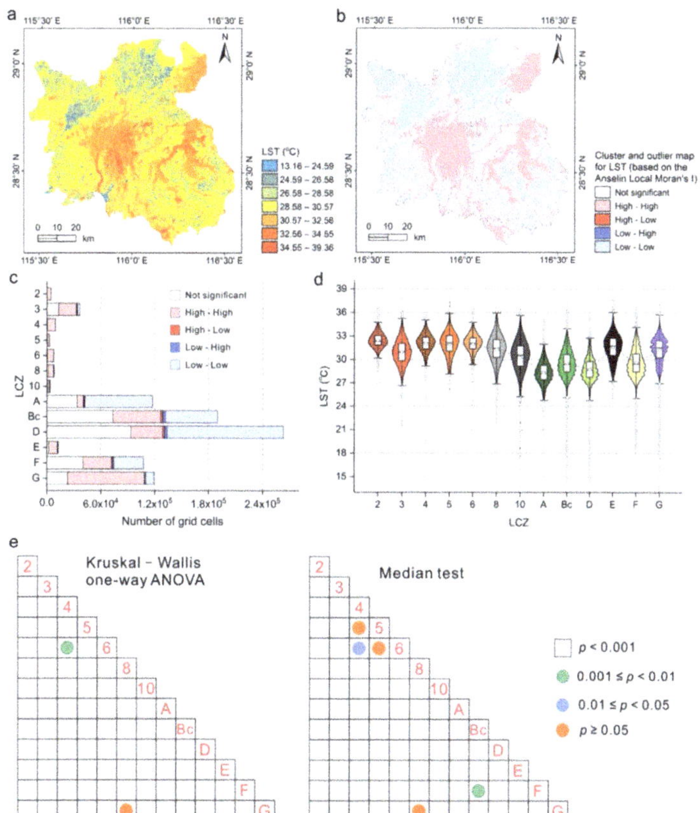

Figure 11. (**a**) Spatial distribution of nighttime LSTs in Nanchang. (**b**) Cluster and outlier map for nighttime LST. (**c**) Number of grid cells for each cluster/outlier type. (**d**) Nighttime LSTs for each LCZ class. The violin density plot displays the probability density of the LST data, with a boxplot of the mean (hollow circle), median (center horizontal line), interquartile range (black rectangle), and upper and lower whiskers (vertical lines between upper and lower horizontal lines). (**e**) Pairwise multiple comparison results of the Kruskal–Wallis one-way ANOVA test and median test, respectively. Blank cells indicate pairs of LCZs with significantly different LSTs ($p < 0.001$).

As shown in Figure 11d, different LST variations within a single LCZ class were observed. In general, there were large differences in nighttime LSTs between LCZ classes, especially between land cover LCZ types. The nighttime LST was generally higher for the built LCZ types than for the land cover LCZ types. Residential buildings (LCZs 2–6) had higher nighttime LSTs than nonresidential buildings (LCZs 8 and 10), except for LCZ 3 (compact low-rise). Both Kruskal–Wallis one-way ANOVA test and median tests showed statistically significant differences ($p < 0.001$). Overall, there were statistically significant LST differences for most LCZ classes (Figure 11e).

5. Discussion

5.1. LCZ Classification Using Sentinel-2 Imagery and PALSAR-2 Data

The fact that the accuracies of the LCZ map obtained by the D6 dataset were acceptable for further studies (as expected) revealed the potential of combining optical and SAR data for LCZ classification in urban areas. The LCZ classification using only PALSAR-2 was not satisfactory, especially for land cover LCZ types, such as those exhibiting many LCZ G (water) in the D3 dataset and LCZ E-bare rock or paved (stripes) in the D4 dataset. This reveals the limitations of using only SAR data for land cover classification in complex urban and peri-urban environments [10,15,66]. Comparing the LCZ classification maps obtained from the four datasets (D1, D2, D5, and D6), we found that the basic patterns of these LCZ maps were generally similar. It can be concluded that optical data are still dominant in LCZ classification compared to SAR data [10].

Compared to the study of La et al. [16] that combined Sentinel-2 with full-polarized PALSAR-2, this study introduced textural features derived from Sentinel-2 and dual-polarized PALSAR-2 but did not consider the contribution of polarimetric parameters. When fully polarimetric SAR data are available, adding various types of information obtained by polarimetric target decomposition methods to the classification will help to improve classification accuracy [16,67]. However, the fully polarimetric data are not always available due to its limited swath width [21]. As an alternative, our results showed the attractiveness of dual-polarized SAR data for LCZ classification. In addition, we showed the advantages of a majority rule-based grid-cells process in generating LCZ maps with generalized urban patterns.

For the D6 dataset, there are still limitations in the separability between LCZ classes with similar spectral characteristics, such as LCZs B and C; LCZ E (bare rock or paved) and built LCZ types; and LCZs 8 (large low-rise) and 10 (heavy industry). These problems can be solved by adding more discriminatory data in the classification or by improving classification algorithms. The information on building height is beneficial for the discrimination between built LCZ types. Further research on combining height data with other datasets for LCZ classification will be required in the future. It has been shown that the inclusion of LiDAR data in the classification can assist in urban land cover classification [68,69]. Moreover, deep learning methods, such as convolutional neural networks, have recently shown promising performance in LCZ classification [19,70].

Considering that the inherent speckle noise in SAR data makes individual pixels unreliable, the textural features from SAR data can provide attractive information [15]. However, there is a need to further investigate the effect of textural features from both optical and SAR data on LCZ classification. It is important to select the optimal combination of textural features for LCZ classification. Because various combinations of different bands, window sizes, and texture measures will produce many textural features, using these massive features may lead to the curse of dimensionality and reduce the accuracy in the classification using a finite-sized training set [71].

5.2. Implications of the Grid-Cell-Based Method

For the nearest neighbor resampling, the difference between the LCZ maps obtained before and after classification was small, which illustrates the applicability of resampling before classification, as implemented by the WUDAPT method [8]. The effect of the different resampling rules on the LCZ maps is more significant than whether the resampling is implemented before or after classification. As the postprocessing approach for categorical data, the LCZ map obtained by the majority resampling was not as homogeneous as those obtained by the majority rule-based grid-cells method, which may be limited by the built-in default filter window in ArcGIS software.

5.3. Assessment of Interpretability of Features

The significant importance and contribution of features can be explained by the good representation of those features in the characteristics of LCZ classes [11,12]. Not all of the features were equally beneficial to the LCZ classification for the RF models. The fact that the overall importance of the PALSAR-2 derived features was not as prominent as the Sentinel-2 derived features was probably due to the inconsistent dates of PALSAR-2 data and Sentinel-2 imagery. The seasonal variation of vegetation may make polarimetric features contribute little to identifying these changing ground objects. For PALSAR-2 dual-polarization data, the HV polarization band is more conducive to land cover classification than the HH polarization band, which may be due to the unique scattering information about ground objects provided by cross-polarization [15,25]. Our study also indicated that the GLCM textural features have limited ability in land cover classification. This limited ability may reflect the spatial resolution of Sentinel-2 MSI imagery and PALSAR-2 data. This problem will probably be resolved with the improvement of spatial resolution [72].

The fact that most features performed better in land cover LCZ types than built LCZ types indicated that there were still limitations in discriminating built LCZ types for these features. Interestingly, band 1 (coastal aerosol) made a significant contribution to LCZs 8 (large low-rise) and 10 (heavy industry); and band 9 (water vapor) contributed significantly to LCZs 2 (compact mid-rise), D (low plants), and 6 (open low-rise). These results showed that both band 1 (dedicated for aerosol retrieval) and band 9 (dedicated for water vapor correction) in Sentinel-2 imagery were beneficial for LCZ classification, despite their relatively low spatial resolution (60 m). In addition, for LCZ 5 (open mid-rise), S2_GLCM_Mean made the highest contribution. This observation highlights the fact that features that are generally unimportant for the model may be important for a specific class [37].

It is worth noting that there were many features that have very low permutation importance (Figure 9b), probably because of the correlation between features. When features are correlated, permutating one feature has little impact on the model's performance because it can obtain the same information from the correlated features. In the future, it will be necessary to evaluate additional features to provide more information on how to allow LCZ classes to be better differentiated. In addition, it is also important to analyze the correlation in the features extracted from the remote sensing data.

5.4. LST Differentiation of LCZs

In general terms, the fact that there were statistically significant nighttime LST differences between most LCZs indicated that different LCZ classes exhibited thermal environments associated with their surface characteristics [4,49]. For example, built LCZ types (LCZs 2–6, 8, and 10) and LCZ E (bare rock or paved) were clustered as high-high on nighttime LST (Figure 11c), probably because of the thermal properties of impervious surfaces [45]. Compared to other land cover LCZ types, LCZs E and G (water) had relatively high nighttime LSTs, probably because they cool off more slowly during nighttime. The fact that LCZ A (dense trees) had lower nighttime LSTs than LCZ B_C (scattered trees with bush and scrub) indicates that aggregated vegetation cools better than dispersed vegetation [73]. The fact that the nighttime LSTs of LCZ D (low plants) were lower than that of LCZ A was probably because dense trees have greater shading coverage that influences the penetration of solar radiation [74]. For buildings located in urban areas with the same heights, open buildings had lower nighttime LSTs than compact buildings. The former may benefit from the surrounding vegetation and good ventilation [75].

However, the fact that the nighttime LSTs of several LCZ classes were not significantly different statistically from other classes may have been related to the influence of local or regional climate [43]. In addition, the intra-LCZ variability of nighttime LST revealed the potential effects of heterogeneous surrounding environments [76]. For example, the nighttime LSTs of both LCZs 3 (compact low-rise) and B_C were generally higher in urban areas than in rural areas. Similarly, LCZ F (Bare soil or sand) was warmer near water than near dense trees during nighttime. Buildings surrounded by large areas of vegetation tended to have lower nighttime LSTs than buildings surrounded by a small amount of vegetation.

Furthermore, several issues need to be further explored in studies of LSTs or surface urban heat islands using LCZ maps, including seasonal changes in LCZs, the effects of multitemporal (day and night), seasonal, and thermal anisotropy on LST variations [42,43]. Considering that it is not the focus of this study, we only used two dates of LST data in summer to analyze the relationship between LCZ and nighttime LST. Many studies have shown LST differences within LCZs using many dates of LST data [38,40,42,43]. Therefore, the applicability of the inclusion of thermal remote sensing images from different sensors (e.g., Landsat and ASTER) over multiple periods in LCZ classification can be investigated in the future. However, this could potentially lead to methodological bias in LST analysis [44].

6. Conclusions

The combination of Sentinel-2 MSI imagery and dual-polarized (HH + HV) PALSAR-2 data was found to be promising and beneficial for LCZ mapping. The quantitative analysis of input features based on the RF classifier showed that in LCZ classification, band 12-SWIR 2 is crucial for Sentinel-2 imagery, whereas the HV polarization is important for dual-polarized PALSAR-2 data. By using the feature contribution approach based on decision paths, each input feature was found to contribute differently to LCZ classes. These different contributions may not be detected by a standard feature importance analysis. Through this class-based analysis of feature contributions, it is possible to reveal the effective features in distinguishing different LCZ classes. In addition, our comparative results showed that the grid-cell-based method produced more homogeneous LCZ maps than the usual resampling methods.

Spatial analysis of LCZs and summer nighttime LST showed that high LSTs were concentrated mostly in the built LCZ types, LCZ E, and LCZ G, whereas low LSTs were mostly concentrated in LCZs A and D. Statistical analysis showed that the summer nighttime LST differences between most LCZ classes were statistically significant, but this phenomenon needs to be further investigated using more dates of thermal remote sensing images. Considering the thermal differentiation within LCZs, the effect of thermal remote sensing data in LCZ classification can also be further explored.

This study provided insights into the performance of RF classifiers in LCZ mapping and feature assessment that could contribute to future LCZ mapping. In addition, this study highlighted the potential of the LCZ map and the grid-cell-based method for urban climate research that could contribute to a better understanding of the impact of urban morphology defined by LCZs on local climatic conditions.

Author Contributions: Conceptualization, C.C. and H.B.; methodology, C.C. and H.B.; software, C.C.; formal analysis, C.C., H.B. and X.X.; investigation, X.X.; resources and data curation, Y.L.; writing—original draft preparation, C.C.; writing—review and editing, H.B.; visualization, C.C.; supervision, H.B. and Y.Y.; project administration, H.B.; funding acquisition, H.B. All authors have read and agreed to the published version of the manuscript.

Funding: This research was funded by the National Natural Science Key Foundation of China (Grant No. 41730642), and the National Natural Science Foundation of China (Grant No. 41771372).

Institutional Review Board Statement: Not applicable.

Informed Consent Statement: Not applicable.

Data Availability Statement: Not applicable.

Acknowledgments: This research was supported by ALOS-2 RA-6.

Conflicts of Interest: The authors declare no conflict of interest. The funders had no role in the design of the study; in the collection, analyses, or interpretation of data; in the writing of the manuscript, or in the decision to publish the results.

References

1. Voogt, J.; Oke, T. Thermal remote sensing of urban climates. *Remote Sens. Environ.* **2003**, *86*, 370–384. [CrossRef]
2. Mirzaei, P.A. Recent challenges in modeling of urban heat island. *Sustain. Cities Soc.* **2015**, *19*, 200–206. [CrossRef]
3. Parsaee, M.; Joybari, M.M.; Mirzaei, P.A.; Haghighat, F. Urban heat island, urban climate maps and urban development policies and action plans. *Environ. Technol. Innov.* **2019**, *14*, 100341. [CrossRef]
4. Stewart, I.D.; Oke, T.R. Local Climate Zones for Urban Temperature Studies. *Bull. Am. Meteorol. Soc.* **2012**, *93*, 1879–1900. [CrossRef]
5. Zhao, C.; Jensen, J.L.R.; Weng, Q.; Currit, N.; Weaver, R. Use of Local Climate Zones to investigate surface urban heat islands in Texas. *GIScience Remote Sens.* **2020**, *57*, 1083–1101. [CrossRef]

6. Danylo, O.; See, L.; Bechtel, B.; Schepaschenko, D.; Fritz, S. Contributing to WUDAPT: A Local Climate Zone Classification of Two Cities in Ukraine. *IEEE J. Sel. Top. Appl. Earth Obs. Remote Sens.* **2016**, *9*, 1841–1853. [CrossRef]
7. Ching, J.; Mills, G.; Bechtel, B.; See, L.; Feddema, J.; Wang, X.; Ren, C.; Brousse, O.; Martilli, A.; Neophytou, M.; et al. WUDAPT: An Urban Weather, Climate, and Environmental Modeling Infrastructure for the Anthropocene. *Bull. Am. Meteorol. Soc.* **2018**, *99*, 1907–1924. [CrossRef]
8. Bechtel, B.; Alexander, P.J.; Beck, C.; Böhner, J.; Brousse, O.; Ching, J.; Demuzere, M.; Fonte, C.; Gál, T.; Hidalgo, J.; et al. Generating WUDAPT Level 0 data—Current status of production and evaluation. *Urban Clim.* **2019**, *27*, 24–45. [CrossRef]
9. Bechtel, B.; Daneke, C. Classification of Local Climate Zones Based on Multiple Earth Observation Data. *IEEE J. Sel. Top. Appl. Earth Obs. Remote Sens.* **2012**, *5*, 1191–1202. [CrossRef]
10. Bechtel, B.; See, L.; Mills, G.; Foley, M. Classification of Local Climate Zones Using SAR and Multispectral Data in an Arid Environment. *IEEE J. Sel. Top. Appl. Earth Obs. Remote Sens.* **2016**, *9*, 3097–3105. [CrossRef]
11. Xu, Y.; Ren, C.; Cai, M.; Edward, N.Y.Y.; Wu, T. Classification of Local Climate Zones Using ASTER and Landsat Data for High-Density Cities. *IEEE J. Sel. Top. Appl. Earth Obs. Remote Sens.* **2017**, *10*, 3397–3405. [CrossRef]
12. Demuzere, M.; Bechtel, B.; Middel, A.; Mills, G. Mapping Europe into local climate zones. *PLoS ONE* **2019**, *14*, e0214474. [CrossRef]
13. Reba, M.; Seto, K.C. A systematic review and assessment of algorithms to detect, characterize, and monitor urban land change. *Remote Sens. Environ.* **2020**, *242*, 111739. [CrossRef]
14. Amarsaikhan, D.; Ganzorig, M.; Ache, P.; Blotevogel, H. The integrated use of optical and InSAR data for urban land-cover mapping. *Int. J. Remote Sens.* **2007**, *28*, 1161–1171. [CrossRef]
15. Zhu, Z.; Woodcock, C.E.; Rogan, J.; Kellndorfer, J. Assessment of spectral, polarimetric, temporal, and spatial dimensions for urban and peri-urban land cover classification using Landsat and SAR data. *Remote Sens. Environ.* **2012**, *117*, 72–82. [CrossRef]
16. La, Y.; Bagan, H.; Yamagata, Y. Urban land cover mapping under the Local Climate Zone scheme using Sentinel-2 and PALSAR-2 data. *Urban Clim.* **2020**, *33*, 100661. [CrossRef]
17. Qiu, C.; Schmitt, M.; Mou, L.; Ghamisi, P.; Zhu, X.X. Feature Importance Analysis for Local Climate Zone Classification Using a Residual Convolutional Neural Network with Multi-Source Datasets. *Remote Sens.* **2018**, *10*, 1572. [CrossRef]
18. Qiu, C.; Mou, L.; Schmitt, M.; Zhu, X.X. Local climate zone-based urban land cover classification from multi-seasonal Sentinel-2 images with a recurrent residual network. *ISPRS J. Photogramm. Remote Sens.* **2019**, *154*, 151–162. [CrossRef]
19. Rosentreter, J.; Hagensieker, R.; Waske, B. Towards large-scale mapping of local climate zones using multitemporal Sentinel 2 data and convolutional neural networks. *Remote Sens. Environ.* **2020**, *237*, 111472. [CrossRef]
20. Yoo, C.; Lee, Y.; Cho, D.; Im, J.; Han, D. Improving local climate zone classification using incomplete building data and Sentinel 2 images based on convolutional neural networks. *Remote Sens.* **2020**, *12*, 3552. [CrossRef]
21. Ohki, M.; Shimada, M. Large-Area Land Use and Land Cover Classification With Quad, Compact, and Dual Polarization SAR Data by PALSAR-2. *IEEE Trans. Geosci. Remote Sens.* **2018**, *56*, 5550–5557. [CrossRef]
22. Breiman, L. Random Forests. *Mach. Learn.* **2001**, *45*, 5–32. [CrossRef]
23. Rodriguez-Galiano, V.; Ghimire, B.; Rogan, J.; Chica-Olmo, M.; Rigol-Sanchez, J. An assessment of the effectiveness of a random forest classifier for land-cover classification. *ISPRS J. Photogramm. Remote Sens.* **2012**, *67*, 93–104. [CrossRef]
24. Belgiu, M.; Drăguţ, L. Random forest in remote sensing: A review of applications and future directions. *ISPRS J. Photogramm. Remote Sens.* **2016**, *114*, 24–31. [CrossRef]
25. Jin, H.; Mountrakis, G.; Stehman, S.V. Assessing integration of intensity, polarimetric scattering, interferometric coherence and spatial texture metrics in PALSAR-derived land cover classification. *ISPRS J. Photogramm. Remote Sens.* **2014**, *98*, 70–84. [CrossRef]
26. Van Beijma, S.; Comber, A.; Lamb, A. Random forest classification of salt marsh vegetation habitats using quad-polarimetric airborne SAR, elevation and optical RS data. *Remote Sens. Environ.* **2014**, *149*, 118–129. [CrossRef]
27. Mahdianpari, M.; Salehi, B.; Mohammadimanesh, F.; Motagh, M. Random forest wetland classification using ALOS-2 L-band, RADARSAT-2 C-band, and TerraSAR-X imagery. *ISPRS J. Photogramm. Remote Sens.* **2017**, *130*, 13–31. [CrossRef]
28. Merchant, M.A.; Warren, R.K.; Edwards, R.; Kenyon, J.K. An Object-Based Assessment of Multi-Wavelength SAR, Optical Imagery and Topographical Datasets for Operational Wetland Mapping in Boreal Yukon, Canada. *Can. J. Remote Sens.* **2019**, *45*, 308–332. [CrossRef]
29. Abdi, A.M. Land cover and land use classification performance of machine learning algorithms in a boreal landscape using Sentinel-2 data. *GIScience Remote Sens.* **2020**, *57*, 1–20. [CrossRef]
30. Gislason, P.O.; Benediktsson, J.A.; Sveinsson, J.R. Random Forests for land cover classification. *Pattern Recognit. Lett.* **2006**, *27*, 294–300. [CrossRef]
31. Rodriguez-Galiano, V.; Chica-Olmo, M.; Abarca-Hernandez, F.; Atkinson, P.; Jeganathan, C. Random Forest classification of Mediterranean land cover using multi-seasonal imagery and multi-seasonal texture. *Remote Sens. Environ.* **2012**, *121*, 93–107. [CrossRef]
32. Guan, H.; Li, J.; Chapman, M.; Deng, F.; Ji, Z.; Yang, X. Integration of orthoimagery and lidar data for object-based urban thematic mapping using random forests. *Int. J. Remote Sens.* **2013**, *34*, 5166–5186. [CrossRef]
33. Tatsumi, K.; Yamashiki, Y.; Torres, M.A.C.; Taipe, C.L.R. Crop classification of upland fields using Random forest of time-series Landsat 7 ETM+ data. *Comput. Electron. Agric.* **2015**, *115*, 171–179. [CrossRef]

34. Xu, Z.; Chen, J.; Xia, J.; Du, P.; Zheng, H.; Gan, L. Multisource Earth Observation Data for Land-Cover Classification Using Random Forest. *IEEE Geosci. Remote Sens. Lett.* **2018**, *15*, 789–793. [CrossRef]
35. Yokoya, N.; Ghamisi, P.; Xia, J.; Sukhanov, S.; Heremans, R.; Tankoyeu, I.; Bechtel, B.; le Saux, B.; Moser, G.; Tuia, D. Open Data for Global Multimodal Land Use Classification: Outcome of the 2017 IEEE GRSS Data Fusion Contest. *IEEE J. Sel. Top. Appl. Earth Obs. Remote Sens.* **2018**, *11*, 1363–1377. [CrossRef]
36. Palczewska, A.; Palczewski, J.; Robinson, R.M.; Neagu, D. Interpreting random forest models using a feature contribution method. In Proceedings of the 2013 IEEE 14th International Conference on Information Reuse & Integration (IRI), San Francisco, CA, USA, 14–16 August 2013; pp. 112–119.
37. Palczewska, A.; Palczewski, J.; Robinson, R.M.; Neagu, D. Interpreting Random Forest Classification Models Using a Feature Contribution Method. In *Advances in Intelligent Systems and Computing*; Bouabana-Tebibel, T., Rubin, S.H., Kacprzyk, J., Eds.; Springer Science and Business Media LLC: Cham, Switzerland, 2014; Volume 263, pp. 193–218.
38. Bechtel, B.; Demuzere, M.; Mills, G.; Zhan, W.; Sismanidis, P.; Small, C.; Voogt, J. SUHI analysis using Local Climate Zones—A comparison of 50 cities. *Urban Clim.* **2019**, *28*, 100451. [CrossRef]
39. Rahman, M.; Avtar, R.; Yunus, A.P.; Dou, J.; Misra, P.; Takeuchi, W.; Sahu, N.; Kumar, P.; Johnson, B.A.; Dasgupta, R.; et al. Monitoring Effect of Spatial Growth on Land Surface Temperature in Dhaka. *Remote Sens.* **2020**, *12*, 1191. [CrossRef]
40. Du, P.; Chen, J.; Bai, X.; Han, W. Understanding the seasonal variations of land surface temperature in Nanjing urban area based on local climate zone. *Urban Clim.* **2020**, *33*, 100657. [CrossRef]
41. Mushore, T.D.; Odindi, J.; Dube, T.; Matongera, T.N.; Mutanga, O. Remote sensing applications in monitoring urban growth impacts on in-and-out door thermal conditions: A review. *Remote Sens. Appl. Soc. Environ.* **2017**, *8*, 83–93. [CrossRef]
42. Geletič, J.; Lehnert, M.; Dobrovolný, P. Land Surface Temperature Differences within Local Climate Zones, Based on Two Central European Cities. *Remote Sens.* **2016**, *8*, 788. [CrossRef]
43. Geletič, J.; Lehnert, M.; Savić, S.; Milošević, D. Inter-/intra-zonal seasonal variability of the surface urban heat island based on local climate zones in three central European cities. *Build. Environ.* **2019**, *156*, 21–32. [CrossRef]
44. Wang, C.; Middel, A.; Myint, S.W.; Kaplan, S.; Brazel, A.J.; Lukasczyk, J. Assessing local climate zones in arid cities: The case of Phoenix, Arizona and Las Vegas, Nevada. *ISPRS J. Photogramm. Remote Sens.* **2018**, *141*, 59–71. [CrossRef]
45. Khamchiangta, D.; Dhakal, S. Physical and non-physical factors driving urban heat island: Case of Bangkok Metropolitan Administration, Thailand. *J. Environ. Manag.* **2019**, *248*, 109285. [CrossRef] [PubMed]
46. Hartz, D.; Prashad, L.; Hedquist, B.; Golden, J.; Brazel, A. Linking satellite images and hand-held infrared thermography to observed neighborhood climate conditions. *Remote Sens. Environ.* **2006**, *104*, 190–200. [CrossRef]
47. Sekertekin, A.; Bonafoni, S. Sensitivity Analysis and Validation of Daytime and Nighttime Land Surface Temperature Retrievals from Landsat 8 Using Different Algorithms and Emissivity Models. *Remote Sens.* **2020**, *12*, 2776. [CrossRef]
48. Peng, J.; Ma, J.; Liu, Q.; Liu, Y.; Hu, Y.; Li, Y.; Yue, Y. Spatial-temporal change of land surface temperature across 285 cities in China: An urban-rural contrast perspective. *Sci. Total. Environ.* **2018**, *635*, 487–497. [CrossRef]
49. Stewart, I.D.; Oke, T.R.; Krayenhoff, E.S. Evaluation of the 'local climate zone' scheme using temperature observations and model simulations. *Int. J. Clim.* **2014**, *34*, 1062–1080. [CrossRef]
50. Bagan, H.; Yamagata, Y. Landsat analysis of urban growth: How Tokyo became the world's largest megacity during the last 40 years. *Remote Sens. Environ.* **2012**, *127*, 210–222. [CrossRef]
51. Bagan, H.; Millington, A.; Takeuchi, W.; Yamagata, Y. Spatiotemporal analysis of deforestation in the Chapare region of Bolivia using LANDSAT images. *Land Degrad. Dev.* **2020**, *31*, 3024–3039. [CrossRef]
52. Zhang, Y.; Xu, B. Spatiotemporal analysis of land use/cover changes in Nanchang area, China. *Int. J. Digit. Earth* **2014**, *8*, 312–333. [CrossRef]
53. Zhang, X.; Estoque, R.C.; Murayama, Y. An urban heat island study in Nanchang City, China based on land surface temperature and social-ecological variables. *Sustain. Cities Soc.* **2017**, *32*, 557–568. [CrossRef]
54. Statistics Bureau of Nanchang, Nanchang Investigation Team of National Bureau of Statistics. *Nanchang Statistical Yearbook of 2020*; China Statistics Press: Beijing, China, 2020.
55. Drusch, M.; del Bello, U.; Carlier, S.; Colin, O.; Fernandez, V.; Gascon, F.; Hoersch, B.; Isola, C.; Laberinti, P.; Martimort, P.; et al. Sentinel-2: ESA's Optical High-Resolution Mission for GMES Operational Services. *Remote Sens. Environ.* **2012**, *120*, 25–36. [CrossRef]
56. Rosenqvist, A.; Shimada, M.; Suzuki, S.; Ohgushi, F.; Tadono, T.; Watanabe, M.; Tsuzuku, K.; Kamijo, S.; Aoki, E. Operational performance of the ALOS global systematic acquisition strategy and observation plans for ALOS-2 PALSAR-2. *Remote Sens. Environ.* **2014**, *155*, 3–12. [CrossRef]
57. Shimada, M. *Imaging from Spaceborne and Airborne SARs, Calibration, and Applications*; CRC Press: Boca Raton, FL, USA, 2018.
58. Gillespie, A.; Rokugawa, S.; Matsunaga, T.; Cothern, J.; Hook, S.; Kahle, A. A temperature and emissivity separation algorithm for Advanced Spaceborne Thermal Emission and Reflection Radiometer (ASTER) images. *IEEE Trans. Geosci. Remote Sens.* **1998**, *36*, 1113–1126. [CrossRef]
59. NASA/METI/AIST/Japan Spacesystems; U.S./Japan ASTER Science Team. ASTER Level 2 Surface Temperature Product. 2001. NASA EOSDIS Land Processes DAAC. Available online: https://doi.org/10.5067/ASTER/AST_08.003 (accessed on 18 May 2020).

60. Mellor, A.; Boukir, S.; Haywood, A.; Jones, S. Exploring issues of training data imbalance and mislabelling on random forest performance for large area land cover classification using the ensemble margin. *ISPRS J. Photogramm. Remote Sens.* **2015**, *105*, 155–168. [CrossRef]
61. Congalton, R.G.; Green, K. *Assessing the Accuracy of Remotely Sensed Data*, 3rd ed.; CRC Press; Taylor & Francis Group: Boca Raton, FL, USA, 2019.
62. Green, A.; Berman, M.; Switzer, P.; Craig, M. A transformation for ordering multispectral data in terms of image quality with implications for noise removal. *IEEE Trans. Geosci. Remote Sens.* **1988**, *26*, 65–74. [CrossRef]
63. Haralick, R.M.; Shanmugam, K.; Dinstein, I. Textural Features for Image Classification. *IEEE Trans. Syst. Man, Cybern.* **1973**, *SMC-3*, 610–621. [CrossRef]
64. Rokach, L. Ensemble-based classifiers. *Artif. Intell. Rev.* **2010**, *33*, 1–39. [CrossRef]
65. Pedregosa, F.; Varoquaux, G.; Gramfort, A.; Michel, V.; Thirion, B.; Grisel, O.; Blondel, M.; Prettenhofer, P.; Weiss, R.; Dubourg, V.; et al. Scikit-learn, Machine learning in Python. *J. Mach. Learn. Res.* **2011**, *12*, 2825–2830. Available online: https://jmlr.csail.mit.edu/papers/v12/pedregosa11a.html (accessed on 26 May 2020).
66. Hu, J.; Ghamisi, P.; Zhu, X.X. Feature Extraction and Selection of Sentinel-1 Dual-Pol Data for Global-Scale Local Climate Zone Classification. *ISPRS Int. J. Geo-Inf.* **2018**, *7*, 379. [CrossRef]
67. Bagan, H.; Kinoshita, T.; Yamagata, Y. Combination of AVNIR-2, PALSAR, and Polarimetric Parameters for Land Cover Classification. *IEEE Trans. Geosci. Remote Sens.* **2011**, *50*, 1318–1328. [CrossRef]
68. Gu, Y.; Wang, Q.; Jia, X.; Benediktsson, J.A. A Novel MKL Model of Integrating LiDAR Data and MSI for Urban Area Classification. *IEEE Trans. Geosci. Remote Sens.* **2015**, *53*, 5312–5326. [CrossRef]
69. Yan, W.Y.; Shaker, A.; El-Ashmawy, N. Urban land cover classification using airborne LiDAR data: A review. *Remote Sens. Environ.* **2015**, *158*, 295–310. [CrossRef]
70. Yoo, C.; Han, D.; Im, J.; Bechtel, B. Comparison between convolutional neural networks and random forest for local climate zone classification in mega urban areas using Landsat images. *ISPRS J. Photogramm. Remote Sens.* **2019**, *157*, 155–170. [CrossRef]
71. Laurin, G.V.; Liesenberg, V.; Chen, Q.; Guerriero, L.; del Frate, F.; Bartolini, A.; Coomes, D.; Wilebore, B.; Lindsell, J.; Valentini, R. Optical and SAR sensor synergies for forest and land cover mapping in a tropical site in West Africa. *Int. J. Appl. Earth Obs. Geoinf.* **2013**, *21*, 7–16. [CrossRef]
72. Mishra, V.N.; Prasad, R.; Rai, P.K.; Vishwakarma, A.K.; Arora, A. Performance evaluation of textural features in improving land use/land cover classification accuracy of heterogeneous landscape using multi-sensor remote sensing data. *Earth Sci. Inform.* **2018**, *12*, 71–86. [CrossRef]
73. Estoque, R.C.; Murayama, Y.; Myint, S.W. Effects of landscape composition and pattern on land surface temperature: An urban heat island study in the megacities of Southeast Asia. *Sci. Total. Environ.* **2017**, *577*, 349–359. [CrossRef]
74. Ng, E.; Chen, L.; Wang, Y.; Yuan, C. A study on the cooling effects of greening in a high-density city: An experience from Hong Kong. *Build. Environ.* **2012**, *47*, 256–271. [CrossRef]
75. Yang, J.; Jin, S.; Xiao, X.; Jin, C.; Xia, J.; Li, X.; Wang, S. Local climate zone ventilation and urban land surface temperatures: Towards a performance-based and wind-sensitive planning proposal in megacities. *Sustain. Cities Soc.* **2019**, *47*, 101487. [CrossRef]
76. Shi, Y.; Lau, K.K.-L.; Ren, C.; Ng, E. Evaluating the local climate zone classification in high-density heterogeneous urban environment using mobile measurement. *Urban Clim.* **2018**, *25*, 167–186. [CrossRef]

Article

Assessing the Accuracy of ALOS/PALSAR-2 and Sentinel-1 Radar Images in Estimating the Land Subsidence of Coastal Areas: A Case Study in Alexandria City, Egypt

Noura Darwish [1], Mona Kaiser [1], Magaly Koch [2] and Ahmed Gaber [3,*]

[1] Geology Department, Faculty of Science, Suez Canal University, Ismailia 41522, Egypt; nourahmoustafa123@yahoo.com (N.D.); mona_kaysar@science.suez.edu.eg (M.K.)
[2] Center for Remote Sensing, Boston University, Boston, MA 02215, USA; Mkoch@bu.edu
[3] Geology Department, Faculty of Science, Port Said University, Port Said 42522, Egypt
* Correspondence: gaber@sci.psu.edu.eg

Abstract: Recently, the Differential Interferometric Synthetic Aperture Radar (DInSAR) technique is widely used for quantifying the land surface deformation, which is very important to assess the potential impact on social and economic activities. Radar satellites operate in different wavelengths and each provides different levels of vertical displacement accuracy. In this study, the accuracies of Sentinel-1 (C-band) and ALOS/PALSAR-2 (L-band) were investigated in terms of estimating the land subsidence rate along the study area of Alexandria City, Egypt. A total of nine Sentinel-1 and 11 ALOS/PALSAR-2 scenes were used for such assessment. The small baseline subset (SBAS) processing scheme, which detects the land deformation with a high spatial and temporal coverage, was performed. The results show that the threshold coherence values of the generated interferograms from ALOS-2 data are highly concentrated between 0.2 and 0.3, while a higher threshold value of 0.4 shows no coherent pixels for about 80% of Alexandria's urban area. However, the coherence values of Sentinel-1 interferograms ranged between 0.3 and 1, with most of the urban area in Alexandria showing coherent pixels at a 0.4 value. In addition, both data types produced different residual topography values of almost 0 m with a standard deviation of 13.5 m for Sentinel-1 and −20.5 m with a standard deviation of 33.24 m for ALOS-2 using the same digital elevation model (DEM) and wavelet number. Consequently, the final deformation was estimated using high coherent pixels with a threshold of 0.4 for Sentinel-1, which is comparable to a threshold of about 0.8 when using ALOS-2 data. The cumulative vertical displacement along the study area from 2017 to 2020 reached −60 mm with an average of −12.5 mm and mean displacement rate of −1.73 mm/year. Accordingly, the Alexandrian coastal plain and city center are found to be relatively stable, with land subsidence rates ranging from 0 to −5 mm/year. The maximum subsidence rate reached −20 mm/year and was found along the boundary of Mariout Lakes and former Abu Qir Lagoon. Finally, the affected buildings recorded during the field survey were plotted on the final land subsidence maps and show high consistency with the DInSAR results. For future developmental urban plans in Alexandria City, it is recommended to expand towards the western desert fringes instead of the south where the present day ground lies on top of the former wetland areas.

Keywords: Sentinel-1; ALOS/PALSAR-2; land subsidence; accuracy assessment; Alexandria City; Egypt

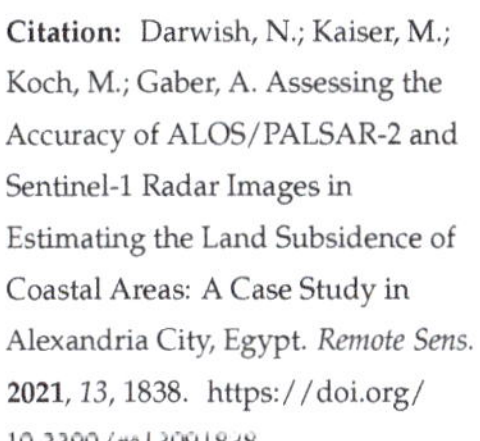

Citation: Darwish, N.; Kaiser, M.; Koch, M.; Gaber, A. Assessing the Accuracy of ALOS/PALSAR-2 and Sentinel-1 Radar Images in Estimating the Land Subsidence of Coastal Areas: A Case Study in Alexandria City, Egypt. *Remote Sens.* **2021**, *13*, 1838. https://doi.org/10.3390/rs13091838

Academic Editor: Takeo Tadono

Received: 28 March 2021
Accepted: 5 May 2021
Published: 9 May 2021

1. Introduction

Coastal cities along the Nile Delta encompass most of the social and economic activities in Egypt. Meanwhile, the combined impact of sea level rise and land subsidence causes serious risks and problems [1]. Together, subsidence and lack of sediment supply along the depositional zone of the Nile Basin (downstream Nile Delta) caused by the intensive construction of dams could potentially cause a relative rise in sea level over time [2]. Additionally, coastal erosion is predicted to increase hazard risks in coastal cities [3,4]. Therefore, there is an urgent need for regular monitoring, with high accuracy, the deformation in coastal cities in order to mitigate the expected hazards. Traditional survey instruments are point observations and thus have a spatially limited coverage. Remote sensing techniques are more cost-effective and provide better spatial coverage for coastal and land observations [5–11].

Numerous studies have been conducted to measure land subsidence in coastal areas across the Nile Delta of Egypt using SAR remote sensing data [12–25]. Information on the spatial distribution of natural hazards needs to be generated as quickly as possible in order to be useful for emergency response efforts [26]. Such information is usually generated from the analysis of optical satellite imagery [27,28]. However, relying only on optical satellite imagery for natural hazards assessment is problematic as the mapping processes can be significantly delayed by cloud cover and bad weather conditions [29]. Synthetic Aperture Radar (SAR) satellite imagery provides an alternative method to generate information under all-weather conditions. SAR is most widely used to measure the downslope velocity of slow-moving landslides and to provide subtle measurements of coastal subsidence at a significantly improved spatial resolution and over large areas [30–41].

Achache et al. 1996 [42] have demonstrated the ability of InSAR to monitor small displacements at the required scale for large landslide monitoring and their work shows similar trends to those acquired from ground-based measurements. Many studies are adopting InSAR technology for land subsidence tracking, and the derived patterns and results are comparable to ground-based measurements [43–46]. In this context, high-accuracy monitoring methods, including persistent scatterer InSAR [47] and small baseline subset DInSAR (SBAS-InSAR) [48], were proposed and applied to monitor land subsidence with high accuracy. Indeed, the SBAS approach allows for the detection of land deformation at high spatial and temporal coverage. This research work aims to improve the SBAS DInSAR methodology and to determine the best SAR data (L-band or C-band) for estimating the land-subsidence rate with high accuracy in an economically important and densely populated part of the Nile Delta of Egypt, which has suffered from many subsidence events in recent years [49]. In addition, a systematic comparison of two SAR sensors (ALOS/PALSAR-2 and Sentinel-1A) is presented to determine which dataset provided the most suitable and accurate results of land subsidence measure in locations showing high coherence within the Alexandrian urban area (Figure 1).

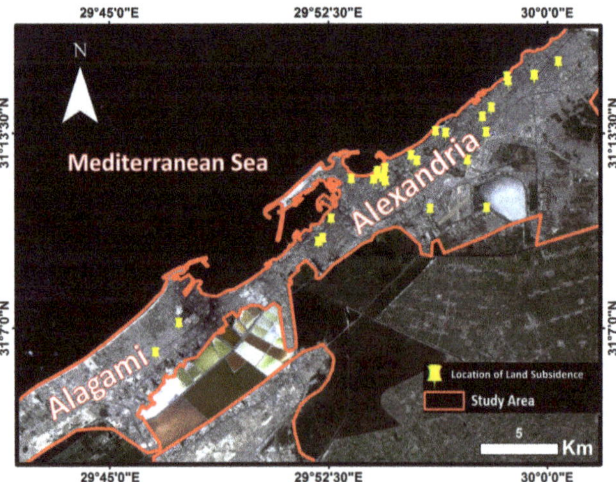

Figure 1. The urban study area in Alexandria City is outlined with a red border plotted on a Sentinel-2 image of October 2019. The yellow dots represent the recently recorded subsidence events.

2. Study Area

Alexandria was chosen as a research area due to its complicated pattern of urban subsidence. The city lies on the Mediterranean Sea at the western edge of the Nile Delta, about 183 km northwest of Cairo City and encloses an area of approximately 2679 km^2. It is Egypt's second-largest city and a major economic center. It hosts 40% of the Egyptian industrial and commercial activities, as well as the largest harbor in the country [50,51]. It is located in a moderately tectonically stable plate in North East Africa. The periodic instability has been caused by the readjustment to downwrapping (sediment compaction, faulting, isostatic lowering) of the sedimentary sequences (locally exceeding 4000 m) [52,53].

Generally, the low-lying region of the Nile Delta is subjected to significant differential subsidence. The reduction in annual Nile water and sediment discharge over the last two centuries associated with dams constructions along the Nile River has resulted in an environmental imbalance along the depositional zone of the Nile watershed, and thus, has increased the impact of sea level rise across the city [54].

Many subsidence events were recorded along Alexandria City in recent years [49]. Land subsidence in Alexandria has not been thoroughly documented in the literature; however, subsidence was partially studied using traditional geological measurements, mostly in the eastern and southern parts of the city [55,56]. The DInSAR technique has been used to study land subsidence in the central part of the city and its relation to sea level rise [57]. It is worth mentioning that the previous DInSAR studies in Alexandria mainly focused on the period before 2015, using a variety of radar images with different wavelengths and producing results with different levels of accuracy.

3. Datasets

In this study, a total of 11 ALOS/PALSAR-2 L-band images acquired from 8 March 2015 to 31 March 2019, as well as 9 Sentinel-1A C-band images acquired from 7 August 2017 to 24 January 2020 were utilized to map the ground deformation over Alexandria City, Egypt. The ALOS-2 acquisitions were captured in ScanSAR mode, ascending with single-look complex format (SLC), right-looking and HH polarization. All the Sentinel-1 images were acquired in the descending path, SLC format and Interferometric Wide (IW) swath mode. The short sampling rate of the Sentinel-1 images has the potential to maintain the coherence, even if the wavelength is short (~5.5 cm). The ALOS-2 images with a longer wavelength (~24 cm) are less affected by temporal decorrelation of the SAR signal. Table 1 lists the details of the used SAR data.

Table 1. Characteristics of ALOS/PALSAR-2 and Sentinel-1A SAR data used in this study.

Satellite	ALOS/PALSAR-2	Sentinel-1A
Band	L	C
Orbit	Ascending	Descending
Master image	27 November 2016	27 June 2018
Number of scenes	11	9
Acquisition period	2015–2019	2017–2020
λ (cm)	23.6	5.6
Polarization	HH	VV
Revisit cycle	14 days	12 days
Mode	ScanSAR	IW

Sentinel-1 products consist of three main sub-swaths (Figure 2, red labels) for each polarization channel, with a total of three (single polarizations) or six (dual polarization) images in an IW product. Each sub-swath image consists of a number of bursts (Figure 2, white labels), where each burst has been processed as a separate SLC image. The individually focused complex burst images are combined, in azimuth-time order, into a single sub-swath image with black-fill demarcation in between. The data used in evaluating the land subsidence in Alexandria City lies in the sub-swath IW2 in the western part of Nile Delta.

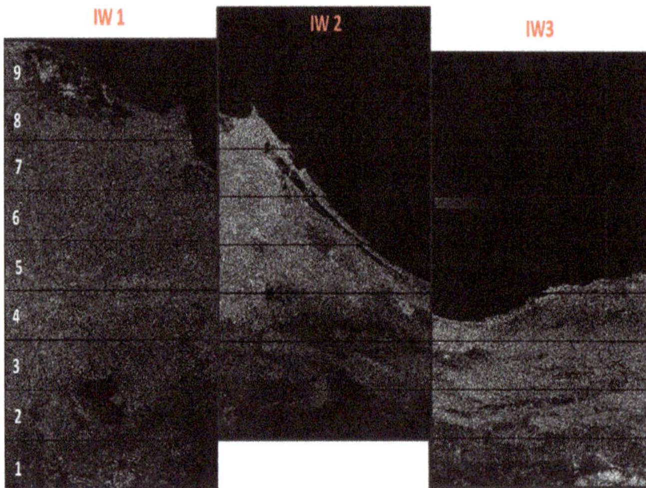

Figure 2. Sub-swaths (red) and bursts (white) of Sentinel-1 products of the study area.

4. SBAS Processing

The Interferometric analysis of ALOS/PALSAR-2 and Sentinel-1A sensors was carried out through the SBAS method [58]. All acquired steps to calculate the rates of displacement affecting the study area are shown in Figure 3. Firstly, the specifications of the perpendicular and temporal baselines thresholds for processing were calculated based on the conditions of the study area and the used SAR data. If the study area is urban, the time baseline can be set for a longer period, as several stable points can stay without changing their coherence over a long time period. An interferogram can still be generated over an urban area, even if the temporal baseline is up to four years [59]. Based on these considerations, the connection graphs were produced for each group (Figure 4). A comparison was made between the spatial and temporal baselines of the radar images. Subsequently, the links between radar images were built to show which baselines are small enough to be comparable [60].

The result is a network of connections, defined as a connection graph (Figure 4a,b), where the yellow and green dots in these graphs represent the super master and the slave of the SAR images, respectively. The red lines represent the interferograms that pass the SBAS minimal requirements. Figure 4c,d indicate the distance between the various images, depending on the date of acquisition. These graphs make it possible to conduct a quick visual evaluation of the relations between the images and the time distribution of the SAR data being used. Consequently, pairs of images were generated and later used to create interferograms [61,62]. For the ALOS/PALSAR-2 data (2015–2019) about 39 pairs of interferograms were generated and 22 pairs for Sentinel-1A data (2017–2020).

Each pair of compatible radar images identified in the previous step is used to generate an interferogram. In addition to the radar images and the connecting graph, this process requires the use of an accurate digital terrain model [61]. The digital elevation model SRTM 1-arcsecond provided by NASA with a resolution of 30 m was used in this study. During the first step of co-registration, all SAR images were geometrically adjusted by resampling each image with the master, which was chosen during the correlation step to give them the same geometry [62,63]. During the second stage, a stack of interferograms formed, followed by a process of flattening.

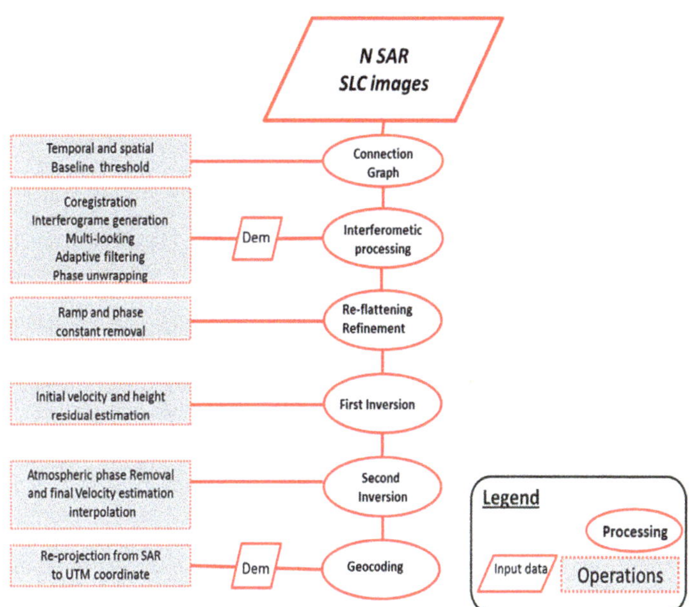

Figure 3. Flowchart of SBAS-DInSAR processing steps.

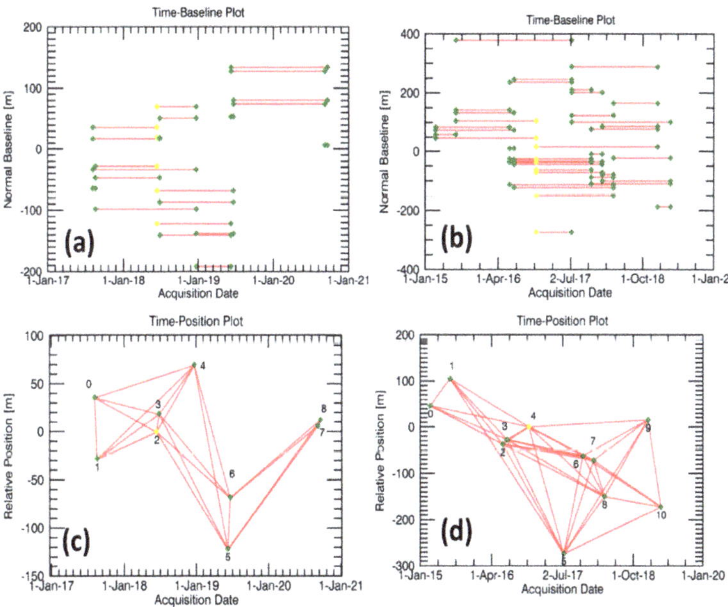

Figure 4. The temporal and spatial baseline distributions of the SAR interferograms from the Sentinel-1A and ALOS/PALSAR-2 data sets (**a**–**d**), where each acquisition is represented by a diamond associated to an ID number; the green diamonds represent the valid acquisitions and the yellow diamonds represent the super master image of the small baseline subset (SBAS). (**a**) Time–baseline plot of SBAS interferograms generated by the Sentinel-1A data, with 27 June 2018 as the super master image (**b**) time–baseline plot of SBAS interferograms generated by the ALOS/PALSAR-2 data, with 27 November 2016 as the super master image; (**c**) time–position plot generated by the Sentinel-1A and (**d**) time–position plot generated by the ALOS/PALSAR-2.

In this study, the multi-looking process was performed by setting the number of looks for both SAR data (ALOS/PALSAR-2 and Sentinel-1A) as 1 in the range direction and 4 in azimuth direction. Such multi-looking process increases Signal to Noise Ratio (SNR) of the interferograms and thus improves the quality of coherence estimation [64]. Next, interferograms were generated after meeting the temporal, geometric baseline and Doppler difference criteria. The formed interferograms are considered as Stripmap like interferograms, thus phase filtering and unwrapping were implemented. There are several methods for unwrapping the interferograms; however, in this study, the Minimum Cost Flow (MCF) method was used for both ALOS/PALSAR-2 and Sentinel-1A data. The other methods were tested, compared visually, and were found to be less suited for the study area than the MCF method.

In order to use the possible maximum number of coherent pixels a compromise value of 0.35 is suggested, when using the minimum cost flow method for the unwrapping stage. Indeed, all areas with coherence lower than the 0.35 threshold value were eliminated. As low coherence values can lead to particularly noisy areas in the analysis, it reduces the reliability of the final results [65]. Besides unwrapping interferograms, coherence maps for each interferogram were generated. Therefore, the Goldstein filters were applied to the generated interferograms in order to minimize the amount of noise [66].

In order to re-flatten the interferograms, ground control points (GCPs) are required as input, which should be positioned on areas that are thought to be stable or with pre-known deformation values. The height values of these GCPs were estimated from the input DEM [67]. In this study a total number of 150 points were used and manually placed only on the persistent scatterers with very high coherence pixels to ensure that each interferogram has received the necessary amount of control points to correct any inconsistencies caused by orbital fluctuations and re-flattening the interferograms to make phase data more accurate [68].

Following the previous processing scheme, the first inversion step was applied to the generated and re-flattened interferograms in order to measure the residual height and the velocity of the displacement using the linear model [69]. In the same context, the second unwrapping process was performed to improve the SAR data for the next step. In addition, the second inversion step was applied to provide more accurate estimation of the final velocity displacement. The high and low pass atmospheric filter was applied to remove the noise through temporal and spatial filtering operations [70,71]. The displacement observed by DInSAR is one dimensional along the Line of Sight (LOS). In order to convert the LOS displacement to the vertical direction (subsidence), an additional equation was used [72]. This operation suggests that the displacement is primarily caused by the subsidence (vertical displacement) and that the horizontal displacement, which is in the same direction of the LOS is very small compared to the vertical subsidence; thus, it can be ignored. However, it is very difficult to convert the measured phase change along the LOS into the perpendicular horizontal displacement. Therefore, the geocoding was applied to the different outputs from the previous steps to convert the slant range format to the geocoded images with the required coordinate system. Finally, the data was exported in the Geotiff format to ArcGIS, where statistics have been calculated.

5. Results

After assessing the accuracy of both ALOS-2 and Sentinel-1 in estimating the vertical displacement, the final land subsidence rate of Alexandria City and its surroundings was monitored from August 2017 to September 2020 by using nine Sentinel-1 SAR images with 5×20 m spatial resolution and VV polarization. A total of 22 interferometric pairs were generated using the image of 27 June 2018 as the super master. All slant ranges of the nine images were co-registered with this super master, which was used as the reference image.

Selecting the proper perpendicular and temporal baselines is considered to be a very important step to discard the unsuitable SAR images and to examine the validity of the generated interferograms. For Sentinel-1 interferograms, the mean absolute normal baseline was 78 m with minimum and maximum absolute baseline of 13 m and 191 m, respectively. The mean absolute temporal baseline was 285 days with minimum and maximum value equal to 12 days and 504 days, respectively. Table 2 illustrates each master and slave image of Sentinel-1, which was combined to generate 22 interferograms. While, for ALOS/PALSAR-2 interferograms, the mean absolute normal baseline was 110 m with minimum and maximum absolute baseline of 18 m and 287 m, respectively. The mean absolute temporal baseline was 329 days with minimum and maximum values equal to 28 days and 600 days, respectively.

Table 2. Sentinel-1 data pairs for SBAS processing.

Master	Slave	Normal Baseline (m)	Temporal Baseline (Days)
7/8/2017	19/8/2017	−62	12
	15/6/2018	−36	312
	27/6/2018	−20	324
	24/12/2018	34	504
19/8/2017	15/6/2018	28	300
	26/6/2018	44	312
	24/12/2018	95	492
15/6/2018	27/6/2018	18	12
	24/12/2018	69	192
	10/6/2019	−121	360
	22/6/2019	−68	372
27/6/2018	28/12/2018	53	180
	10/6/2019	−138	348
	22/6/2019	−84	360
24/12/2018	22/6/2019	−191	168
	10/6/2019	−138	180
10/6/2019	22/6/2019	54	12
	8/9/2020	126	456
	20/9/2020	134	468
22/6/2019	8/9/2020	72	444
	20/9/2020	80	456
8/9/2020	20/9/2020	13	12

Table 3 illustrates the temporal and normal baseline for each master and slave ALOS/PALSAR-2 interferogram pairs, which were combined to generate 33 interferograms. The mean absolute value of the normal and absolute baseline for ALOS-2 interferograms is higher than the value of Sentinel-1 interferograms, with approximately 32 m and 44 days, respectively. The increase in baselines has a negative effect on the reliability and accuracy of the generated interferograms. The correlation between the pixels of the used SAR pair in terms of power and phase is recognized as coherence. The zero coherence means there is no matching between the pixels with high changes on the ground and coherence value of 1 means complete matching with no change. The coherence decreases due to an increase in the normal and temporal baseline, as well as in situ anthropogenic activities. Figure 5 illustrates the coherence values of the generated interferograms from Sentinel-1 and ALOS-2 data. The histograms in Figure 5a show the coherence of ALOS-2 interferograms, with values concentrating mostly in the range between 0.2 and 0.3. Whereas, Figure 5b represents the coherence value of Sentinel-1 interferograms, with values ranging between 0.3 and 1. Based on these histograms and the distribution of coherence values, the coherence thresholds were selected as 2 and 4 for ALOS/PALSAR-2 and Sentinel-1A, respectively.

Table 3. ALOS/PALSAR-2 data pairs for SBAS processing.

Master	Slave	Normal Baseline (m)	Temporal Baseline (Days)
8/3/2015	12/7/2015	84	126
	12/6/2016	−56	462
	10/7/2016	−34	490
12/7/2015	12/6/2016	−134	336
	10/7/2016	−111	364
	27/11/2016	−105	504
12/6/2016	10/7/2016	22	28
	27/11/2016	37	168
	9/7/2017	−242	392
	12/11/2017	−35	518
	21/1/2018	−36	588
10/7/2016	21/1/2018	−58	560
	12/11/2017	−57	490
	9/7/2017	−264	384
	27/11/2016	27	140
27/11/2016	1/4/2018	−149	490
	21/1/2018	−71	420
	12/11/2017	−62	350
	9/7/2017	−273	224
9/7/2017	6/1/2019	287	546
	1/4/2018	128	266
	21/1/2018	206	196
	12/11/2017	211	126
12/11/2017	31/3/2019	−109	504
	6/1/2019	76	420
	1/4/2018	−87	140
	21/1/2018	−18	70
21/10/2018	31/3/2019	−102	434
	6/1/2019	83	390
	1/4/2018	134	468
1/4/2018	6/1/2019	−79	70
	31/3/2019	−23	364
6/1/2019	31/3/2019	−185	84

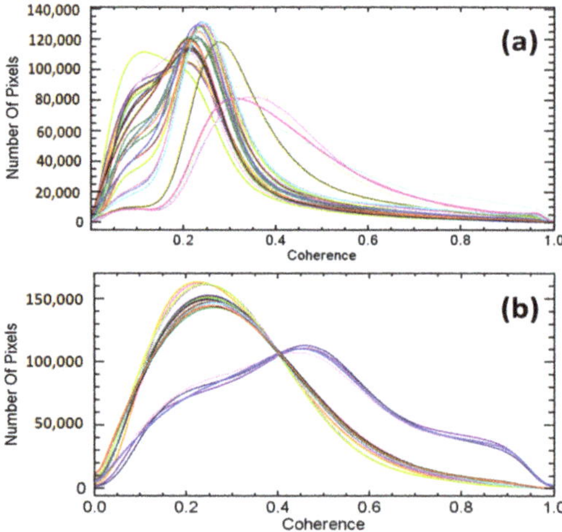

Figure 5. Illustrates the coherence values distribution of interferograms produced from the ALOS-2 data (**a**) and coherent interferogram values produced from the Sentinel-1 data (**b**).

The generated interferograms, which show strong residual phase ramps and jumps originated from the orbital inaccuracies, together with large atmospheric artifacts, were corrected by removing the residual phase frequency. Since Alexandria City lies in a coastal area with a dense cloud cover during the winter season, the atmospheric artifact was expected to cause a negative effect on the interferograms quality. The very large temporal or normal baseline between the two acquisitions resulted in the generation of wrapped interferograms with very low coherence; thus, these interferograms were discarded (Figure 6).

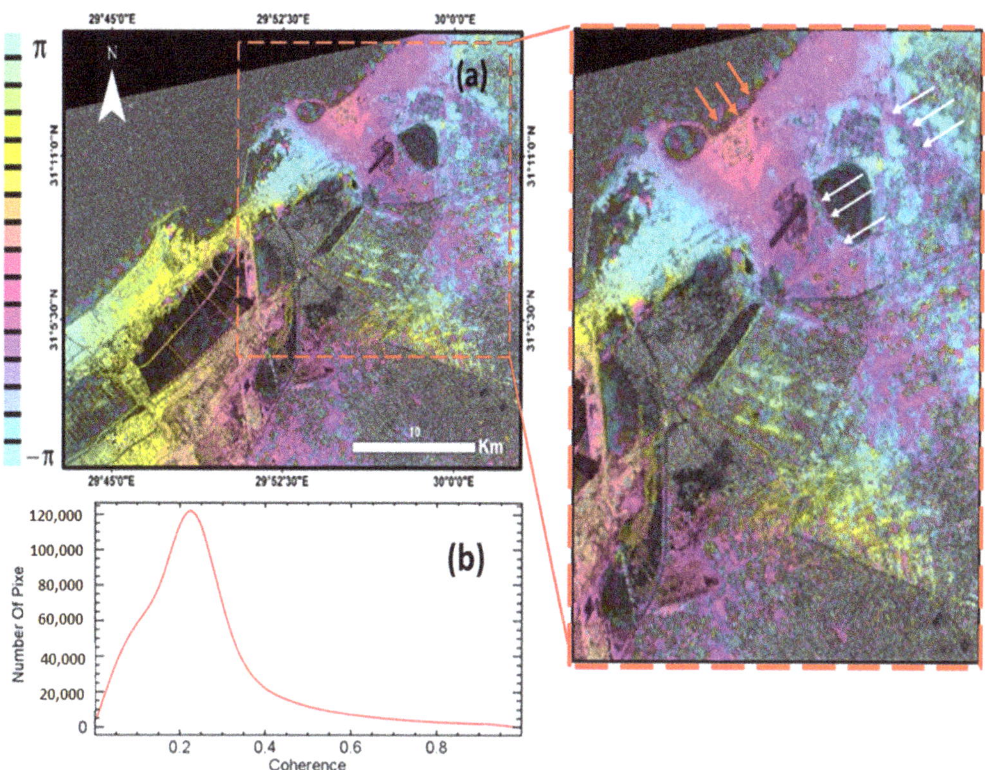

Figure 6. (**a**) Wrapped ALOS-2 low coherent interferogram showing errors during the flattening sub-step; white arrow indicates systematic residual fringes that could have been caused by strong orbital inaccuracy or issues with some parameter settings, while red arrows indicate strong atmospheric artifacts and phase jumps. (**b**) Histogram representing the coherence value of interferogram with a mean coherent value of 0.15.

However, the highest coherent Sentinel-1 wrapped interferogram (Master 15 December 2018 and Slave 10 June 2019) was considered to be acceptable. The urban areas have shown high coherent pixels without any phase jump, unlike the agricultural cover, which shows low coherent pixels (Figure 7). Wrapped interferograms were subsequently filtered and used together with the coherence data to calculate the phase unwrapping. The unwrapped interferogram were refined and re-flattened by using the residual phase method to estimate and remove the remaining phase constants and phase ramps, in order to relate the change in slant range to the deformation only (due to subsidence). A total of 150 GCPs were selected for both Sentinel-1 and ALOS-2 data, where the unwrapped phase value was close to zero and the flat areas were identified.

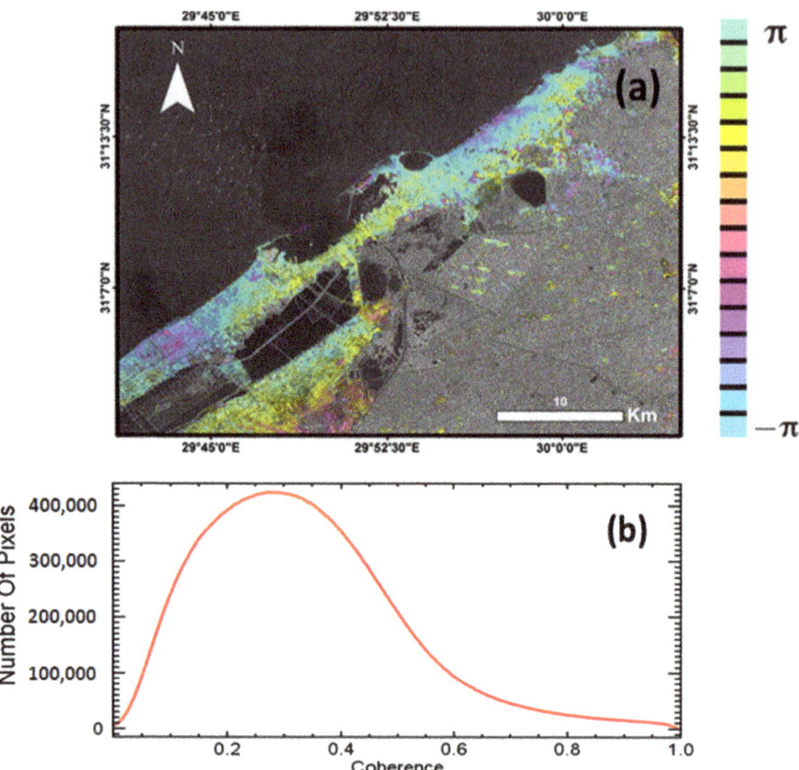

Figure 7. Wrapped Sentinel-1 interferogram with low coherence. The cause is the very large temporal and normal baseline between the two acquisitions used to make the interferogram (**a**). The histogram illustrates the coherence value between two interferogram pairs (master 15 December 2018 and Slave 10 June 2019) (**b**).

The linear model was used to estimate the residual height and the displacement velocity [73]. An incorrect residual topography calculation will cause horizontal shifts in the final SBAS geocoding results. The accuracy of the residual topography calculation depends on the vertical and horizontal accuracy of the used DEM, as well as the pixel spacing of SAR data [73,74]. In this study, the freely available SRTM data was used for estimating the residual topography. But due to the high spatial resolution of Sentinel-1 interferograms, its sensitivity is large enough to estimate the average residual topography to almost 0 m with a standard deviation equal to 13.5 m using 1 as a wavelet number (Figure 8c). Moreover, the accuracy of the ALOS-2 interferograms was also checked using the same SRTM data and the same wavelet number and showed average residual topography equal to −100 m and high standard deviation equal to 70 m (Figure 8a). Meanwhile, by using two wavelet numbers, the ALOS-2 interferograms showed average residual topography equal to −20.5 m and standard deviation equal to 33.24 m (Figure 8b).

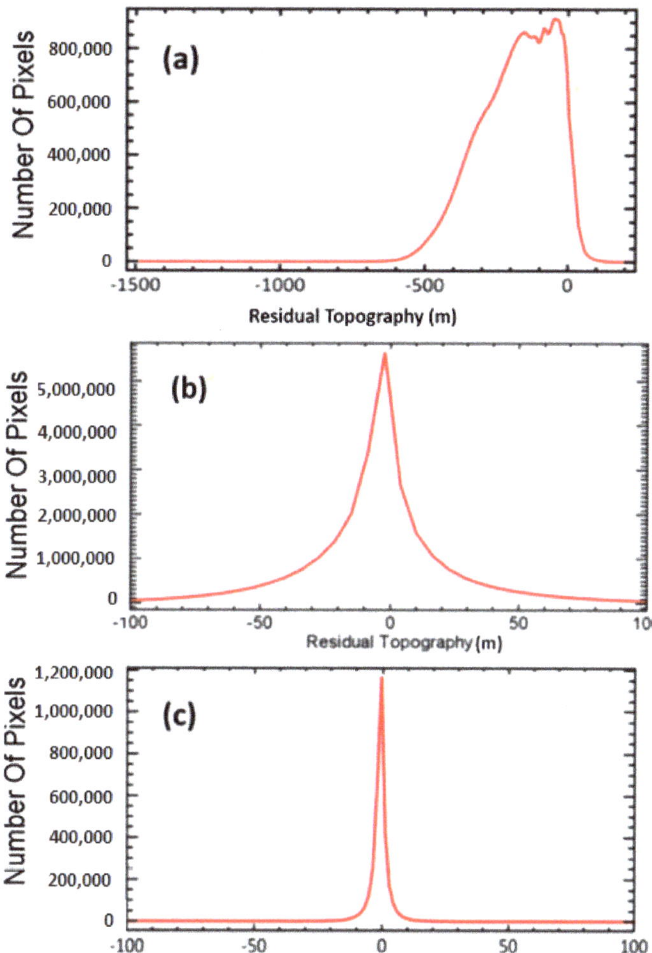

Figure 8. Statistics of estimated residual topography: (**a**) for ALOS-2 interferograms by using the SRTM data, which showed average residual topography equal to −100 m and high standard deviation equal to 70 m; (**b**) the average residual topography and standard deviation reduced by increasing the wavelet number; (**c**) for Sentinel-1 interferogram with average residual topography to almost 0 m with a standard deviation equal to 13.5 m.

Increasing the value of the wavelet to two means that the information that is coarser than 200 m is removed, while the information, which is finer than 200, is preserved. Such a result is highly unlikely because the estimation produced is extremely imprecise. It is clear that even after using a large wavelet number with the ALOS-2 SAR data, the Sentinel-1 interferograms with a zero wavelet showed an accurate and improved estimation of the average residual topography. After removing the residual topography, the phase information characterizes the displacement along the line of sight (LOS) direction. The sensitivity of displacement rate depends on the system wavelength. Thus, sensors with longer wavelength will have a sensitivity smaller than the sensors with a shorter wavelength.

Finally, all the obtained results were geocoded to adopt two constraints: the velocity and height precisions thresholds. To estimate meaningful thresholds, in terms of coverage and precision, the statistic tool of the ArcGIS was used (Figure 9). Consequently, the velocity precision threshold value of 9 mm/year and the height precision threshold value of 35 m were used for Sentinel-1 data, while the velocity precision threshold value used for ALOS-2 data was 40 mm/year and the height precision threshold of 100 m.

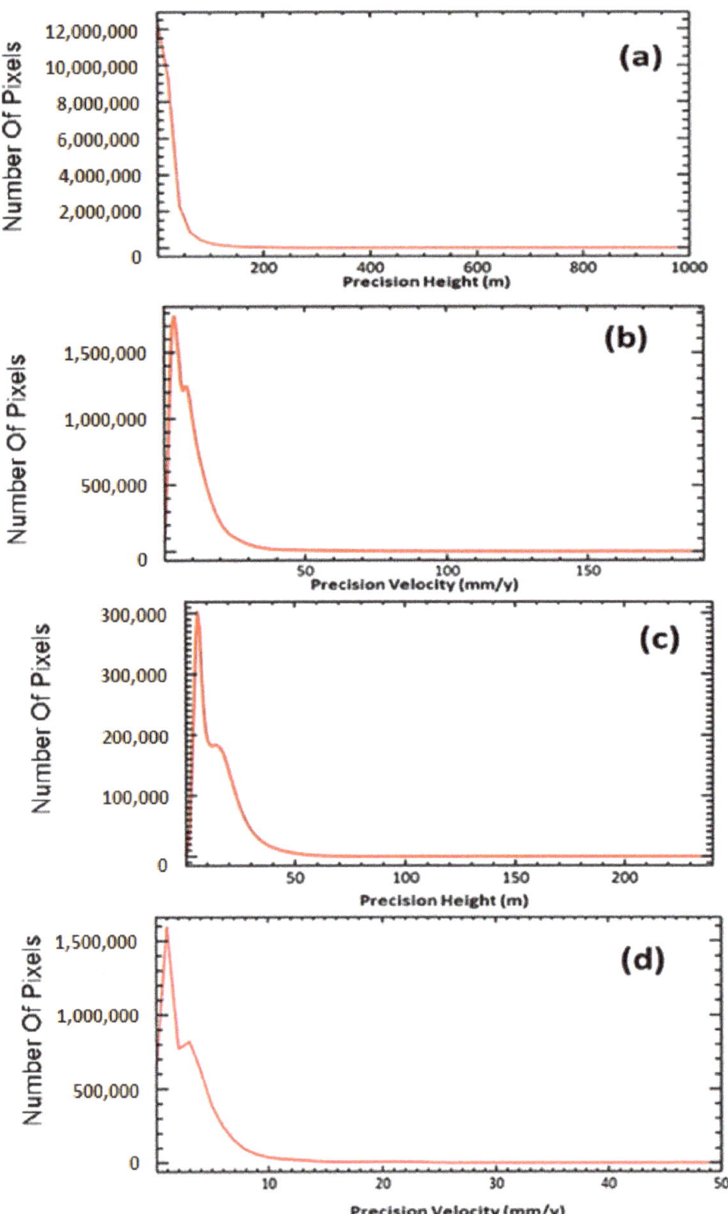

Figure 9. (**a**) Statistics histograms of precision height and (**b**) the corresponding height precision for ALOS/PALSAR-2 data. (**c**) Representative histograms of Sentinel-1 data precision height and (**d**) the corresponding precision velocity.

The deformation component can be isolated from the non-deformation component by addressing phase noise due to changing the properties of scattering over time. This has been achieved by using the phase behavior of radar signals to select pixels with minimal decorrelation [75]. Thus, the accuracy of the final results depends on the final coherence and the wavelength of the used data. Accordingly, coherence of 0.2 in C-band comparable to a coherence of about 0.6 in L-band to obtain the same precision, but with less pixels. The final coherence result of Sentinel-1 data ranges from 0.2 to 0.75 with an average coherence value of about 0.4 and standard deviation of 0.11. Coherent pixels represent good coverage of all urban area of Alexandria City, which has been used in calculating the land subsidence (Figure 10). However, the final coherence coverage of ALOS-2 data is very poor with an average coherence value of about 0.6 and standard deviation of 0.04 and about 80% of Alexandria City showed no coherent data (Figure 11).

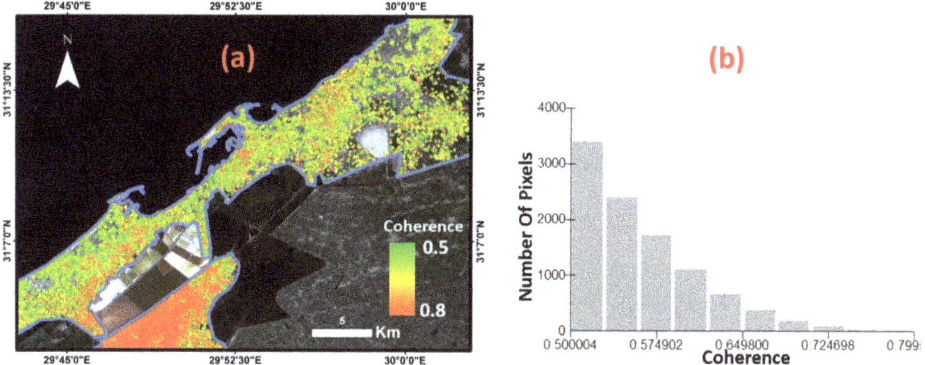

Figure 10. An excellent final coherence map result is shown based on the C-band Sentinel-1 data with values ranging from 0.5 to 0.8 (**a**) and its associated histogram (**b**).

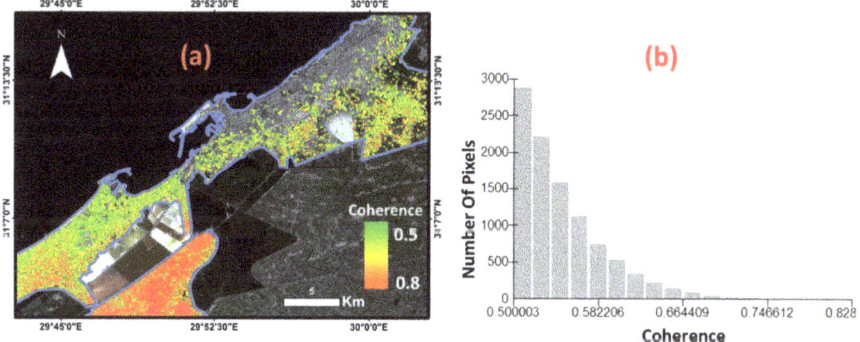

Figure 11. The final coherence coverage of ALOS-2 data which is very poor with average coherence values of about 0.6, standard deviation of 0.04 and about 80% of Alexandria City show no coherent data (**a**) and its associated histogram (**b**).

The measured average vertical displacement of Alexandria City during the period from 2015 to 2019 using a coherence threshold value of 0.2 for ALOS/PALSAR-2 data is about 5 cm for the maximum uplift rate and about −15 cm for the maximum vertical subsidence rate (Figure 12). Since ALOS-2 data do not have enough coherent pixels to be used to estimate the deformation along the urban areas, the final deformation results are discussed by using Sentinel-1 data with a threshold value of 0.4, which corresponds to a threshold value of 0.8 using ALOS-2 data.

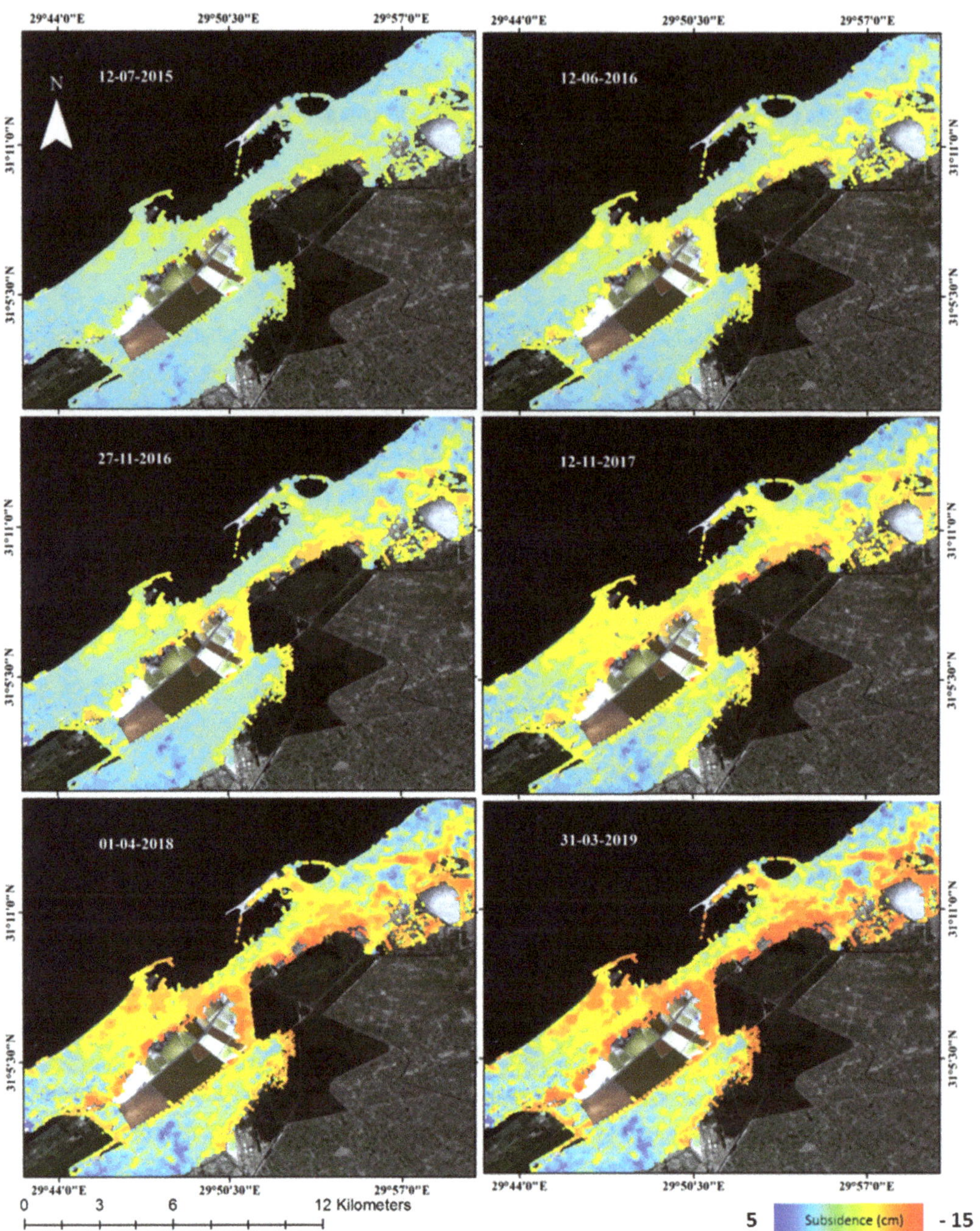

Figure 12. Vertical displacement of Alexandria City using coherence threshold value of 0.2 during the period from 2015 to 2019 using ALOS/PALSAR-2 images.

The deformation in the vertical direction along the study area can be clearly observed in Figure 13 during the period from 2017 to 2020. The color ramp from red to blue indicates the negative to positive velocities in the vertical direction. The negative values indicate the surface is moving away from the satellite (subsidence), while the positive values indicate the opposite direction of movement (uplift). Vertical deformation from 2017 to 2018 varies between −30 mm and 20 mm (Figure 13a) with an average of −4 mm. Figure 13c,d presents the vertical motion from 2017 to 2019, which ranged between −40 mm and 20 mm with an average about −4.3 mm with a standard deviation of 8 mm. The cumulative displacement in vertical direction along the study area from 2017 to 2020 reached −60 mm away from the sensor with an average of −12.5 mm and standard deviation of −10 mm (Figure 13e,f). These calculated subsidence rates show the subsidence rate is not constant over the years.

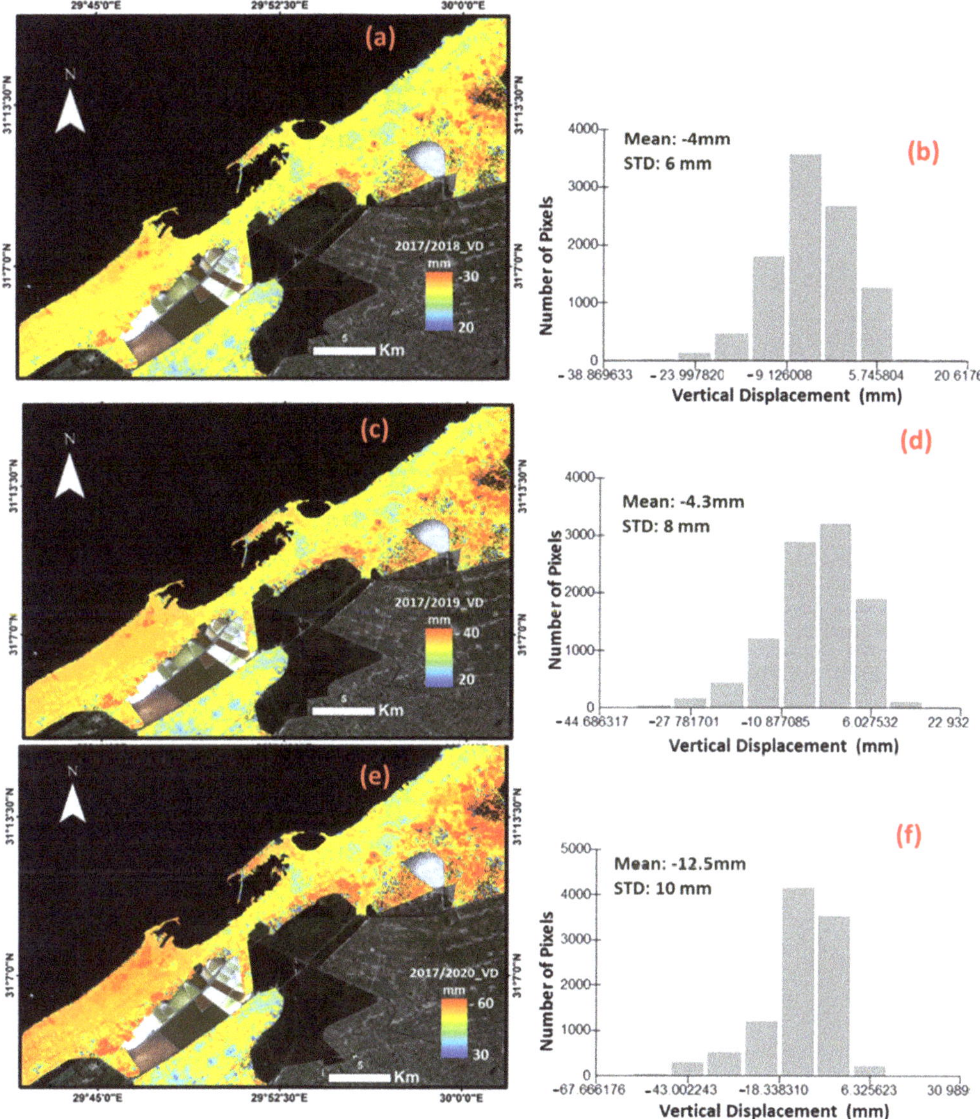

Figure 13. Vertical displacement of Alexandria City from: (**a**) 2017/2018; (**c**) 2017/2019; (**e**) 2017/2020 and the corresponding histograms of each period (**b**,**d**,**f**).

The mean velocity (mm/year) maps of the final geocoded vertical displacements generated from the Sentinel-1 data are shown in Figure 14. The histograms of the estimated displacement velocities along the study area are shown in Figure 14b, with an average displacement rate and a standard deviation of −1.73 and 4 mm/year, respectively.

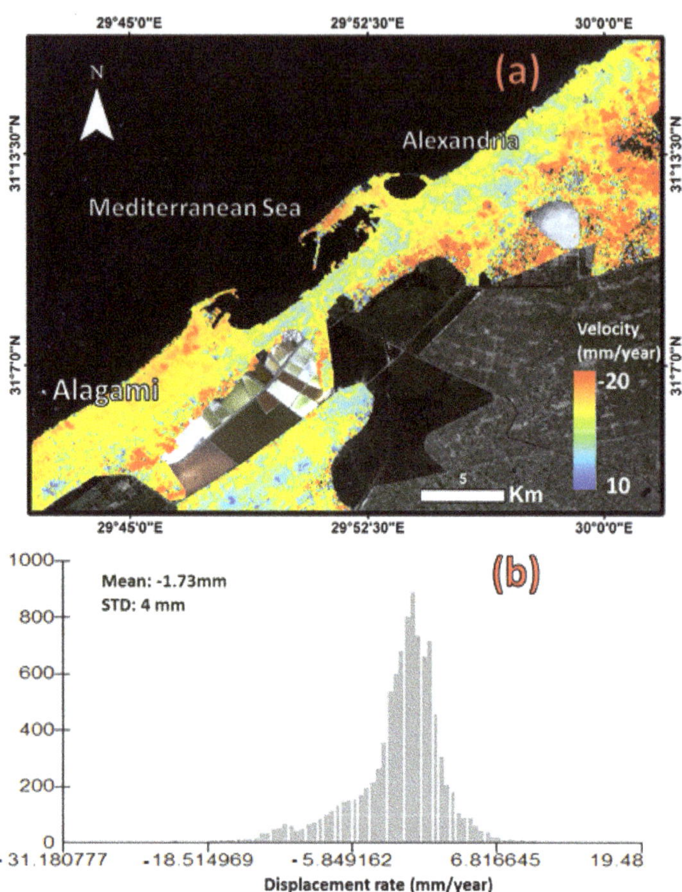

Figure 14. (**a**) The estimated mean displacement velocity of Alexandria and Alagami Cities using the Sentinel-1 data from August 2017 to September 2020 with coherence threshold of 0.4; (**b**) the corresponding histogram distribution of the derived displacement velocity rates.

6. Discussion

Selecting the proper master and slave SAR images with acceptable perpendicular and temporal baselines is a very important issue to avoid generating noisy interfeorograms. Such a selection should also consider the types of the surface features along the study area, where the urban areas can maintain a large temporal baseline, unlike the vegetation cover. Moreover, the coherence threshold value should be determined in a professional way so that a large number of pixels is used in order to achive high coherence with no surface changes and good matching between the different images.

All phase jumps, phase ramps, orbital inaccuracies, atmospheric artifacts and residual topography were calibrated and corrected before estimating the final vertical displacement to improve the accuracy of the results. The longer wavelength of SAR data shows less sensitivity, as well as less spatial coverage in calculating the land subsidence.

However, Alexandria is located at the western margin of the Nile Delta on a cemented Pleistocene sandstone ridge covered by a thin layer of Holocene sediments [76], except in the paleo island of Pharos and the former southern wetlands. Accordingly, the Alexandrian coastal plain and city center are considered to be relatively stable, with an estimated land subsidence of 0 to −5 mm/year. The maximum average subsidence value reached −20 mm/year, where the high subsidence areas are located mostly in the former Abu Qir Lagoon, the dry and recently reclaimed region of the former Mariout Lake, and parts of the northeast Alagami area.

The port of Alexandria plays an important role in Egypt's economy. Its capacity represents 75% of Egypt's total capacity Ports of the Mediterranean. It accounts for 40% of Egypt's total population industry and 56% of the petroleum industry. Due to new urbanized areas and infrastructure constructed for all of these facilities it has negatively affected the land deformation of Alexandria City. Whereas the city area constructed before 1917 was relatively stable relative to urban expansion. The city has been expanding more than double, mainly along its built-up areas during the last quarter century.

The type of Alexandria substrate rocks has an effect on the surface deformation. The soil substrate of Alexandria City represents carbon ridges, gravel, sand, stabilized sand dunes, oolitic beach and beach ridge, Nile silt and sabkha deposits, as well as refilled materials of the former lagoons. The maximum thickness of sabkha deposits reaches about 35 m in the south-western part of Alexandria where it plays an important role in accelerating the land subsidence compared to other lithological formations. The thickness of the Nile silt is about 30 m in the eastern and central part of the city. All pixels with velocity of less than −20 mm/year represent areas with different degrees of subsidence, as shown in the small-data-frame of Figure 14. Subsided areas were mainly distributed in the southern newly urbanized areas of the city, built on dried grounds from former lakes and lagoons (Figure 15).

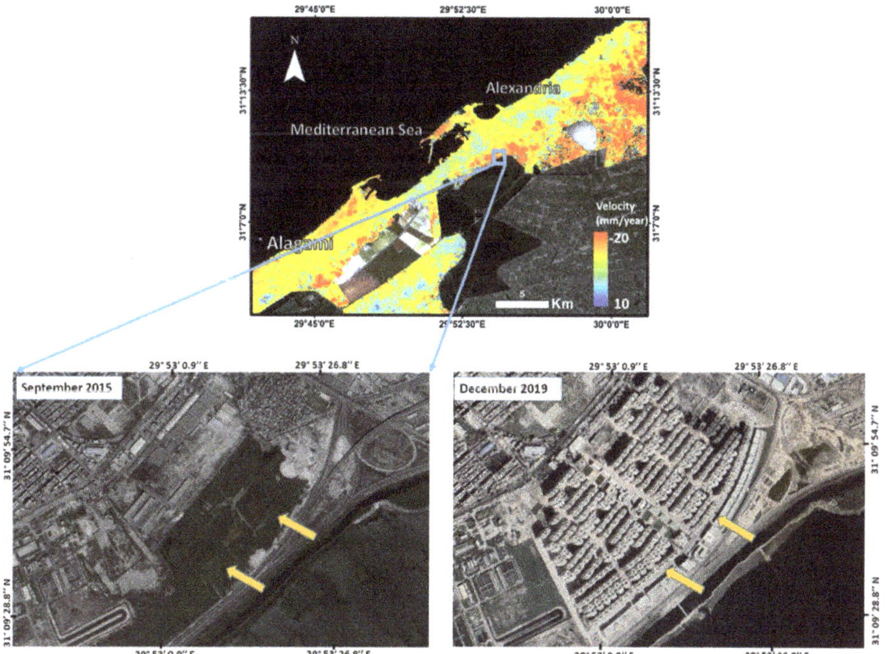

Figure 15. Shows the recent urban expansions along the border of former Mariout Lake causing land subsidence.

Subsidence are found in parts of the northwest (Alagami), newly reclaimed areas in the northeast of palaeo-island of Farous, the sandy tombolo of the old city, some parts of stabilized sand dunes in the eastern part of the city, and newly reclaimed areas in the far eastern side of Abu-Qir area. These subsidence areas are controlled by the subsurface rock type. There are three refilled and reclaimed areas from former lagoons, including the former Alharda Lake, a part of the former Abu-Qir Lagoon and a part of the former Mariout Lake in the south Western section of the study area (Figure 16). The dried and reclaimed areas from Abu-Qir Lagoon in the southeast of the study area have the highest average subsidence when compared to other dried and reclaimed former lakes and lagoon areas, about −20 mm/year. The reclaimed areas of the former Lake Alhadra showed the lowest annual average subsidence among dried and reclaimed former lakes and lagoon areas, −8.5 mm/year. Land subsidence was recorded in the refilled areas from a former lagoon in the southern part of the study area.

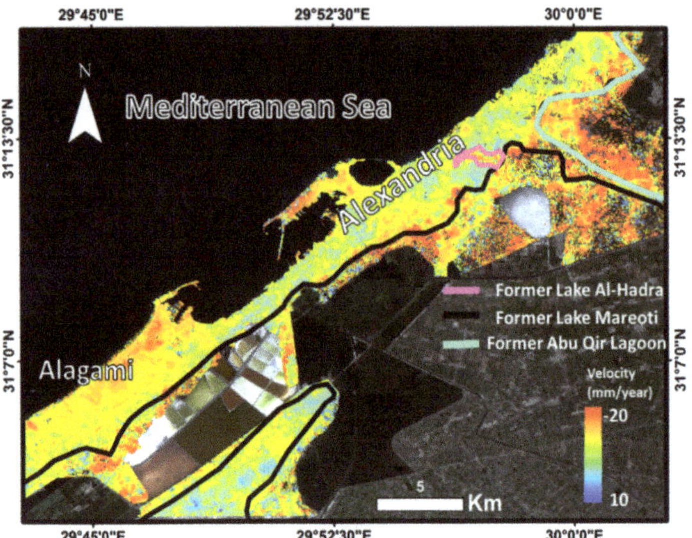

Figure 16. Former lakes and lagoons in Alexandria based on displacement velocity of Alexandria and Alagami cities using the Sentinel-1 data from August 2017 to September 2020.

Different levels of subsidence at different locations along the study area leave visual marks on the surface of the city. The field investigation provided supporting evidence of the results presented in this study. Figure 17 shows the visual cracks on buildings and roads captured at subsided locations, as defined in our vertical displacement results. Different degrees of subsidence can be easily inferred from the road cracks shown in the pictures in Figure 17b–d. The subsidence and field-checked buildings were plotted on the final land subsidence maps derived from Sentinel-1 data using the SBAS technique. The location of the affected buildings shows high consistency with the estimated results of the DInSAR.

The frequently occurred land subsidence events along the Alexandria City, especially along the newly urbanized areas should be carefully considered for future expansion or any other developmental plans. This is to avoid any related hazards that might cause damages. The western desert fringes of Alexandria should be considered for future developmental plans instead of the southern part of the city, which is suffering from the high subsidence rate due to its fragile substrate soil.

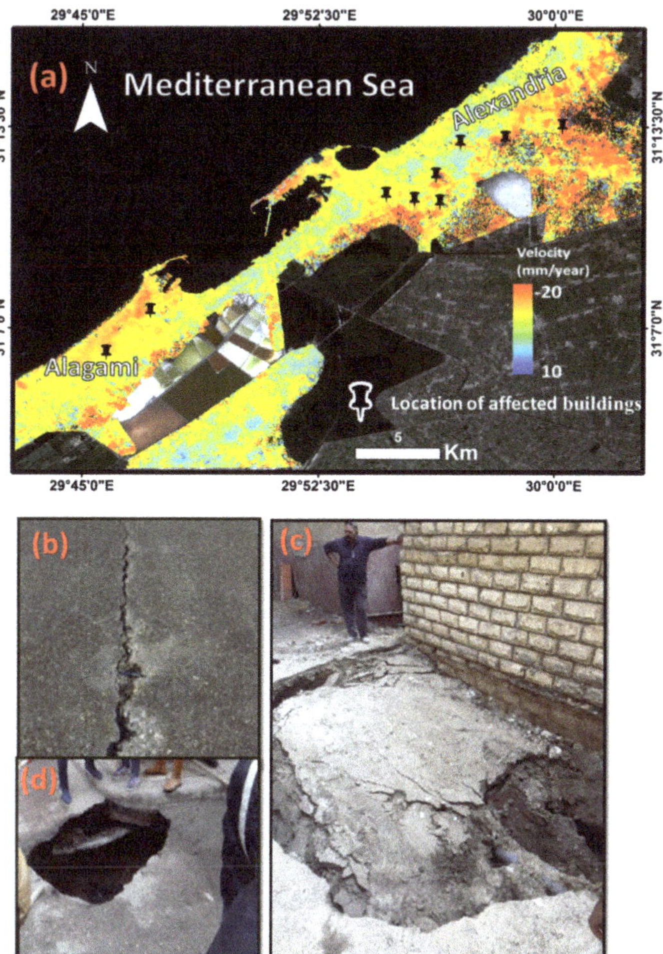

Figure 17. Land-Subsidence map shows the locations of the affected buildings (**a**); samples of cracks on roads and ground deformations located in high subsided areas (**b–d**).

7. Conclusions

Quantifying the vertical land displacement with high accuracy is a very important aspect in many applied science fields, especially those focusing on mitigating the resulting environmental hazards and their impact on various human driven activities and infrastructure. Meanwhile, the DInSAR is considered as one of the best tools to provide such deformation measures with very high accuracy and with good spatial coverage. In this study, different SAR sensors (Sentinel-1 and ALOS/PALSAR-2) were used to assess their relative accuracy in estimating the land subsidence along the coastal city of Alexandria in Egypt, which was selected as a test site. A total of nine Sentinel-1 and 11 ALOS/PALSAR-2 data covering the period of 2017 to 2020 were processed using the SBAS method approach.

The Sentinel-1 C-band data showed higher coherence and less residual topography than the ALOS/PALSAR-2 L-band. Consequently, there are not enough distributed coherent pixels in the ALOS-2 data to be used to accurately represent the deformation for urban along the study area, thus the final deformation result was discussed by using Sentinel-1 data only. The cumulative displacement pattern in vertical direction from 2017 to 2020 recorded −60 mm away from the sensor with an average of −12.5 mm and a standard deviation of −10 mm. These results show that the Alexandria coastal plain and main city center are considered to be relatively stable, with estimates of 0 to −5 mm/year. However, the maximum average subsidence value was estimated as −20 mm/year and located mostly along the dried regions of the former Abu Qir Lagoon and Mariout Lake, as well as parts of the northeast Alagami area. Finally, the results have been validated using field information, which show good correlation. In addition, the western desert fringes of Alexandria should be considered for future developmental plans instead of the southern part, which show a high subsidence rate due to its fragile substrate soil.

Author Contributions: Methodology, writing and editing A.G. and N.D.; Supervision, and reviewing, M.K. (Mona Kaiser) and M.K. (Magaly Koch). All authors have read and agreed to the published version of the manuscript.

Funding: This research received no external funding.

Acknowledgments: The authors would like to thank the JAXA, and ESA for providing the ALOS/PALSAR-2 (through ALOS-2 6th RA PI-3131 agreement) and Sentinel-1 satellite images as well as SNAP software, respectively, free of charge.

Conflicts of Interest: The authors declare no conflict of interest.

References

1. Frihy, O.E. Some proposals for coastal management of the Nile delta coast. *Ocean Coast. Manag.* **1996**, *30*, 43–59. [CrossRef]
2. Warne, A.G.; Stanley, J.D. Late Quaternary evolution of the northwest Nile delta and adjacent coast in the Al-exandria region, Egypt. *J. Coast. Res.* **1993**, *9*, 26–64.
3. Stanley, D.J.; Warne, A.G. Nile Delta in its destruction phase. *J. Coast. Res.* **1998**, *14*, 795–825.
4. Bridge, J.S. *Rivers and Floodplains: Forms, Processes, and Sedimentary Record*; John Wiley & Sons: Hoboken, NJ, USA, 2009.
5. Lu, Z.; Kwoun, O.-I. Radarsat-1 and ERS InSAR Analysis Over Southeastern Coastal Louisiana: Implications for Mapping Water-Level Changes Beneath Swamp Forests. *IEEE Trans. Geosci. Remote Sens.* **2008**, *46*, 2167–2184. [CrossRef]
6. Jiang, L.; Lin, H.; Cheng, S. Monitoring and assessing reclamation settlement in coastal areas with advanced InSAR techniques: Macao city (China) case study. *Int. J. Remote Sens.* **2011**, *32*, 3565–3588. [CrossRef]
7. Liu, P. *InSAR Observations and Modeling of Earth Surface Displacements in the Yellow River Delta (China)*; University of Glasgow: Glasgow, UK, 2012.
8. Wang, H.; Wright, T.J.; Yu, Y.; Lin, H.; Jiang, L.; Li, C.; Qiu, G. InSAR reveals coastal subsidence in the Pearl River Delta, China. *Geophys. J. Int.* **2012**, *191*, 1119–1128. [CrossRef]
9. Oliver-Cabrera, T.; Wdowinski, S. InSAR-based mapping of tidal inundation extent and amplitude in Louisiana coastal wetlands. *Remote Sens.* **2016**, *8*, 393. [CrossRef]
10. Zhang, B.; Wdowinski, S.; Oliver-Cabrera, T.; Koirala, R.; Jo, M.J.; Osmanoglu, B. Mapping the Extent and Magnitude of Sever Flooding Induced by Hurricane Irma with Multi-Temporal Sentinel-1 Sar and Insar Observations. *ISPRS Int. Arch. Photogramm. Remote Sens. Spat. Inf. Sci.* **2018**, *3*, 2237–2244. [CrossRef]
11. Zhao, Q.; Ma, G.; Wang, Q.; Yang, T.; Liu, M.; Gao, W.; Falabella, F.; Mastro, P.; Pepe, A. Generation of long-term InSAR ground displacement time-series through a novel multi-sensor data merging technique: The case study of the Shanghai coastal area. *ISPRS J. Photogramm. Remote Sens.* **2019**, *154*, 10–27. [CrossRef]
12. Aly, M.H. Radar Interferometry for Monitoring Land Subsidence and Coastal Change in the Nile Delta, Egypt. Ph.D. Thesis, Texas A&M University, College Station, TX, USA, 2006.
13. Becker, R.H.; Sultan, M. Land subsidence in the Nile Delta: Inferences from radar interferometry. *Holocene* **2009**, *19*, 949–954. [CrossRef]
14. Hereher, M.E. Vulnerability of the Nile Delta to sea level rise: An assessment using remote sensing. *Geomat. Nat. Hazards Risk* **2010**, *1*, 315–321. [CrossRef]
15. El-Asmar, H.M.; Hereher, M.E. Change detection of the coastal zone east of the Nile Delta using remote sensing. *Environ. Earth Sci.* **2010**, *62*, 769–777. [CrossRef]
16. Aly, M.H.; Klein, A.G.; Zebker, H.A.; Giardino, J.R. Land subsidence in the Nile Delta of Egypt observed by persistent scatterer interferometry. *Remote Sens. Lett.* **2012**, *3*, 621–630. [CrossRef]
17. Bouali, E.; Sultan, M.; Becker, R.; Cherif, O. *Using Persistent Scatterers Interferometry to Create a Subsidence Map of the Nile Delta in Egypt*; American Geophysical Union: Washington, DC, USA, 2013.
18. Hassaan, M.A.; Abdrabo, M.A. Vulnerability of the Nile Delta coastal areas to inundation by sea level rise. *Environ. Monit. Assess.* **2012**, *185*, 6607–6616. [CrossRef] [PubMed]
19. Fugate, J.M. Measurements of Land Subsidence Rates on the North-Western Portion of the Nile Delta Using Radar Interferometry Techniques. Ph.D. Thesis, University of Toledo, Toledo, OH, USA, 2014.

20. Gaber, A.; Darwish, N.; Sultan, Y.; Arafat, S.; Koch, M. Monitoring Building Stability in Port-Said City, Egypt Using Differential SAR Interferometry. *Int. J. Environ. Sustain.* **2014**, *3*. [CrossRef]
21. Gaber, A.; Darwish, N.; Koch, M. Minimizing the Residual Topography Effect on Interferograms to Improve DInSAR Results: Estimating Land Subsidence in Port-Said City, Egypt. *Remote Sens.* **2017**, *9*, 752. [CrossRef]
22. Stanley, J.-D.; Clemente, P.L. Increased Land Subsidence and Sea-Level Rise Are Submerging Egypt's Nile Delta Coastal Margin. *GSA Today* **2017**, *27*, 4–11. [CrossRef]
23. Gebremichael, E.; Sultan, M.; Becker, R.; El Bastawesy, M.; Cherif, O.; Emil, M. Assessing Land Deformation and Sea Encroachment in the Nile Delta: A Radar Interferometric and Inundation Modeling Approach. *J. Geophys. Res. Solid Earth* **2018**, *123*, 3208–3224. [CrossRef]
24. Sataer, G.; Sultan, M.; Emil, M.K.; Palaseanu, M.; Becker, R.; Kehew, A.; Yellich, J.A.; Kincare, K. *Visualizing and Monitoring Bluff Retreat Using Sentinel 1 Radar Interferometry and UAV Imagery*; American Geophysical Union: Washington, DC, USA, 2019.
25. Rateb, A.; Abotalib, A.Z. Inferencing the land subsidence in the Nile Delta using Sentinel-1 satellites and GPS between 2015 and 2019. *Sci. Total. Environ.* **2020**, *729*, 138868. [CrossRef]
26. Williams, J.G.; Rosser, N.J.; Kincey, M.E.; Benjamin, J.; Oven, K.J.; Densmore, A.L.; Milledge, D.G.; Robinson, T.R.; Jordan, C.A.; Dijkstra, T.A. Satellite-based emergency mapping using optical imagery: Experience and reflections from the 2015 Nepal earthquakes. *Nat. Hazards Earth Syst. Sci.* **2018**, *18*, 185–205. [CrossRef]
27. Kargel, J.S.; Leonard, G.J.; Shugar, D.H.; Haritashya, U.K.; Bevington, A.; Fielding, E.J.; Fujita, K.; Geertsema, M.; Miles, E.S.; Steiner, J.; et al. Geomorphic and geologic controls of geohazards induced by Nepals 2015 Gorkha earthquake. *Science* **2015**, *351*, aac8353. [CrossRef] [PubMed]
28. Bessette-Kirton, E.K.; Cerovski-Darriau, C.; Schulz, W.H.; Coe, J.A.; Kean, J.W.; Godt, J.W.; Thomas, M.A.; Hughes, K.S. Landslides triggered by Hurricane Maria: Assessment of an extreme event in Puerto Rico. *GSA Today* **2019**, *29*, 4–10.
29. Robinson, T.R.; Rosser, N.; Walters, R.J. The Spatial and Temporal Influence of Cloud Cover on Satellite-Based Emergency Mapping of Earthquake Disasters. *Sci. Rep.* **2019**, *9*, 12455–12459. [CrossRef] [PubMed]
30. Dai, K.; Li, Z.; Tomás, R.; Liu, G.; Yu, B.; Wang, X.; Cheng, H.; Chen, J.; Stockamp, J. Monitoring activity at the Daguangbao mega-landslide (China) using Sentinel-1 TOPS time series interferometry. *Remote Sens. Environ.* **2016**, *186*, 501–513. [CrossRef]
31. Bonì, R.; Bordoni, M.; Colombo, A.; Lanteri, L.; Meisina, C. Landslide state of activity maps by combining multi-temporal A-DInSAR (LAMBDA). *Remote Sens. Environ.* **2018**, *217*, 172–190. [CrossRef]
32. Handwerger, A.L.; Fielding, E.J.; Huang, M.; Bennett, G.L.; Liang, C.; Schulz, W.H. Widespread Initiation, Reactivation, and Acceleration of Landslides in the Northern California Coast Ranges due to Extreme Rainfall. *J. Geophys. Res. Earth Surf.* **2019**, *124*, 1782–1797. [CrossRef]
33. Hu, X.; Bürgmann, R.; Lu, Z.; Handwerger, A.L.; Wang, T.; Miao, R. Mobility, Thickness, and Hydraulic Diffusivity of the Slow-Moving Monroe Landslide in California Revealed by L-Band Satellite Radar Interferometry. *J. Geophys. Res. Solid Earth* **2019**, *124*, 7504–7518. [CrossRef]
34. Aslan, G.; Foumelis, M.; Raucoules, D.; De Michele, M.; Bernardie, S.; Cakir, Z. Landslide Mapping and Monitoring Using Persistent Scatterer Interferometry (PSI) Technique in the French Alps. *Remote Sens.* **2020**, *12*, 1305. [CrossRef]
35. Bekaert, D.P.; Handwerger, A.L.; Agram, P.; Kirschbaum, D.B. InSAR-based detection method for mapping and monitoring slow-moving landslides in remote regions with steep and mountainous terrain: An application to Nepal. *Remote Sens. Environ.* **2020**, *249*, 111983. [CrossRef]
36. Reyes-Carmona, C.; Barra, A.; Galve, J.P.; Monserrat, O.; Pérez-Peña, J.V.; Mateos, R.M.; Notti, D.; Ruano, P.; Millares, A.; López-Vinielles, J. Sentinel-1 DInSAR for Monitoring Active Landslides in Critical Infrastructures: The Case of the Rules Reservoir (Southern Spain). *Remote Sens.* **2020**, *12*, 809. [CrossRef]
37. Solari, L.; Del Soldato, M.; Raspini, F.; Barra, A.; Bianchini, S.; Confuorto, P.; Casagli, N.; Crosetto, M. Review of Satellite Interferometry for Landslide Detection in Italy. *Remote Sens.* **2020**, *12*, 1351. [CrossRef]
38. Bitelli, G.; Bonsignore, F.; Carbognin, L.; Ferretti, A.; Strozzi, T.; Teatini, P.; Tosi, L.; Vittuari, L. Radar interferome-try-based mapping of the present land subsidence along the low-lying northern Adriatic coast of Italy. Land Subsidence, Associated Hazards and the Role of Natural Resources Development. In Proceedings of the EISOLS 2010, Queretaro, Mexico, 17–22 October 2010; pp. 279–286.
39. Ruiz, A.M.; Caro, M.; Sousa, J.J.; Gil, A.J.; Hanssen, R.F.; Perski, Z.; Galindo-Zaldívar, J.; de Galdeano, C.S. Land subsidence monitoring in the southern Spanish coast using satellite radar interferometry. *Sci. Total Environ. J.* **2018**, *636*, 670–687. [CrossRef]
40. De Macedo, K.A.C.; Ramos, F.L.G.; Gaboardi, C.; Moreira, J.R.; Vissirini, F.; Da Costa, M.S. A Compact Ground-Based Interferometric Radar for Landslide Monitoring: The Xerém Experiment. *IEEE J. Sel. Top. Appl. Earth Obs. Remote Sens.* **2017**, *10*, 975–986. [CrossRef]
41. Ruiz-Armenteros, A.M.; Lazecky, M.; Ruiz-Constán, A.; Bakoň, M.; Delgado, J.M.; Sousa, J.J.; Galindo-Zaldívar, J.; de Galdeano, C.S.; Caro-Cuenca, M.; Martos-Rosillo, S. Monitoring continuous subsidence in the Costa del Sol (Málaga province, southern Spanish coast) using ERS-1/2, Envisat, and Sentinel-1A/B SAR interferometry. *Procedia Comput. Sci.* **2018**, *138*, 354–361. [CrossRef]
42. Achache, J.; Fruneau, B.; Delacourt, C. Applicability of SAR interferometry for monitoring of landslides. In Proceedings of the ERS Applications Workshop, London, UK, 6–8 December 1995.
43. Fruneau, B.; Achache, J.; Delacourt, C. Observation and modelling of the Saint-Étienne-de-Tinée landslide using SAR interferometry. *Tectonophysics* **1996**, *265*, 181–190. [CrossRef]

44. Vietmeier, J.W.; Wagner, W.; Dikau, R. Monitoring moderate slope movements (landslides) in the southern French Alps using differential SAR interferometry. In Proceedings of the Second International Workshop on ERS SAR Interferometry, FRINGE'99, Liège, Belgium, 10–12 November 1999.
45. Rizo, V.; Tesauro, M. SAR interferometry and field data of Randazzo landslide (Eastern Sicily, Italy). *Phys. Chem. Earth Part B Hydrol. Oceans Atmos.* **2000**, *25*, 771–780. [CrossRef]
46. Squarzoni, C.; Delacourt, C.; Allemand, P. Nine years of spatial and temporal evolution of the La Valette landslide observed by SAR interferometry. *Eng. Geol.* **2003**, *68*, 53–66. [CrossRef]
47. Ferretti, A.; Prati, C.; Rocca, F. Permanent scatterers in SAR interferometry. *IEEE Trans. Geosci. Remote Sens.* **2001**, *39*, 8–20. [CrossRef]
48. Berardino, P.; Fornaro, G.; Lanari, R.; Sansosti, E. A new algorithm for surface deformation monitoring based on small baseline differential SAR interferograms. *IEEE Trans. Geosci. Remote Sens.* **2002**, *40*, 2375–2383. [CrossRef]
49. Mohamed, A. A Decision from the Governor of Alexandria Regarding Land Subsidence and the Neighborhood: The Geb Factory. *Egypt News*. Available online: https://www.eg24.news (accessed on 28 March 2021).
50. Agrawala, S.; Moehner, A.; El Raey, M.; Conway, D.; Van Aalst, M.; Hagenstad, M.; Smith, J. Development and climate change in Egypt: Focus on coastal resources and the Nile. *Organ. Econ. Coop. Dev.* **2004**, *1*, 1–68.
51. El-Raey, M. Impacts and implications of climate change for the coastal zones of Egypt. In *Coastal Zones and Climate Change*; The Henry L. Stimson Center: Washington, DC, USA, 2010; pp. 31–49, ISBN 978-0-9821935-5-6.
52. Sampsell, B.M.; van Hasselt, B. *Geology of Egypt*; The American University in Cairo Press: Cairo, Egypt, 2003.
53. Said, R. *The Geological Evolution of the River Nile*; Springer: Berlin/Heidelberg, Germany, 1981.
54. Stanley, J.-D.; Clemente, P.L. Clay distributions, grain sizes, sediment thicknesses, and compaction rates to in-terpret subsidence in Egypt's northern Nile Delta. *J. Coast. Res.* **2014**, *30*, 88–101.
55. Stanley, J.-D. Growth faults, a distinct carbonate-siliciclastic interface and recent coastal evolution, NW Nile Delta, Egypt. *J. Coast. Res.* **2005**, *42*, 309–318.
56. Foumelis, M.; Blasco, J.M.D.; Desnos, Y.-L.; Engdahl, M.; Fernández, D.; Veci, L.; Lu, J.; Wong, C. ESA SNAP-StaMPS integrated processing for Sentinel-1 persistent scatterer interferometry. In Proceedings of the International Geoscience and Remote Sensing Symposium—IGARSS 2018, Valencia, Spain, 22–27 July 2018; pp. 1364–1367.
57. Wöppelmann, G.; Le Cozannet, G.; De Michele, M.; Raucoules, D.; Cazenave, A.; Garcin, M.; Hanson, S.; Marcos, M.; Santamaría-Gómez, A. Is land subsidence increasing the exposure to sea level rise in Alexandria, Egypt? *Geophys. Res. Lett.* **2013**, *40*, 2953–2957. [CrossRef]
58. Liosis, N.; Marpu, P.R.; Pavlopoulos, K.; Ouarda, T.B. Ground subsidence monitoring with SAR interferometry techniques in the rural area of Al Wagan, UAE. *Remote Sens. Environ.* **2018**, *216*, 276–288. [CrossRef]
59. Lanari, R.; Casu, F.; Manzo, M.; Zeni, G.; Berardino, P.; Manunta, M.; Pepe, A. An Overview of the Small Baseline Subset Algorithm: A DInSAR Technique for Surface Deformation Analysis. *Pure Appl. Geophys.* **2007**, *164*, 637–661. [CrossRef]
60. Diao, X.; Bai, Z.; Wu, K.; Zhou, D.; Li, Z. Assessment of mining-induced damage to structures using InSAR time series analysis: A case study of Jiulong Mine, China. *Environ. Earth Sci.* **2018**, *77*, 166. [CrossRef]
61. Hu, B. Monitoring of Ground Deformation due to Excessive Withdrawal of Natural Gas Using SBAS. *Math. Probl. Eng.* **2014**, *2014*, 1–6. [CrossRef]
62. Chaabani, A.; Deffontaines, B. Application of the SBAS-DInSAR technique for deformation monitoring in Tunis City and Mornag plain. *Geomat. Nat. Hazards Risk* **2020**, *11*, 1346–1377. [CrossRef]
63. Hooper, A. A multi-temporal InSAR method incorporating both persistent scatterer and small baseline approaches. *Geophys. Res. Lett.* **2008**, *35*, 302. [CrossRef]
64. Gierull, C. Statistical analysis of multilook SAR interferograms for CFAR detection of ground moving targets. *IEEE Trans. Geosci. Remote Sens.* **2004**, *42*, 691–701. [CrossRef]
65. Woodhouse, I.H. *Introduction to Microwave Remote Sensing*; CRC Press: Boca Raton, FL, USA, 2017.
66. Sarmap. PS Tutorial: Version 0.9. 2014. Available online: http://sarmap.ch/page.php?page=tutorials (accessed on 16 February 2016).
67. Gama, F.F.; Cantone, A.; Mura, J.C.; Pasquali, P.; Paradella, W.R.; dos Santos, A.R.; Silva, G.G. Monitoring subsidence of open pit iron mines at Carajás Province based on SBAS interferometric technique using TerraSAR-X data. *Remote Sens. Appl. Soc. Environ.* **2017**, *8*, 199–211. [CrossRef]
68. Zhang, Y.; Meng, X.; Jordan, C.; Novellino, A.; Dijkstra, T.; Chen, G. Investigating slow-moving landslides in the Zhouqu region of China using InSAR time series. *Landslides* **2018**, *15*, 1299–1315. [CrossRef]
69. Loesch, E.; Sagan, V. SBAS Analysis of Induced Ground Surface Deformation from Wastewater Injection in East Central Oklahoma, USA. *Remote Sens.* **2018**, *10*, 283. [CrossRef]
70. Ferretti, A.; Prati, C.; Rocca, F. Nonlinear subsidence rate estimation using permanent scatterers in differential SAR interferometry. *IEEE Trans. Geosci. Remote Sens.* **2000**, *38*, 2202–2212. [CrossRef]
71. Pendry, L.F.; Salvatore, J. Individual and social benefits of online discussion forums. *Comput. Hum. Behav.* **2015**, *50*, 211–220. [CrossRef]
72. Zebker, H.A.; Goldstein, R.M. Topographic mapping from interferometric synthetic aperture radar observations. *J. Geophys. Res. Solid Earth* **1986**, *91*, 4993–4999. [CrossRef]

Mora, O.; Lanari, R.; Mallorqui, J.J.; Berardino, P.; Sansosti, E. A new algorithm for monitoring localized deformation phenomena based on small baseline differential SAR interferograms. In Proceedings of the IEEE International Geoscience and Remote Sensing Symposium, Toronto, ON, Canada, 24–28 June 2002.
Rao, X.; Tang, Y. Small baseline subsets approach of DInSAR for investigating land surface deformation along the high-speed railway. In *SPIE Asia-Pacific Remote Sensing*; SPIE International Society for Optics and Photonics: Beijing, China, 2014; p. 92601C.
Bekaert, D.; Walters, R.; Wright, T.; Hooper, A.; Parker, D. Statistical comparison of InSAR tropospheric correction techniques. *Remote Sens. Environ.* **2015**, *170*, 40–47. [CrossRef]
Stanley, J.-D.; Toscano, M.A. Ancient archaeological sites buried and submerged along Egypt's Nile delta coast: Gauges of Holocene delta margin subsidence. *J. Coast. Res.* **2009**, *25*, 158–170. [CrossRef]

MDPI
St. Alban-Anlage 66
4052 Basel
Switzerland
Tel. +41 61 683 77 34
Fax +41 61 302 89 18
www.mdpi.com

Remote Sensing Editorial Office
E-mail: remotesensing@mdpi.com
www.mdpi.com/journal/remotesensing

www.ingramcontent.com/pod-product-compliance
Lightning Source LLC
LaVergne TN
LVHW070058170726
843515LV00007B/1565

* 9 7 8 3 0 3 6 5 6 1 0 6 6 *